Weather and Climate Science

Weather and Climate Science

Editor: Jose Wells

R CALLISTO
REFERENCE

www.callistoreference.com

Callisto Reference,
118-35 Queens Blvd., Suite 400,
Forest Hills, NY 11375, USA

Visit us on the World Wide Web at:
www.callistoreference.com

ISBN: 978-1-64116-135-0 (Hardback)

Cataloging-in-Publication Data

Weather and climate science / edited by Jose Wells.
    p. cm.
Includes bibliographical references and index.
ISBN 978-1-64116-135-0
1. Weather. 2. Climatology. 3. Meteorology. 4. Climatic changes. I. Wells, Jose.
QC981 .W43 2019
551.6--dc23

Table of Contents

Preface

The main aim of this book is to educate learners and enhance their research focus by presenting diverse topics covering this vast field. This is an advanced book which compiles significant studies by distinguished experts in the area of analysis. This book addresses successive solutions to the challenges arising in the area of application, along with it; the book provides scope for future developments.

Climate and weather studies are pursued by the respective branches of climatology and meteorology, which fall in the domain of atmospheric sciences. Climatology is the qualitative and quantitative study of long-term weather conditions and climate trends. Meteorology involves weather predictions and an analysis of atmospheric physics and chemistry. Different analog techniques can be used in climatology for the prediction of short-term weather forecasting. Some of these techniques are El Nino-Southern Oscillation, Madden-Julian oscillation, North Atlantic oscillation, etc. Weather and climate prediction is useful in agriculture, for determining the distribution of radioactive aerosols and gases in the atmosphere, for analyzing industrial pollution dispersion, etc. This book unfolds the innovative aspects of weather and climate studies, which will be crucial for the progress of these fields in the future. Also included in this book is a detailed explanation of the various principles and applications of meteorology and climatology. With state-of-the-art inputs by acclaimed experts of this field, this book targets students and professionals alike.

It was a great honour to edit this book, though there were challenges, as it involved a lot of communication and networking between me and the editorial team. However, the end result was this all-inclusive book covering diverse themes in the field.

Finally, it is important to acknowledge the efforts of the contributors for their excellent chapters, through which a wide variety of issues have been addressed. I would also like to thank my colleagues for their valuable feedback during the making of this book.

Editor

Detection and attribution analysis of annual mean temperature changes in China

Ying Xu[1,*], Xuejie Gao[2], Ying Shi[1], Zhou Botao[1]

[1]National Climate Center, China Meteorological Administration, 100081 Beijing, China
[2]Climate Change Research Center, Institute of Atmospheric Physics, Chinese Academy of Sciences, 100029 Beijing, China

ABSTRACT: Using an optimal fingerprinting method and the Coupled Model Intercomparison Project Phase 5 (CMIP5) multi-model simulations, we attempted to quantify the human contribution to the observed annual mean temperature change that occurred over China between 1961 and 2005. Results indicate that the combined effects of greenhouse gases and sulfate aerosol forcing are clearly detectable in the observed annual mean temperature change. Effects of anthropogenic and natural forcings are separately detectable, and the climate response to greenhouse gas forcing can be identified clearly and robustly. Our results also show that only when anthropogenic forcing is involved can the observed changes in China's mean temperature from 1961 to 2005 be explained.

KEY WORDS: Detection and attribution · Climate change · CMIP5 · Models

1. INTRODUCTION

The Fifth Assessment Report (AR5) of the Intergovernmental Panel on Climate Change (IPCC 2013) illustrates the latest advances in climate change research since the publication of AR4 (IPCC 2007). It assesses the science of climate change, impacts, adaptation and mitigation in the present and future at both regional and global scales. Understanding the causes of observed changes in climate is an important focus of this assessment, since this provides not only an explanation for the observed changes, but also confidence in model performance and projection. AR5 states that 'Human influence has been detected in warming of the atmosphere and the ocean, in changes in the global water cycle, in reductions in snow and ice, in global mean sea level rise, and in changes in some climate extremes' (IPCC 2013, p. 17).

Detection and attribution of climate change in terms of the relative contribution of anthropogenic and natural factors is a hot topic, and one of the key issues in the global climate change debate. It provides an important scientific basis to answer the question of whether, and to what extent, human activities have influenced climate change. The detection and attribution of climate change was initially conducted on the annual mean temperature at a global scale. Now, it extends to other components of the climate system such as precipitation, sea level pressure, climate extremes, and other variables that are more impact-relevant and occur on various space–time scales (IPCC 2013).

The use of climate model simulations is a very important approach in climate change detection and attribution analyses, as reviewed by Hegerl & Zwiers (2011). The recently completed Coupled Model Intercomparison Project Phase 5 (CMIP5) provides historical climate simulations under the combined effect of external forcings as well as individual forcings, with a compilation of over 50 models from different modeling centers around the world (Taylor et al. 2012). These simulations provide an important foundation for the detection and attribution of climate change as it enables climatic responses to various forcings to be assessed. Using the Hadley Centre new Global Envi-

*Corresponding author: xuying@cma.gov.cn

ronmental Model 2 Earth System (HadGEM2-ES) model simulations, Stott & Jones (2012) detected the effect of greenhouse gas forcing on the global mean temperature. Gareth et al. (2013) conducted a more comprehensive analysis by using simulations from multiple models, trying to attribute observed historical near-surface temperatures to anthropogenic and natural external forcings. The results indicated that anthropogenic emissions of greenhouse gases are the dominate causes of the observed global warming that has occurred since the mid-20th century. Of the estimated observed warming trend of ~0.6 K, between 0.6 and 1.2 K can be attributed to greenhouse gases, balanced by a counteracting cooling of between 0 and −0.5 K from other anthropogenic forcings such as anthropogenic aerosols.

While it is important to understand the causes of warming at a global scale, the interests and concerns of policy makers are aimed more towards regional and even local scales. Thus, considerable efforts have been made towards attributing observed changes at regional scales (e.g. Zwiers & Zhang 2003, Zhang et al. 2006, Hegerl et al. 2007, Stott et al. 2010). As in other parts of the world, significant warming has been observed over China in the last several decades (e.g. Zhai & Pan 2003, Tang & Ren 2005, Qian et al. 2011, Ren & Zhou 2014).

A limited number of studies have been conducted using the optimal fingerprinting method, to understand the causes of these observed temperature changes. Zhou & Yu (2006) analyzed temperature changes in the 20th century, simulated by 19 models from CMIP3. They found a high correlation between mean temperature in China and model simulated responses under the combined effects of anthropogenic and natural forcing, indicating an influence from external forcings. External forcing explained 32.5 % of the annual mean temperature change in the 20th century over China, while internal variability was as high as 67.5 %. This suggests that attribution of the changes at a regional scale is much more complex compared to that at a global scale. Using the optimal fingerprinting method, Zhang et al. (2006) quantitatively compared temperature changes in the observations and responses to external forcings simulated by 2 models over different regions, including China. They detected the effect of anthropogenic forcing from greenhouse gases and sulfate aerosols in the annual mean temperature over China. Using an optimal detection method, Wen et al. (2013) compared spatio-temporal patterns of 4 extreme indices (namely annual maxima of daily maximum and daily minimum temperatures and annual minima of daily

maximum and daily minimum temperatures) with those observed and simulated by the Canadian Earth System Model 2 (CanESM2) over China between 1961 and 2007. They established a clear connection between human emission of greenhouse gases and extreme temperature changes in China. More recently, Sun et al. (2014) compared the observed summer temperature changes with those simulated by CMIP5 models, with a focus on the 2013 heat event in eastern China, and found contributions from anthropogenic influences. Zhou et al. (2014) reported that both anthropogenic influences and the internal variability of the Pacific decadal oscillation/interdecadal Pacific oscillation (PDO/IPO) contributed to the heat event in eastern China in 2013. Other attribution studies include the applications of a single global model, the Bergen Climate Model (BCM; Wang et al. 2013) and a single regional model, the Regional Climate Model version 4 (RegCM4; Zhang et al. 2015) to investigate the possible human influences on the rainfall patterns observed in China over the last several decades.

The objective of the study was to try to further understand the surface temperature response over China to external forcings during the period of 1961–2005 based on the CMIP5 multi-model ensemble. We focused on annual mean temperature only in the present period.

2. DATA

2.1 Observational data

The observational dataset employed in this study is an updated version of the gridded daily scale dataset, CN05.1, developed by Wu & Gao (2013). The difference between the updated dataset and the original CN05.1 is that station data employed in the interpolation has been homogeneously adjusted by China's National Meteorological Information Center using RHtest of Wang et al. (2007). This homogenization resulted in some differences between the 2 versions, similar to that found by Li et al. (2015) using a different method, Multiple Analysis of Series for Homogenization (MASH).

CN05.1 is the further improvement of CN05 (Xu et al. 2009), but is based on interpolation from more station observations (760 for CN05 vs. 2416 for CN05.1), comprises more variables and has a higher spatial resolution (0.25° latitude × 0.25° longitude). The dataset was constructed following the commonly used 'anomaly approach', in which a gridded climatology is first calculated, and then a gridded daily anomaly is added to the climatology to obtain the final dataset. The density of stations used in CN05.1 is quite high over eastern China, but not so high in the west where population density and urban establishments are much lower, in particular over the Tibetan Plateau. More detailed information concerning the data construction and comparison with other datasets can be found in Xu et al. (2009) and Wu & Gao (2013). CN05 and the updated CN05.1 are becoming more popular models for validation analysis over the Chinese region (e.g. Gao et al. 2011, Wu et al. 2012, Guo & Wang 2013, Sui et al. 2014).

2.2. Model data

A typical detection and attribution analysis requires the use of model simulations to estimate climate responses (or signals) to external forcing as well as to internal variability. For this purpose, we used 296 simulations from the 20 models participating in CMIP5 under various external forcings (Table 1). These models produced different ensembles for the period from 1850–2005, including (1) 118 historical forcing runs that included both anthropogenic and natural external forcing (ALL); (2) 48 natural forcing runs that only contain changes in solar irradiance and volcanic activity (NAT); and individual forcing runs, including (3) 42 runs under greenhouse gases (GHG); (4) 33 runs under anthropogenic aerosol (AA); (5) 15 runs under land use change (LU); and (6) 40 anthropogenic runs (ANT) (see Table 1). Each ensemble from a model contained at least 3 member runs. Detailed information on forcing data can be found on the CMIP5 website (http://cmip-pcmdi.llnl.gov/cmip5/forcing.html). These models were also conducted using different years of pre-industrial control simulation (CTL): a total of 13 008 yr in the 20 models. Values of annual mean temperature were extracted from the monthly outputs of these simulations for the estimation of internal variability. All model data were bilinearly interpolated to the common 1° × 1° grids.

Table 1. The 20 models included in the Coupled Model Intercomparison Project Phase 5 (CMIP5). Atm. Res: atmospheric model resolution (in no. of grid squares); ALL: includes anthropogenic and natural external forcings; NAT: only solar irradiance and volcanic activity; GHG: greenhouse gases; ANT: anthropogenic influences; LU: land use changes; AA: anthropogenic aerosols; (–) not included

Models	Institution and country	Atm. Res (lon × lat)	ALL	NAT	GHG	ANT	LU	AA
BCC-CSM1-1	Beijing Climate Center, China Meteorological Administration, China	128 × 64	3	–	–	–	–	–
BCC-CSM1-1-m	Beijing Climate Center, China Meteorological Administration, China	320 × 160	3	–	–	–	–	–
CCSM4	National Center for Atmosphere Research, United States	288 × 192	6	4	3	–	–	–
CESM1-CAM5	National Center for Atmosphere Research, United States	288 × 192	3	–	–	–	–	–
CNRM-CM5	CNRM and CERFACS, France	256 × 128	10	6	6	10	–	–
CSIRO-Mk3-6-0	CSIRO and QCCCE, Australia	192 × 96	10	5	5	5	–	5
CanESM2	Canadian Centre for Climate Modeling and Analysis, Canada	128 × 64	5	5	5	–	5	5
FGOALS-g2	Institute of Atmospheric Physics, Chinese Academy of Sciences and Tsinghua University, China	128 × 60	5	–	–	–	–	–
FIO-ESM	The First Institution of Oceanography, China	128 × 64	3	–	–	–	–	–
GFDL-CM3	NOAA Geophysical Fluid Dynamics Laboratory, United States	144 × 90	5	3	3	3	–	3
GISS-E2-H(p1,p2,p3)	NASA, GISS (Goddard Institute for Space Studies), United States	144 × 90	3 × 5	5	5	10	5	10
GISS-E2-R(p1,p2,p3)	NASA, GISS (Goddard Institute for Space Studies), United States	144 × 90	3 × 6	10	5	10	5	10
HadGEM2-ES	MOHC (Met Office Hadley Centre), UK	192 × 145	4	4	4	–	–	–
IPSL-CM5A-LR	Institute Pierre-Simon Laplace, France	96 × 96	6	3	3	2	–	–
MIROC-ESM	NIES, JAMSTEC, Japan	128 × 64	3	3	3	–	–	–
MIROC5	AORI, NIES, JAMSTEC, Japan	256 × 128	5	–	–	–	–	–
MPI-ESM-LR	Max Planck Institute for Meteorology, Germany	192 × 96	3	–	–	–	–	–
MPI-ESM-MR	Max Planck Institute for Meteorology, Germany	192 × 96	3	–	–	–	–	–
MRI-CGCM3	Meteorological Research Institute, Japan	320 × 160	5	–	–	–	–	–
NorESM1-M	Norwegian Climate Centre, Norway	144 × 96	3	–	–	–	–	–
Sum Models (runs)			20 (118)	10 (48)	10) (42	6 (40)	3 (15)	5 (33)

3. METHODS AND DATA PROCESSING

3.1. Detection and attribution method

Climate change detection is defined as the process of demonstrating that climate (or a system affected by climate) has changed in some defined statistical sense, without providing a reason for that change. Climate change attribution is defined as the process of evaluating the relative contributions of multiple causal factors to a change or event with an assignment of statistical confidence. In the context of IPCC, detection also occasionally refers to the discovery of an influence from external forcings, and thus is closely linked to attribution (IPCC 2013). Casual factors usually refer to external influences, which may be anthropogenic (e.g. greenhouse gases, aerosols, ozone precursors, land use) and/or natural (e.g. volcanic eruptions, solar cycle modulations).

Detection and attribution have 4 core elements: observations of climate indicators from which climate responses to external forcing might be detected; an estimate of external forcing; a quantitative physically-based understanding of how external forcing might affect these climate indicators, usually based on climate model simulations; and an estimate of climate internal variability, also typically based on climate model simulations. Additionally, it is important that the key external forces are identified, that signals and noise are additive, and that the large-scale patterns of response are correctly simulated by climate models (Bindoff et al. 2013).

A typical detection and attribution analysis uses an optimal fingerprint method based on generalized multivariate linear regression (e.g. Allen & Tett 1999, Allen & Stott 2003). The optimal fingerprinting method assumes that the climate response to different forcings superimposes linearly, and that the forced response also superimposes linearly on unforced climate variability. Consequently, detection and attribution relies heavily on climate models.

The optimal fingerprint method assumes that signals are linearly additive. This was found to be the case even for a highly non-linear system (Allen & Stott 2003). The linearity of the climate response to different forcings has been evaluated in numerous studies and found to work well, at least for the transient climate response. The larger equilibrium temperature responses to aerosol and greenhouse gas emissions do not necessarily combine linearly (Ming & Ramaswamy 2009). However, as the changes we examined in this study were small relative to the equilibrium state, we would assume such non-linearity to have little impact on our analysis.

The optimal fingerprint method assumes the observations are the sum of expected changes (scaled signals) and internal variability (residual), expressed as following equation:

$$y = \beta X + \varepsilon$$

Here, the vector y is a filtered version of the observed temperature record. Matrix X represents model-simulated signal patterns consisting of 1 or 2 vectors (i.e. either the estimated ANT signal in our 1-signal analysis or the estimated GHG and AA signals in our 2-signal analysis). Vector ε represents natural (residual) variability that is not explained by the signals. The scaling factor β, a vector of 1 or 2 elements, adjusts the signal amplitude so that the scaled signal best matches the observations.

We conducted a single-signal analysis involving only 1signal vector at a time. The detection of a signal would indicate the presence of response to a particular forcing (or combined forcings) in the observations. As the observations may be influenced by multiple forcing factors, regression models with multiple predictors provide a better fit. Therefore, we also conducted a 2-signal analysis in which X had 2 signal vectors. The use of a 2-signal analysis allows the separation of responses to different forcings, so a clearer attribution to individual forcing can be obtained.

3.2. Data processing

3.2.1. Methods

The detection analysis was conducted based on temperature evolution over both space and time. For both observations and model simulations, we computed a regional time series of annual mean temperature anomalies over 8 sub-regions of China using area weighting. These regions were defined according to administrative boundaries and societal and geographical conditions, and were used in China's National Assessment Report on Climate Change (CCNARCC 2007). The 8 sub-regions consisted of northeastern China (NEC: 39–54° N, 119–134° E), north China (NC: 36–46° N, 111–119° E), eastern China (EC: 27–36° N, 116–122° E), central China (CC: 27–36° N, 106–116° E), southern China (SC: 20–27° N, 106–120° E), southwestern China (SWC1: 27–36° N, 77–106° E; SWC2: 22–27° N, 98–106° E) and northwestern China (NWC: 36–46° N, 75–111° E), as outlined in Fig. 1 by the red rectangles.

Detection and attribution analysis must usually be conducted within relatively small dimensions, due to a lack of sample data for the estimation of noise covariance, or because the estimated noise covariance matrix is not 'full rank' if the dimension is too large. We used 2 approaches to obtain small dimensions from annual mean temperatures: (1) we used multi-year non-overlapping means to reduce the time dimension; and (2) we further reduced the dimension by projecting the space–time series onto leading empirical orthogonal functions (EOFs) of the model-simulated natural variability. The use of shorter time averages may provide a better chance to detect climate response to short-time period forcings such as volcanic activities. However, the use of shorter time averages also requires estimation of a larger covariance matrix, which results in larger estimation error with the limited data sample. According to Wen et al. (2013), if an analysis is conducted on different multi-year non-overlapping mean series, the detection results are not sensitive to the use of time averaging. For this reason, we based our results on the analysis of a 9 yr mean series. The number of

EOFs retained was determined by a residual consistency check (Allen & Tett 1999). Further details on the data processing are described in the next section.

3.2.2. Observations and model data processing

The original resolution of CN05.1 is 0.25° latitude × 0.25° longitude, and was re-grid to the 1° × 1° grids in the present study using the interpolation method of conservative remapping in order to facilitate comparison with model simulations. Observed daily temperatures were averaged to obtain monthly values over China. Monthly anomalies relative to the base period of 1961–1990 were then computed for each grid. Finally, regional mean series were computed based on available grid values weighted by the grid box area for the 8 sub-regions. The 9 yr non-overlapping averages of regional annual series were used for subsequent detection and attribution analyses.

Due to the different spatial resolution of multi-model data, we re-grid all model data into the common 1° latitude × 1°longitude grid. Also, prior to

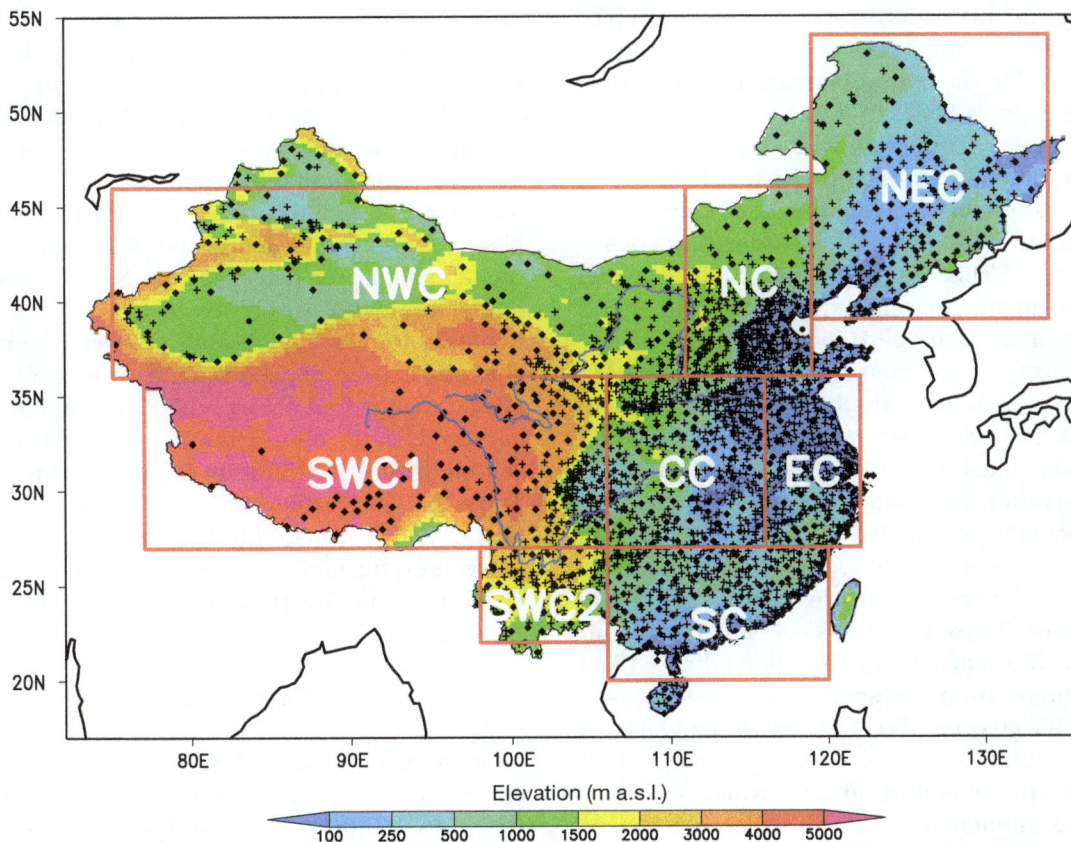

Fig. 1. Topography of China showing the spatial distribution of 2416 observation stations (dots: major stations; crosses: other stations). Red rectangles identify the 8 sub-regions used in this study: NEC, northeastern China; NC, north China; EC, eastern China; CC, central China; SC, southern China; SWC1 and SWC2, southwestern China; NWC, northwestern China

analysis, we masked them to the same space–time coverage as the observations. Signals were estimated as multiple model ensemble means by first computing individual model ensemble means and then averaging the utilized models. These included signals for the effects of ANT and NAT and their combined effects, as well GHG, AA, and LU.

Many different methods have been employed to estimate the climate response to anthropogenic forcing. One method is to compute multi-model ensemble means for simulations under ANT forcing. The resulting signal is referred to as ANT1. Another method is to use the difference between the ALL and NAT signals (i.e. ANT = ALL − NAT). The resulting signal is referred to as ANT2. The third method is to accumulate responses from individual anthropogenic forcings, including GHG, AA, and LU. The resulting signal is referred to as ANT3. Note that the latter 2 estimates assume that climate response to various forcing agents may be additive (e.g. Cubasch et al. 2001, Meehl et al. 2004). This assumption has been validated for temperature at the global scale (Meehl et al. 2004), although it has not been examined over China. Each method has its own advantages and disadvantages. In this paper, we only used the ANT1 (multi-model ensemble mean under ANT forcing) as the estimate for the anthropogenic forcing signal (hereinafter referred to as 'ANT').

Model data were further processed to produce data for noise estimation as detailed below. In order to make the best use of the available data, we included inter-ensemble differences when estimating covariance structure. This was done by dividing model simulations under different external forcings into 2 chunks: for 1961–2005 and 1901–1945. Respective ensemble means were then removed. To make the noise data comparable with observational coverage, simulations for 1901–1945 were masked with the observational data for 1961–2005, with the year 1961 in the observation matching the year 1901 in the simulation. Secondly, we divided the 13 008 yr CTL from the 20 models into non-overlapping, 45 yr chunks and obtained 2 independent samples, each having 130 chunks of data. These 130 chunks of data, along with those from different forcing runs (including 118 of the 45 yr chunks from ALL simulations, 42 from NAT, 48 from GHG, 40 from ANT, 15 from LU, and 33 from AA), made up 2 sets of 426 45-yr model output chunks, plus the preindustrial CTL, which are available for the estimation of natural variability (noise1 and noise2). The 45 yr chunks of noise data were split into 2 independent sets, with one set used for optimization and the other for testing.

4. RESULTS

4.1. Observations and simulations

Fig. 2 shows linear trends in annual mean temperatures for the observed trend as well as for model simulated temperature responses to different forcings. An increase in the observed mean temperature occurred almost everywhere in China. Increases in northeastern and western China were much stronger than in other regions. Some cooling trends occurred in the Sichuan Basin area (Fig. 2a). These findings are consistent with results reported in previous studies (e.g. Ren et al. 2012). Mean temperature responses to ALL, GHG, ANT, and NAT were characterized by warming trends. NAT (Fig. 2c) showed the weakest trend; GHG (Fig. 2d) exhibited the strongest trend. ALL (Fig. 2b) and ANT (Fig. 2g) trends were of similar magnitudes, which were generally weaker than the observed trend. In general, the response to LU forcing (Fig. 2e) exhibited a negative trend, with a similar negative trend appearing in most areas of China, especially on the Tibetan Plateau.

Fig. 3 displays a time series of annual mean temperatures averaged over China, expressed as temperature anomalies relative to the 1961–1990 mean for both observations and model simulations. Visual inspection suggests a good match in the long-term changes in annual mean temperature between the observed trend and the responses to ALL, or between observations and responses to GHG. However, the observed trend was clearly above the upper range of model simulated responses to NAT. Additionally, model simulated responses to LU and AA showed negative trends. Clearly, the observed temperature changes can be explained by a response to ALL or GHG forcings, but not by NAT, LU or AA alone. Linear trends in the observations and in multi-model ensemble means for different forcings are provided in Table 2. The results indicate that trends of ALL, GHG, ANT were close to that of the observed trend (especially for GHG), while the trend of NAT was very weak, and the trend of LU and AA was negative.

4.2. One-signal analyses

Fig. 4a shows the best estimates of scaling factors and their 90% confidence intervals for ALL, GHG, ANT, NAT, and LU based on 1-signal optimal analyses for the annual mean temperature. The detection results are not sensitive to the number of EOFs retained. The ALL, GHG and ANT signals were ro-

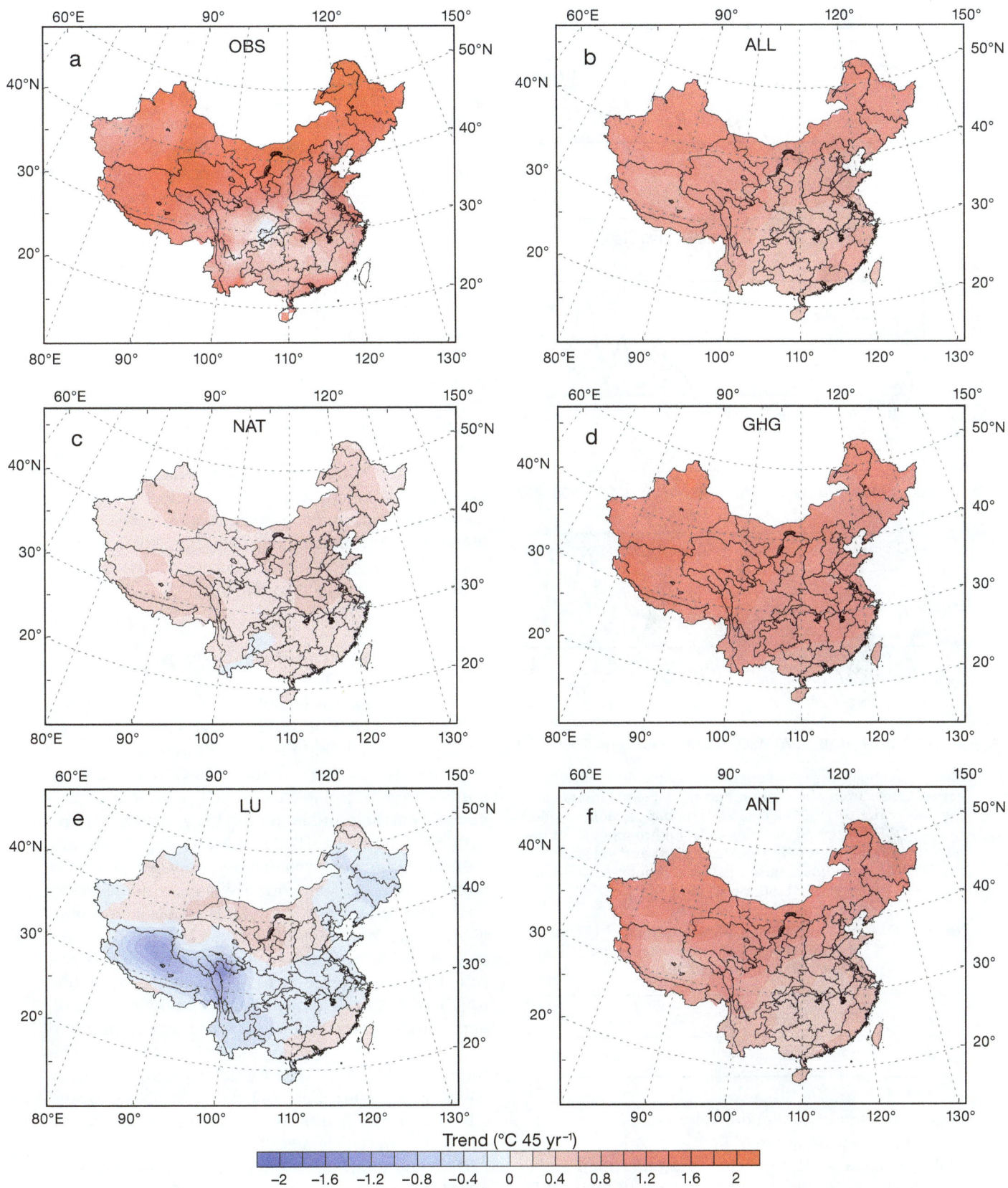

Fig. 2. Estimated trend of annual mean temperatures (°C change over 45 yr) over China under different forcings during the period 1961–2005. OBS: observed trend; ALL: includes anthropogenic and natural external forcings; NAT: includes only solar irradiance and volcanic activity; GHG: greenhouse gases; LU: land use change; ANT: anthropogenic influences

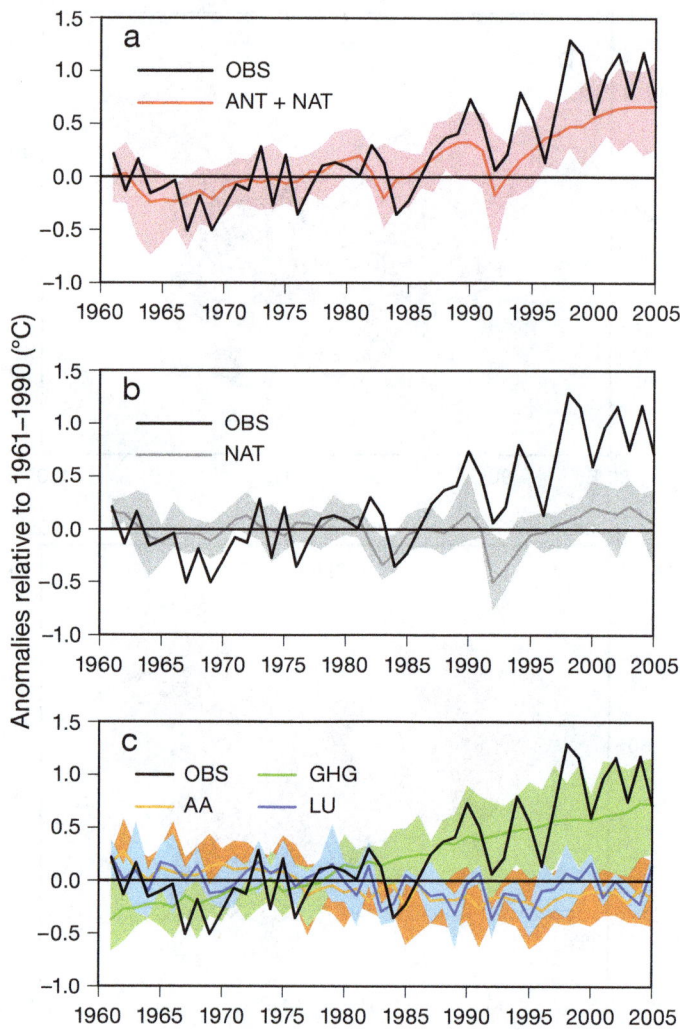

Fig. 3. Annual mean temperature anomalies averaged across China from observation and model simulations during the period of 1961–2005. Shaded bands: multi-model range. OBS: observed trend; ANT: anthropogenic influences; NAT: includes only solar irradiance and volcanic activity; GHG: greenhouse gases; AA: anthropogenic aerosols; LU: land use change

Table 2. Observed and model simulated trends (°C change over 10 yr) of annual mean temperature over China. See Fig. 2 for abbreviations

OBS	ALL	GHG	NAT	LU	AA	ANT
0.28	0.18	0.25	0.09	−0.04	−0.08	0.17

bustly detected in annual mean temperature individually across a wide range of EOF truncations as their scaling factors were significantly greater than zero. Results shown in Fig. 4a are based on 40 EOFs 'full rank', which passed a residual consistency test. Model simulated responses to the effect of combined

anthropogenic and natural forcing were smaller (though not significantly) than the observed change. The NAT-only and LU signals were not detected in the observation. This suggests that GHG, or ANT including GHG, or ALL including ANT alone can explain observed temperature changes, but NAT forcing or LU forcing alone cannot. Changes in land cover can have substantial impacts on extreme temperatures due to land–atmosphere interactions (Seneviratne et al. 2006). Christidis et al. (2013) showed that land use changes may have a cooling effect on temperature extremes at a global scale (especially on extremely warm days) due to the increase of albedo accompanied with deforestation. Wen et al. (2013) also found a detectable effect of LU on annual maximum daily temperatures, suggesting that the impact of land use changes on extremely warm days might be detectable even at a regional scale. However, the LU effect is hard to detect in annual mean temperature changes on a global scale (Christidis et al. 2013). Although there is a cooling effect of the mean temperature response to LU over China as simulated by multi-models (Fig. 2e), the signal was not detectable in our research. Consequently, we conclude that LU is not a dominant forcing.

4.3. Two-signal analyses

A 2-signal analysis was also carried out with different combinations of forcing responses, including ANT (individual anthropogenic forcing) and NAT (only natural forcing), GHG (only GHG) and other anthropogenic forcing (ANT-GHG, including anthropogenic aerosols and land use), LU and other anthropogenic forcing (ANT-LU, representing GHG and anthropogenic aerosols). These analyses shed more light on the relative importance of individual forcing.

Fig. 5 displays 90% confidence regions and marginal confidence intervals for the 2-signal detection analysis using ANT and NAT. The origin (0, 0) is outside the 90% confidence region, suggesting that ANT and NAT are jointly detected at the 90% confidence level. The marginal 90% confidence intervals for ANT are above 0, but for NAT, include 0. This suggests that ANT is clearly detected in the annual mean temperature but NAT is not. It also indicates that the effects of ANT on annual mean temperatures can be separated from NAT.

The magnitude of scaling factors and their 90% marginal confidence intervals for ANT are also comparable to those from the 1-signal analysis. The 2-signal analysis results are also very robust to differ-

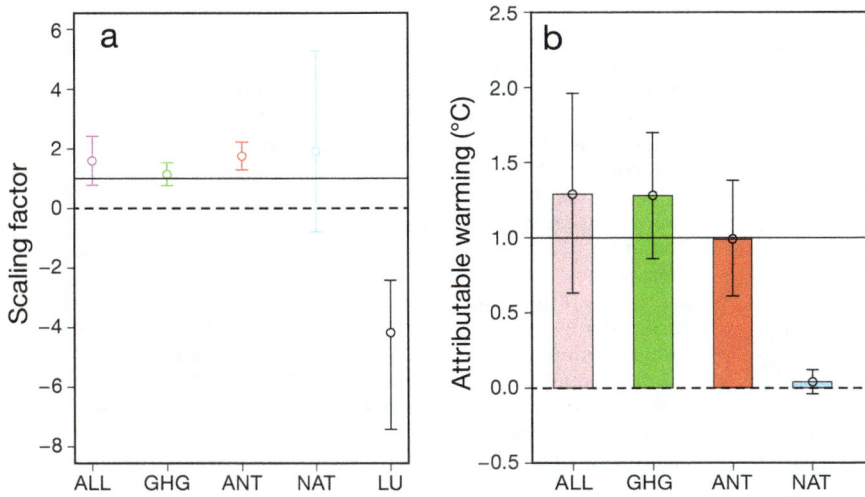

Fig. 4. (a) Scaling factors and their 5–95 % uncertainty ranges for anthropogenic and natural external forcings (ALL); greenhouse gases (GHG); anthropogenic influences (ANT); solar irradiance and volcanic activity (NAT); and land use changes (LU) from 1-signal; (b) attributable warming and their 5–95 % uncertainty ranges from 1-signal (ALL and GHG) and 2-signal analysis (ANT and NAT)

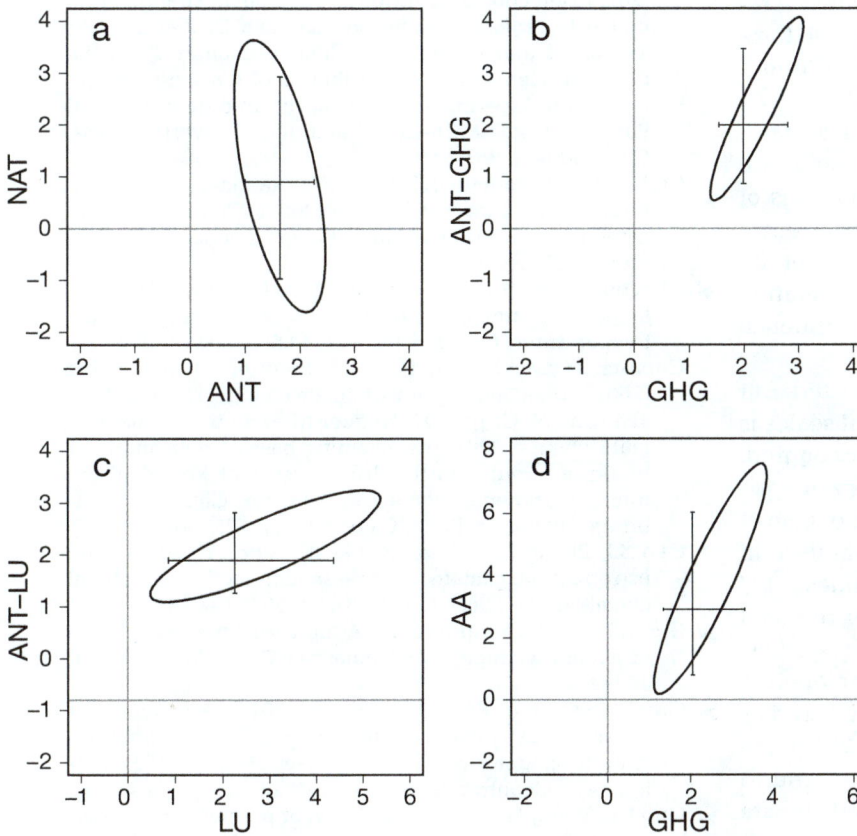

Fig. 5. Scaling factors, their 90 % joint confidence regions (bars), and marginal 90 % confidence intervals (ellipses) obtained from 2-signal analyses of (a) anthropogenic (ANT) and natural (NAT) forcings, (b) greenhouse gas emissions (GHG) and the combination of ANT-GHG, (c) land use (LU) and the combination of ANT-LU, and (d) GHG and anthropogenic aerosols (AA)

ent EOF truncations. ALL and ANT were clearly detectable but NAT was clearly not detectable in the 1-signal analyses. That ANT and NAT can be jointly detected, and the effects of ANT can be separated from those of NAT in the 2-signal analyses suggests that only anthropogenic forcing can explain the observed annual mean temperature changes in China from 1961 to 2005.

A 2-signal analysis conducted on a combination of individual anthropogenic forcings suggests that the effect of GHG could be separated from those of other anthropogenic forcings for annual mean temperature. Analysis for annual mean temperature conducted with GHG against ANT-GHG (Fig. 5b) suggests that the effect of GHG could be separated from the combined effect of anthropogenic aerosols and land use changes. The uncertainty range of the ANT-GHG scaling factor was much larger than that for GHG. A comparison of LU against ANT-LU (Fig. 5c) suggests that the effect of land use change could be separated from the combined effect of GHG and anthropogenic aerosol, but the effect of LU was not detected in the 1-signal analysis.

The 2-signal analysis for AA and GHG indicates that the effect of GHG can be separated from that of AA, and that the separate detection of AA from GHG is very robust (Fig. 5d). Analysis for GHG and LU failed to separate the GHG contribution from that of LU, suggesting that simulated responses to GHG and LU may be highly correlated (not shown).

Fig. 4b shows the attributable warming of different forcings from 1- and 2-signal detection. Attributable warming was estimated as the linear least-square trend of the relevant time series multiplied by the corresponding scaling factors. From the 1-signal detection results, the attributable warming of ALL and GHG to the mean temperature of the observed trend was 1.29°C (90% CI 0.63~1.96°C) and 1.28°C (90% CI 0.86~1.70°C), respectively (Fig. 4b). The warming attributable to ANT was 0.99°C (90% CI 0.61~1.38°C), while the contribution of NAT was 0.04°C (90% CI –0.04~0.12°C) from the 2-signal analysis (Fig. 4b).

5. DISCUSSION AND CONCLUSIONS

Using an optimal detection technique and CMIP5 multi-model simulations, we have shown that the effect of anthropogenic forcing on climate change is detectable in China. The combined effect of greenhouse gas emissions and sulfate aerosol forcing is clearly detectable in the observed annual mean temperature change over China from 1961–2005. The effects of ANT and NAT are separately detectable, and the effect of GHG can be separated from other anthropogenic forcings for annual mean temperature. The effect of LU failed to separate. The climate response to greenhouse gas forcing can be clearly and robustly identified in the annual mean temperature, and results show that only anthropogenic forcing can explain the observed changes in China's mean temperature from 1961 to 2005.

Our detection results were based on the most state-of-the-art multiple CMIP5 models. Our results are consistent with, and further confirm the findings of Stott (2003) — who used the single Hadley Centre Coupled Model (HadCM3), and of Zhang et al. (2006), who used 2 versions of 2 general circulation models (GCMs) — that the anthropogenic influence on warming can be detected in Asia.

In this study, the effect of LU was not detected. But the influence of LU at the regional and local scales is likely not negligible, and should not be discounted, in particular over China with its long history of agricultural activities and its high population (e.g. Gao et al. 2007). Perhaps the global LU forcing data used in our simulation did not fully reflect Chinese LU changes, due to a lack of sufficient local or regional details. Future regional climate detection studies, employing the LU and aerosol datasets developed in China based on both satellite and field based studies (e.g. Liu et al. 2003, Zhang et al. 2012, 2013) are needed in order to increase our knowledge and to better understand the attribution of observed climate change in this region.

Acknowledgements. We acknowledge the World Climate Research Program's Working Group on Coupled Modeling, which is responsible for CMIP, and thank climate modeling groups for producing and making model output available. For CMIP, the US Department of Energy's Program for Climate Model Diagnosis and Intercomparison provided coordinating support and led the development of software infrastructure in partnership with the Global Organization for Earth System Science Portals. This research is a contribution to a collaborative project of the Joint Working Group XIII between the Meteorological Service of Canada and the China Meteorological Administration. We are grateful to X. Zhang, H. Wan and Q. H. Wen of Environment Canada for their support and assistance in this research. We are grateful to Dr. Y. X. Zhang for assistance with English. This research received support from R&D Special Fund for Public Welfare Industry (meteorology) (GYHY201306019), National Natural Science Foundation of China (41275078), Grant Projects of China Clean Development Mechanism Fund (121312), and Climate Change Foundation of China Meteorological Administration (CCSF201339).

LITERATURE CITED

▶ Allen MR, Stott PA (2003) Estimating signal amplitudes in optimal fingerprinting. I. Theory. Clim Dyn 21:477–491

▶ Allen MR, Tett SFB (1999) Checking for model consistency in optimal fingerprinting. Clim Dyn 15:419–434

Bindoff NL, Stott PA, AchutaRao KM, Allen MR and others (2013) Detection and attribution of climate change: from global to regional. In: Stocker TF, Qin D, Plattner GK, Tignor M and others (eds) Climate change 2013: the physical science basis. Contribution of Working Group I to the 5th Assessment Report of the Intergovernmental Panel on Climate Change. Cambridge University Press, Cambridge, p 867–952

CCNARCC (Committee for China's National Assessment Report on Climate Change) (2007) China's national assessment report on climate change. Science Press, Beijing (in Chinese)

▶ Christidis N, Stott PA, Hegerl GC, Betts R (2013) The role of land use change in the recent warming of daily extreme temperatures. Geophys Res Lett 40:589–594

Cubasch U, Meehl GA, Boer GJ, Stouffer RJ and others (2001) Projections of future climate change. In: Houghton JH, Ding Y, Griggs DJ, Noguer M and others (eds) Climate change 2001: the scientific basis. Contribution of Working Group I to the 3rd Assessment Report of the Intergovernmental Panel on Climate Change. Cambridge University Press, Cambridge, p 525–582

Gao XJ, Zhang DF, Chen ZX, Pal JS, Giorgi F (2007) Land use effects on climate in China as simulated by a regional climate model. Sci China Earth Sci 50:620–628

▶ Gao XJ, Shi Y, Giorgi F (2011) A high resolution simulation of climate change over China. Sci China Earth Sci 54: 462–472

▶ Gareth SJ, Stott PA, Christidis N (2013) Attribution of observed historical near-surface temperature variations to anthropogenic and natural causes using CMIP5 simulations. J Geophys Res 118:4001–4024

▶ Guo DL, Wang HJ (2013) Simulation of permafrost and seasonally frozen ground conditions on the Tibetan Plateau, 1981–2010. J Geophys Res Atmos 118:5216–5230

Hegerl GC, Zwiers FW, Braconnot P, Gillett NP and others (2007) Understanding and attributing climate change. In: Solomon S, Qin D, Manning M, Chen Z and others (eds) Climate change 2007: the physical science basis. Contribution of Working Group I to the 4th Assessment Report of the Intergovernmental Panel on Climate Change. Cambridge University Press, Cambridge, p 665–745

Hegerl GC, Zwiers F (2011) Use of models in detection and attribution of climate change. WIREs Clim Change 2: 570–591

IPCC (2007) Climate change 2007: the physical science basis. In: Solomon S, Qin D, Manning M, Chen Z and others (eds) Contribution of Working Group I to the 4th Assessment Report of the Intergovernmental Panel on Climate Change. Cambridge University Press, Cambridge

IPCC (2013) Climate change 2013: the physical science basis. In: Stocker TF, Qin D, Plattner GK, Tignor M and others (eds) Contribution of Working Group I to the 5th Assessment Report of the Intergovernmental Panel on Climate Change. Cambridge University Press, Cambridge

Li Z, Yan ZW, Wu HY (2015) Updated homogenized Chinese temperature series with physical consistency. Atmos Ocean Sci Lett 8:17–22

Liu JY, Liu ML, Zhuang DF, Zhang ZX, Deng XZ (2003) Study on spatial pattern of land-use change in China during 1995–2000. Sci China Earth Sci 46:373–384

Meehl GA, Washington WM, Amman C, Arblaster JM, Wigley TML, Tebaldi C (2004) Combinations of natural and anthropogenic forcings and 20th century climate. J Clim 17:3721–3727

Ming Y, Ramaswamy V (2009) Nonlinear climate and hydrological responses to aerosol effect. J Clim 22:1329–1339

Qian C, Fu CB, Wu ZH (2011) Changes in the amplitude of the temperature annual cycle in China and their implication for climate change research. J Clim 24:5292–5302

Ren GY, Zhou YQ (2014) Urbanization effect on trends of extreme temperature indices of national stations over Mainland China, 1961–2008. J Clim 27:2340–2360

Ren GY, Ding YH, Zhao ZC, Zheng JW, Wu TW, Tang GL, Xu Y (2012) Recent progress in studies of climate change in China. Adv Atmos Sci 29:958–977

Seneviratne SI, Luthi D, Litschi M, Schar C (2006) Land atmosphere coupling and climate change in Europe. Nature 443:205–209

Stott PA (2003) Attribution of regional-scale temperature changes to anthropogenic and natural causes. Geophys Res Lett 30:1728, doi:10.1029/2003GL017324

Stott PA, Jones GS (2012) Observed 21st century temperatures further constrain likely rates of future warming. Atmos Sci Lett 13:151–156

Stott PA, Gillett NP, Hegerl GC, Karoly DJ, Stone DA, Zhang XB, Zwiers F (2010) Detection and attribution of climate change: a regional perspective. WIREs Clim Chang 1: 192–211

Sui Y, Lang XM, Jiang DB (2014) Time of emergence of climate signals over China under the RCP4.5 scenario. Clim Change 125:265–276

Sun Y, Zhang XB, Zwiers FW, Song L and others (2014) Rapid increase in the risk of extreme summer heat in Eastern China. Nat Clim Change 4:1082–1085

Tang GL, Ren GY (2005) Reanalysis of surface air temperature change of the last 100 years over China. Clim Environ Res 10:791–798 (in Chinese)

Taylor KE, Stouffer RJ, Meehl GA (2012) An overview of CMIP5 and the experiment design. Bull Am Meteorol Soc 93:485–498

Wang XL, Wen QH, Wu Y (2007) Penalized maximal t test for detecting undocumented mean change in climate data series. J Appl Meteorol Climatol 46:916–931

Wang T, Wang HJ, Otter OH, Gao YQ, Suo LL, Furevik T, Yu L (2013) Anthropogenic forcing of shift in precipitation in Eastern China in late 1970s. Atmos Chem Phys Discuss 13:11997–12032

Wen HQZ, Zhang XB, Xu Y, Wang B (2013) Detecting human influence on extreme temperatures in China. Geophys Res Lett 40:1171–1176

Wu J, Gao XJ (2013) A gridded daily observation dataset over China region and comparison with the other datasets. Chin J Geophys 56:1102–1111 (in Chinese with English abstract)

Wu J, Gao XJ, Giorgi F, Chen ZhH, Yu DF (2012) Climate effects of the Three Gorges Reservoir as simulated by a high resolution double nested regional climate model. Quart Int 282:27–36

Xu Y, Gao XJ, Shen Y, Xu CH, Shi Y, Giorgi F (2009) A daily temperature dataset over China and its application in validating a RCM simulation. Adv Atmos Sci 26: 763–772

Zhai P, Pan X (2003) Trends in temperature extremes during 1951–1999 in China. Geophys Res Lett 30:1913, doi: 10.1029/2003GL018004

Zhang XB, Zwiers FW, Stott PA (2006) Multimodel multisignal climate change detection at regional scale. J Clim 19: 4294–4307

Zhang XY, Wang YQ, Niu T, Zhang XC, Gong SL, Zhang YM, Sun JY (2012) Atmospheric aerosol compositions in China: spatial/temporal variability, chemical signature, regional haze distribution and comparisons with global aerosols. Atmos Chem Phys 12:779–799

Zhang XY, Sun JY, Wang YQ, Li WJ and others (2013) Factors contributing to haze and fog in China. Chin Sci Bull 58:1178–1187 (in Chinese)

Zhang DF, Gao XJ, Luo Y, Xia J, Giorgi F (2015) The attribution of anthropogenic forcings and natural variability on the observed climate change over China and its main river basins as simulated by RegCM4.0. Chin Sci Bull (in press)

Zhou TJ, Yu RC (2006) Twentieth century surface air temperature over China and the globe simulated by coupled climate models. J Clim 19:5843–5858

Zhou TJ, Ma SM, Zou LW (2014) Understanding a hot summer in central eastern China: summer 2013 in context of multi-model trend analysis. In: Herring SC, Hoerling MP, Peterson TC, Stott PA (eds) Explaining extreme events of 2013 from a climate perspective. Bull Am Meteorol Soc (Spec Suppl) 95:S54–S57

Zwiers FW, Zhang XB (2003) Toward regional-scale climate change detection. J Clim 16:793–797

Rapid urbanization effect on local climate: intercomparison of climate trends in Shenzhen and Hong Kong, 1968–2013

Lei Li[1], Pak Wai Chan[2,*], Deli Wang[1], Mingyan Tan[1]

[1]Shenzhen National Climate Observatory, Meteorological Bureau of Shenzhen Municipality, Shenzhen Key Laboratory of Severe Weather in South China, Shenzhen 518040, PR China
[2]Hong Kong Observatory, 998000 Kowloon, Hong Kong SAR

ABSTRACT: An intercomparison of climate trends between 2 adjacent large cities, Shenzhen and Hong Kong, in the past 46 yr (1968–2013) suggests that the rate of urbanization is clearly reflected in the rate of the change in local climate. Since becoming a special economic zone in 1980, Shenzhen has experienced a very rapid urbanization process. In only 30 yr, Shenzhen has transformed from a small town into a metropolis rivaling Hong Kong. The gross domestic product and urban built-up area in Shenzhen have increased rapidly. Climate data analysis over the last 46 yr showed a greater rate of change in climate in Shenzhen than in Hong Kong over the same period. In Shenzhen, average annual temperature warmed by $1.63 \pm 0.18°C$ (~$0.35 \pm 0.04°C$ per decade), average maximum temperature increased by $0.90 \pm 0.19°C$, average minimum temperature increased by $2.09 \pm 0.23°C$ and the diurnal temperature range (DTR) decreased by $1.18 \pm 0.23°C$. Over the same period in Hong Kong, average temperature increased by $0.47 \pm 0.20°C$ (~$0.10 \pm 0.04°C$ per decade), maximum temperature increased by $0.12 \pm 0.28°C$, minimum temperature increased by $0.55 \pm 0.20°C$ and DTR decreased by $0.43 \pm 0.27°C$. In addition, relative humidity in Shenzhen decreased by $13.13 \pm 1.78\%$ in the last 46 yr, while there were no significant changes in Hong Kong. Finally, data analysis showed that urbanization has no significant effect on total rainfall for both cities.

KEY WORDS: Rapid urbanization · Local climate · Climate change · Shenzhen · Hong Kong

1. INTRODUCTION

Knowledge of the urbanization effect on climate is essential in designing climate-friendly cities (Eliasson 2000, Masson et al. 2013). The study of the effect of urbanization on local climate can be traced back to the early 19th century. Howard (1833) reported the discovery of an urban heat island, where air temperature is higher in urban areas than in rural areas, in London. Howard's study was generally taken as the start of the study of the urban effect on local climate, in which air temperature is always the focus, and some studies show that the urban–rural air temperature difference can reach as high as 12°C in clear,

calm nights (Oke 1981). The air temperature difference between urban and rural areas can be attributed to the different thermal processes over different land surfaces; a comprehensive summary on the physical mechanism of the urban warming process can be found in Grimmond (2007). However, air temperature is not the only climate element influenced by urbanization. Because of changes in the physical properties of land surfaces (such as heat capacity, albedo and roughness) and the activities of humans (such as industrial processes and transportation of people and goods), wind characteristics and exchanges of energy and substances between the atmosphere and land surface in urban areas are quite

*Corresponding author: pwchan@hko.gov.hk

different from those in rural areas, which significantly influences the wind (i.e. Bornstein & Johnson 1977, Goldreich & Surridge 1988, Cheng & Hu 2005), humidity (i.e. Chandler 1967, Lee 1991) and precipitation (i.e. Oke 1987, Kaufmann et al. 2007) over urban areas.

In the last decade, China has become the focus of attention in the field of urbanization effect on climate (Li et al. 2004, Zhou et al. 2004, Kaufmann et al. 2007, Chen et al. 2011, Ge et al. 2013, Yu et al. 2013). Benefiting from the 'reform and opening' strategy implemented in the early 1980s, China's economy developed rapidly over the last 3 decades. Along with this rapid economic growth, the urbanization process in China was also rapid. Official statistics show that the urban population in China in 2011 was around 691 million, more than 3 times the figure in 1980 (Chinese Academy of Social Sciences 2011).

Among all of the cities in China, Shenzhen is especially notable because it was designated as the first special economic zone (SEZ) in China. Shenzhen is famous for its drastically rapid urbanization after establishment of the SEZ in 1980, and is possibly one of the fastest growing urban areas in the world. Over the last 35 yr, Shenzhen transformed from a small town to a megacity, its population increased by around 30 times and its gross domestic product (GDP) by more than 1000 times. At the same time, the landscape also experienced significant changes. Thus, a study on the impact of urbanization on local climate in Shenzhen will further enhance understanding in this field. Some previous studies (Zhou et al. 2004, Chen et al. 2011) of the urbanization effect on climate in China mentioned the rapid urbanization of Shenzhen but did not provide a detailed analysis of its climate trends. Hong Kong, another large city, is located directly adjacent to Shenzhen and has a quite different urbanization process from that of Shenzhen in the same period from the 1970s onwards; therefore, an intercomparison of climate trends between Shenzhen and Hong Kong would be especially helpful to study the impact of rapid urbanization on local climate.

2. BRIEF DESCRIPTION OF DATA

2.1. Meteorological data

Although there are currently numerous automatic weather stations in Shenzhen and Hong Kong, the meteorological data used herein were collected from the Shenzhen Caiwuwei Weather Station (CWW)

and Hong Kong Observatory (HKO), which have the longest continuous climate data records for each city. The locations of the 2 stations are shown in Fig. 1. The elevations above sea level for the CWW and HKO are 18.5 and 32 m, respectively, and all observations were performed at fixed heights during the period of study. According to the 'local climate zone' (LCZ) classification system by Stewart & Oke (2012), both stations are currently located in LCZ1 (compact high-rise).

Data ranged from 1968 to 2013 and included elements such as 6 h temperature (daily at 02:00, 08:00, 14:00 and 20:00 h local sidereal time), daily relative humidity and daily precipitation. Data collection at the CWW is in accordance with ground-based observation specifications issued by the China Meteorological Administration, and the HKO is in accordance with HKO operational standards. Data quality from both sources was carefully scrutinized.

Yearly average data during 1968–2013 were calculated from the aforementioned 6 h and daily data, which include annual average temperature, annual average maximum temperature (at 14:00 h LST), annual average minimum temperature (at 02:00 h LST), annual average diurnal temperature range (DTR, difference in temperatures at 02:00 and 14:00 h), annual average relative humidity and total annual rainfall.

2.2. Urbanization data

Population and GDP data from 1979 to 2012 were used to quantify the urbanization process in Shenzhen and Hong Kong and were sourced from the Statistics

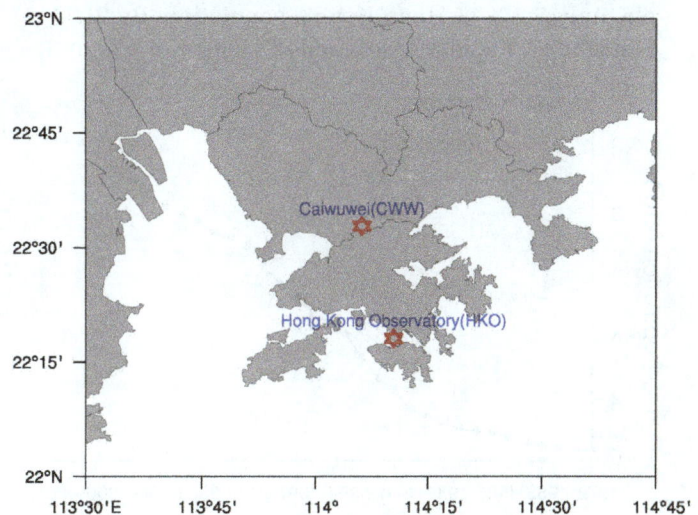

Fig. 1 Locations of the 2 meteorological stations in Hong Kong and Shenzen

Bureau of Shenzhen Municipality and the Hong Kong Census and Statistics Department, respectively. This data range was chosen because the Shenzhen SEZ was established in 1980, and 1979 was selected as the base year for urbanization data; also, official urban development statistics in Shenzhen only began in 1979. In addition to population and GDP, this study also collected data on the size of the built-up area in both cities for some of the years, and data were also sourced from the Urban Planning and Design Institute of Shenzhen and the Hong Kong Census and Statistics Department. To further identify the urbanization process, Landsat satellite data in 1986, 1995 and 2005 were collected to analyze vegetation degradation and urban expansion in Shenzhen.

3. URBAN PROCESSES IN THE LAST 33 YEARS

At the end of 2012, Shenzhen's administrative area was 1996.85 km² and Hong Kong's was 1108 km². In this section, data on population growth, GDP and built-up area of the 2 cities are used to demonstrate the different courses of development in the past 33 yr.

3.1. Population

Fig. 2 shows the population growth in Shenzhen and Hong Kong between 1979 and 2012. As shown, Hong Kong's population variation was relatively flat and only grew from 5.02 million to 7.17 million in 33 yr, a 1.12% compound annual growth rate (CAGR). In contrast, Shenzhen experienced a population explosion; in 1979, there were 0.37 million people, only 7.3% of Hong Kong's population. By 2000, Shenzhen's population exceeded Hong Kong's, and

by 2012, official data showed 10.54 million people, 1.47 times Hong Kong's population. Shenzhen's population grew at 11.03% CAGR in these 33 yr. Even more astonishing is that some unofficial statistics considered 10.54 million a serious underestimate, and if migrant workers are taken into account, Shenzhen's actual population might exceed 14 million.

3.2. GDP

Gross domestic product (GDP) is another indicator used to reflect the pace of city development, and Fig. 3 shows the GDP trends of both cities between 1979 and 2012. Both cities experienced rapid growth over the past 33 yr. Hong Kong's GDP grew from 112.7 billion Hong Kong dollars (HKD) in 1979 to HKD 2 trillion in 2012, a CAGR of 9.47%; Shenzhen's GDP grew from 196 million Chinese yuan (CNY) to CNY 1.3 trillion over the same period, a CAGR of 31.64%. Even as Hong Kong went into a recession and GDP declined after the Asian financial crisis in 1997, Shenzhen was unaffected and maintained rapid growth in GDP. The GDP data illustrated here have not been adjusted for inflation because of the limited data available for Shenzhen. However, the comparison of nominal GDP data is still meaningful, since they reflect the continuous development of economies in the 2 cities year by year.

3.3. Built-up area

Compared to population and GDP data, consistent and continuous data on the size of built-up area is more difficult to obtain. Table 1 lists the size of built-

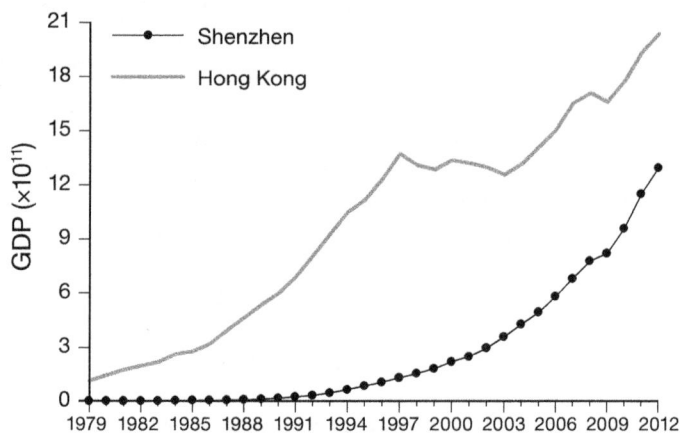

Fig. 2. Population during 1979–2012 in Shenzhen and Hong Kong

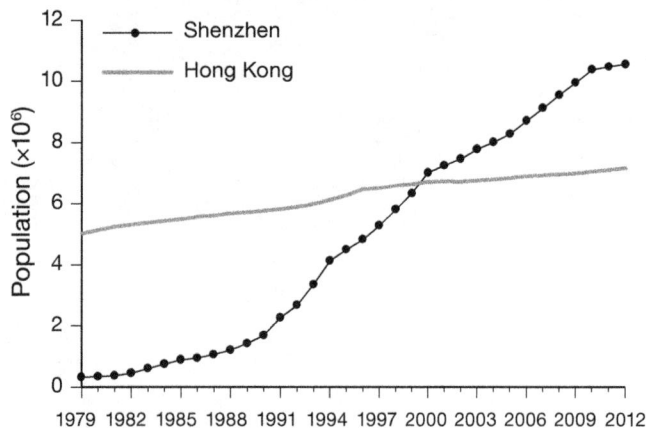

Fig. 3. Gross domestic product (GDP) during 1979–2012 in Shenzhen (in Chinese yuan) and Hong Kong (in Hong Kong dollars)

Table 1. Built-up area (any land on which buildings, non-building structures, roads and/or squares are present) of Shenzhen and Hong Kong in different years (km^2)

Year	Shenzhen	Hong Kong
1980	12.46	166.58
1994	375.05	213.44[a]
2005	629.01	264.00
2012	862.80	265.00

[a]No data for 1994, therefore 1993 data used for comparison

up areas for both cities in 4 separate years. The trend for urban built-up area was relatively flat in Hong Kong, while Shenzhen's growth was very rapid.

In addition to statistical data, urbanization and changes in vegetation can also be described by the normalized difference vegetation index (NDVI) retrieved from satellite data (Lenney et al. 1996, Chen et al. 2006, Stellmes et al. 2013). Fig. 4 shows the change in NDVI over Shenzhen in 3 different years and depicts the evolution of vegetation degradation and urban expansion in Shenzhen. The NVDI data in Fig. 4 were retrieved from Landsat thematic mapper data, in which a negative value indicates ground covered by clouds or water, a larger number represents greater green coverage and a value of 0 to 0.2 denotes bare soil and urban land. Since cloud cover is more intense over southern China, it is difficult to locate satellite images in which the entire Shenzhen area had clear sky; thus, the images in Fig. 4 are made up of images from 2 temporally close days in the same year. For instance, the 1986 image combined those from November 3 and 28, 1986; the 1995 image combined those from December 7 and 30, 1995; and the 2005 image combined those from November 16 and 23, 2005.

Fig. 4 clearly shows the process of urban expansion in Shenzhen over the years. In 1985, when the SEZ had been established for just 5 yr, only the southwestern coastal areas had a dense built-up area, whereas northwestern and northeastern areas were still in their early developmental stages and therefore had relatively higher vegetation coverage. By 1995, the urban built-up area expanded greatly, and northeastern inland and northwestern coastal areas were gradually transformed into visible built-up areas. By 2005, the built-up area had further expanded in northwestern Shenzhen. Throughout the process of urban expansion, not only was vegetation cover reduced but reclamation along the western coastline had transformed part of the ocean into construction sites.

The evolution of land use has significantly changed the surroundings of the 2 weather stations. Based on the LCZ classification system, during the last 35 yr, the HKO had moved from LCZ2 (compact mid-rise) to LCZ1 (compact high-rise), while the CWW of Shenzhen had moved from LCZ6 (open low-rise) to LCZ1.

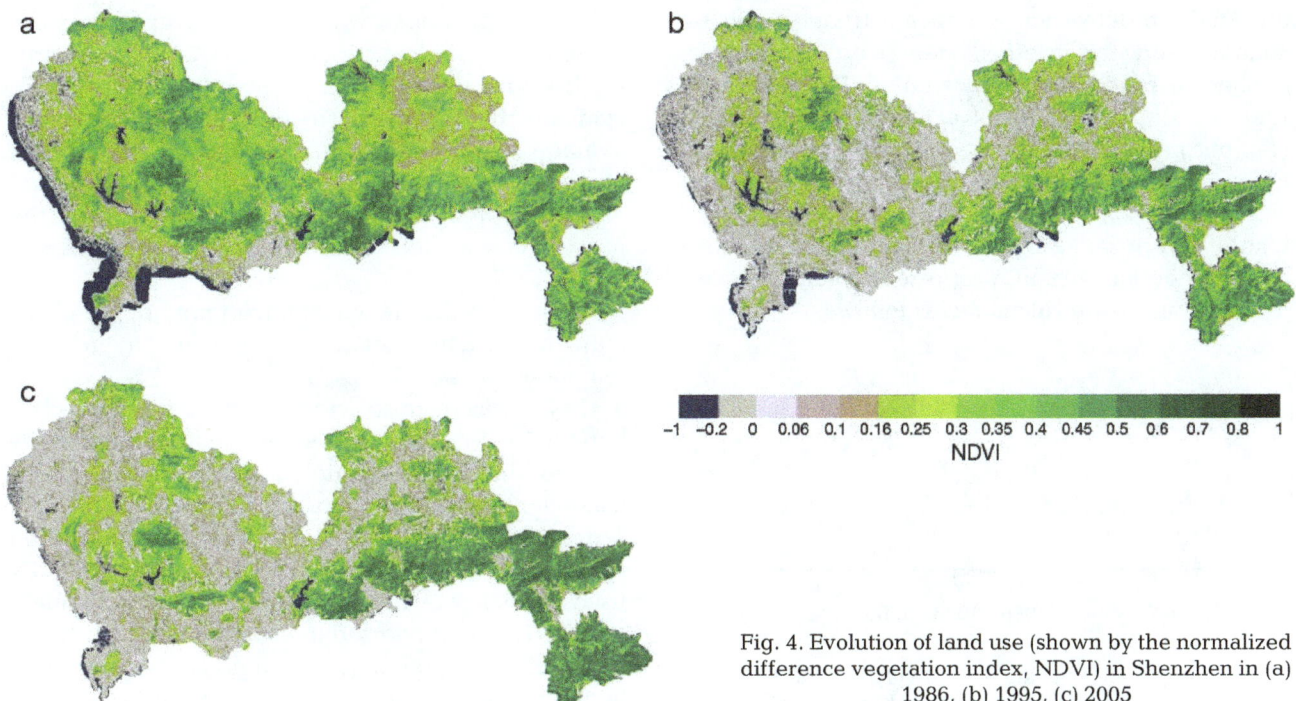

Fig. 4. Evolution of land use (shown by the normalized difference vegetation index, NDVI) in Shenzhen in (a) 1986, (b) 1995, (c) 2005

In terms of the changes of economic, population and built-up area, Shenzhen and Hong Kong experienced very different processes of urbanization in the last 33 yr. In 1979, Hong Kong was already a developed economy and was known as one of the 4 Asian Tigers, whereas Shenzhen was an underdeveloped small town, with an economy and population that were incomparable with those of Hong Kong. The built-up area in Shenzhen in 1979 was less than 10% of Hong Kong's and accounted for less than 1% of the entire Shenzhen administrative area. By 2012, although Hong Kong remained one of the most dynamic economies in the world, Shenzhen's population and built-up area had already exceeded that of Hong Kong, and its economy had reached a level comparable with that of Hong Kong.

4. INTERCOMPARISON OF CLIMATE TRENDS

4.1. Average air temperature (T_a)

Variations in average annual T_a in Shenzhen and Hong Kong are illustrated in Fig. 5. During 1968–1978, Shenzhen's average T_a was always lower than Hong Kong's by about 1°C. From 1978 onwards, the gap between average T_a in these 2 cities began to shrink and became very similar by 1990. After 2006, Shenzhen's average T_a regularly exceeded Hong Kong's by 0.1 to 0.4°C.

Linear least-squares regression (LLSR) was used in this study to detect any significant trend in climate change, and this method has been successfully applied in several previous studies (Li et al. 2004, Zhou et al. 2004, Chen et al. 2011). According to Hamburg (1977), the LLSR equation is expressed as:

$$Y = a + bX \quad (1)$$

where Y represents meteorological elements (such as air temperature), and X represents year. The coefficients a and b are calculated as follows:

$$a = \bar{y} - bX \quad (2)$$

$$b = \frac{\sum_{i=1}^{n} x_i y_i - n\bar{x}\bar{y}}{\sum_{i=1}^{n} x_i^2 - n\bar{x}^2} \quad (3)$$

where $\bar{y}$ and $\bar{x}$ are the average value of samples x_i and y_i, and n is the sample size.

To evaluate the confidence level (CL) from the linear trend derived from the LLSR method, the correlation coefficient of the linear regression is calculated as follows:

$$r = \frac{\sum_{i=1}^{n} (x_i - \bar{x})(y_i - \bar{y})}{\sqrt{(\sum_{i=1}^{n} (x_i - \bar{x})^2)(\sum_{i=1}^{n} (y_i - \bar{y})^2)}} \quad (4)$$

The value of r^2 represents the proportion of variance in Y in Eq. (1) and is called the sample coefficient of determination. Then, the test statistic, t, can be calculated as follows:

$$t(r,n) = \frac{|r|\sqrt{n-2}}{\sqrt{1-r^2}} \quad (5)$$

where, $t(r,n)$ is the point on the Student's t distribution for the degrees of freedom $(n-2)$ and correlation coefficient r. Subsequently, the CL (= $1 - \alpha$) is calculated as follows:

$$1 - \alpha = \int_{-t(r,n)}^{t(r,n)} f(t,n)\,dt \quad (6)$$

where $f(t,n)$ is the probability density function of the test, and α is the significance level.

For T_a, the linear warming trend in both cities passed the significance test at 0.01 CL; thus, warming in both cities was significant in the past 46 yr. To estimate the magnitude of warming in both cities, average temperatures were calculated for the first and last 10 yr of the time period, $T_{1968-1977}$ and $T_{2004-2013}$. Then, the magnitude of warming over the past 46 yr was determined by subtracting $T_{1968-1977}$ from $T_{2004-2013}$.

To estimate the statistical uncertainty range of climate trends, a time series of 10 yr moving average air temperature, say $T_{1968-1977}$, $T_{1969-1978}$, $T_{1970-1979}$, ..., $T_{2004-2013}$, was reconstructed. A new regression equation for 10 yr moving average air temperature can be derived, and the uncertainty range of the regression coefficients, namely, b, in Eq. (3) for the new time series at 0.01 CL can be calculated by using a method introduced by Lane (2014), which was consequently used to estimate the uncertainty range of climate trends. The method used to estimate the uncertainty range of the air temperature trend was also applied to other meteorological elements hereafter.

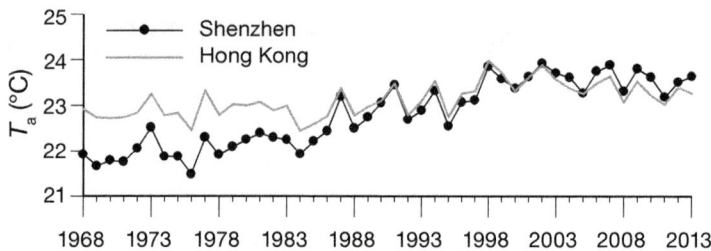

Fig. 5. Yearly average air temperature (T_a) during 1968–2013 in Shenzhen and Hong Kong

Considering the uncertainty range, the air temperature in Hong Kong rose by 0.47 ± 0.20°C in the past 46 yr at a rate of 0.10 ± 0.04°C per decade, while temperature in Shenzhen rose by 1.63 ± 0.18°C at a rate of 0.35 ± 0.04°C per decade. The increased value of Shenzhen's air temperature is larger than the global average value provided in the Fifth Assessment Report of the IPCC, which says that the average surface air temperature rose by around 0.85°C, with an uncertainty range of around 0.2°C, from 1880 to 2012 (IPCC 2013). Furthermore, in a recent study on air temperature series in China, Wang et al. (2014) estimated that the average temperature trend in the period 1956–2005 is around 0.252°C per decade, with an uncertainty range of around 0.095°C per decade. Thus, it can be concluded that Shenzhen's warming rate is also faster than the average warming rate of China.

In addition to calculating the rate of warming, estimating the contribution of urbanization on warming is also important. Chan et al. (2012) used Reanalysis 1 data from the US National Centers for Environment Prediction-National Center for Atmospheric Research (NCEP-NCAR), and meteorological observations and sounding data from HKO stations, leading to estimates that urbanization contributed roughly 42.3 % of warming in Hong Kong during 1971–2010. Therefore, using this conclusion, the contribution of 'background factors' other than urbanization (i.e. global warming) for Hong Kong's warming rate is 100.0 % minus 42.3 %, i.e. 57.7 %. This study assumes that the contribution to warming by the background factors calculated in Chan et al. (2012) is also applicable during 1968–2013, and this assumption is reasonable given that the time periods between the studies are similar. Thus, during 1968–2013, background factors contributed approximately 0.47°C multiplied by 57.7 %, or 0.27°C, to warming in Hong Kong. Given that Shenzhen and Hong Kong are neighbors and that the straight-line distance between the CWW and HKO is only around 30 km, it can be assumed that background factors also contributed 0.27°C to warming in Shenzhen. Therefore, contribution by urbanization to warming in Shenzhen is 1.63 ± 0.18°C minus 0.27°C, or about 1.36 ± 0.18°C, with a contribution rate of 81.4 to 85.1 %, which is quite large, and shows that urbanization is the most important factor driving local warming in Shenzhen.

4.2. Maximum air temperature, minimum air temperature and diurnal temperature range

Fig. 6a shows the trends for maximum temperature (Tmax). Over the past 46 yr, Tmax in Shenzhen also exhibited a clear uptrend; Tmax rose by 0.90 ± 0.19°C, and the linear regression on this warming trend passed the significance test at 0.01 CL. In contrast, trends for Tmax in Hong Kong are not obvious; Tmax rose by only 0.12 ± 0.28°C, and its linear regression failed the significance test at 0.01 CL.

Fig. 6b shows the trends for minimum temperature (Tmin). Over the past 46 yr, Tmin in Shenzhen rose more than Tmax; Tmin rose by 2.09 ± 0.23°C, and the linear regression on this warming trend passed the significance test at 0.01 CL. Similarly, Tmin in Hong Kong also exhibited a more obvious uptrend; Tmin rose by 0.55 ± 0.20°C over the past 46 yr. Despite the smaller magnitude of increase in Hong Kong than in Shenzhen, its linear regression also passed the significance test at 0.01 CL.

Fig. 6c shows the trends for average diurnal temperature range (DTR). A declining trend for average DTR is exhibited in both Shenzhen and Hong Kong over the past 46 yr, and their linear regressions

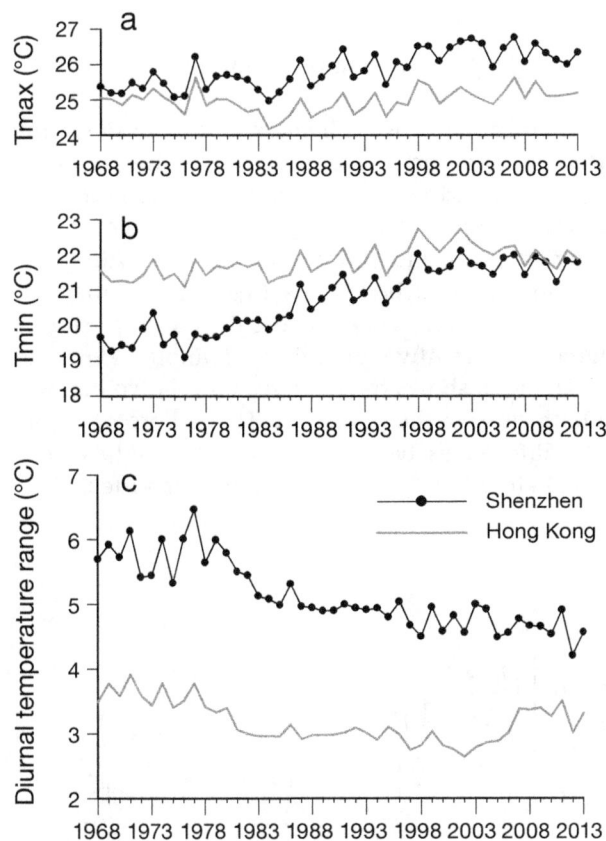

Fig. 6. Yearly average (a) maximum air temperature, (b) minimum air temperature and (c) diurnal temperature range during 1968–2013 in Shenzhen and Hong Kong

passed the significance test at 0.01 CL. However, Shenzhen saw a greater decrease in average DTR than Hong Kong; Shenzhen's DTR dropped by 1.18 ± 0.23°C during the period 1968–2013, while Hong Kong's average DTR dropped by 0.43 ± 0.27°C over the same period. Comparison between Figs. 5 & 6c shows that although annual average temperatures in both cities were similar since the late 1980s, DTRs between the two were still far apart. The average DTR was always lower in Hong Kong, indicating that the climate in Hong Kong exhibited greater oceanic characteristics than that in Shenzhen. Comparison of Figs. 5 & 6 shows that the declining trends of average DTRs in both cities were mainly contributed by rising minimum temperature, which was consistent with the conclusion of Zhou et al. (2004).

4.3. Hot days and cold days

In Shenzhen, 33°C is the threshold for issuing a hot weather warning. If the maximum air temperature exceeds 33°C, a hot weather warning is issued by the Meteorological Bureau of Shenzhen Municipality. Fig. 7a shows the number of days each year with temperatures at 14:00 h LST exceeding 33°C in both cities and is hereinafter defined as the metric measuring the change in number of hot days. Fig. 7a shows that the number of hot days rose very quickly in Shenzhen in the past 46 yr. There was an average of only 7.2 d of hot days annually in 1968–1977, but by 2004–2013, the average increased by 29.1 ± 2.3 d, and its linear regression passed the significance test at 0.01 CL. During the same period, Hong Kong saw an increase of only 2.1 d in the number of hot days. However, the statistical uncertainty range is ±33.4 d, and the linear regression failed the significance test at 0.01 CL. As the number of hot days increased in Shenzhen, energy consumption increased because of

more cooling demand in summer; thus, to some extent, it increased anthropogenic heat emissions and provided positive feedback for further local temperature rises.

Once minimum air temperature falls below 10°C, the Meteorological Bureau of Shenzhen Municipality issues a cold weather warning urging the public to take precautions against cold weather. Fig. 7b shows the number of cold days each year in both cities in the past 46 yr, which are derived from the air temperature at 02:00 h LST. The figure shows that the number of cold days has dropped significantly. In 1968–1977, the average annual number of cold days in Shenzhen was around 31.6 d, and in 2004–2013, the number had dropped by 15.5 ± 5.2 d, and the linear regression for this downtrend passed the significance test at 0.01 CL. During the same period, Hong Kong saw a decrease of only 2.5 ± 2.2 d in the number of cold days. Not only was the rate of decrease much smaller than in Shenzhen, but its variation trend also failed the significance test at 0.01 CL, even though the rising trend of average minimum temperature passed the significance test at 0.01 CL.

4.4. Relative humidity

Urbanization causes changes in the physical properties of land surfaces and increases the amount of impervious surfaces within built-up areas. Grimmond (2007) pointed out that impervious surfaces reduce water seepage and change the surface hydrograph, leading to more urban floods. In addition, impervious surfaces reduce water content, thereby causing changes in relative humidity (Chandler 1967, Lee 1991). Fig. 8 shows the changes for relative humidity in both cities in the past 46 yr. There were no significant differences between relative humidity in both cities before 1983, but they began to diverge in 1983,

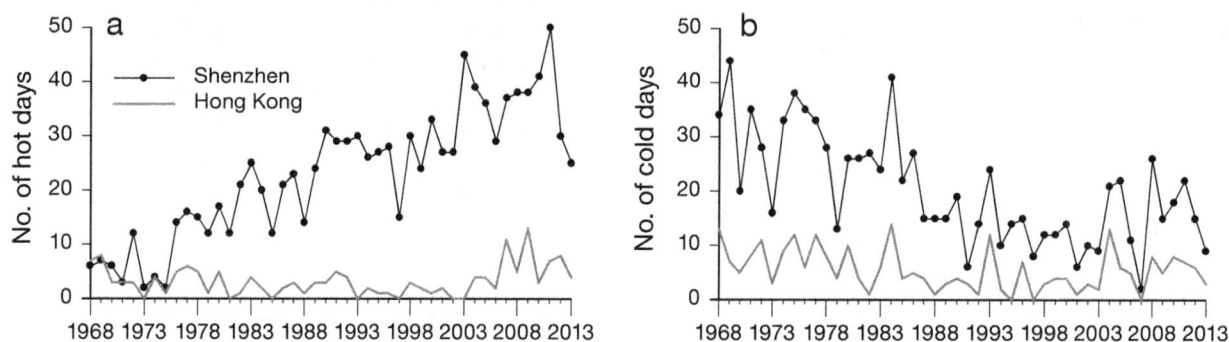

Fig. 7. Number of (a) hot days and (b) cold days during 1968–2013 in Shenzhen and Hong Kong

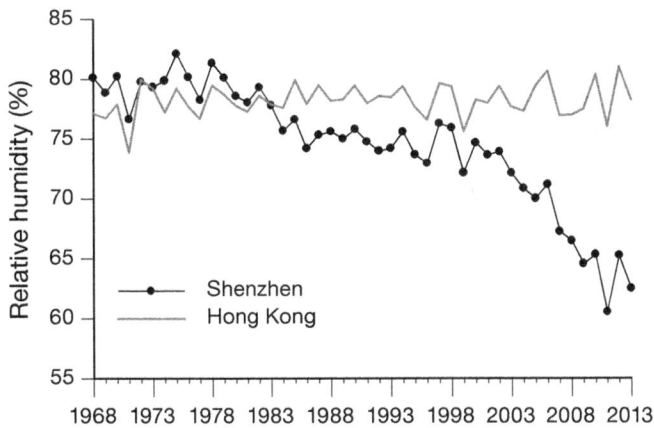

Fig. 8. Relative humidity during 1968–2013 in Shenzhen and Hong Kong

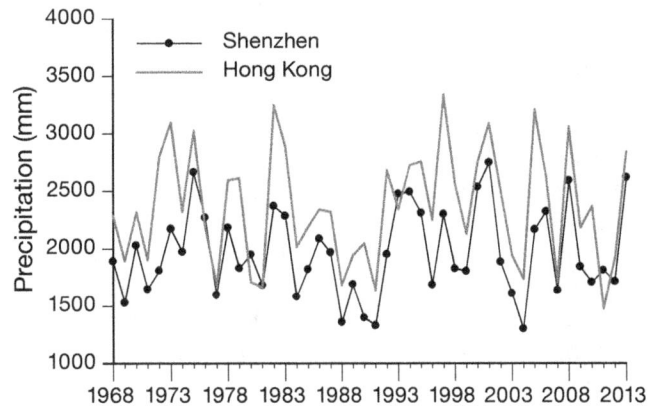

Fig. 9. Yearly precipitation during 1968–2013 in Shenzhen and Hong Kong

with Hong Kong basically showing no changes and Shenzhen showing a significant decline. In Shenzhen, average relative humidity in 2004–2013 was 13.13 ± 1.78 % lower than the average value in 1968–1977. By the 2010s, Shenzhen's average annual relative humidity was around 12 % lower than the average in Hong Kong, which shows that rapid urbanization had a significant influence on relative humidity in Shenzhen. Two possible explanations why relative humidity in Shenzhen is currently around 12 % lower than it is in Hong Kong are that (1) the ocean has greater influence on Hong Kong's weather, similar to the reason behind the difference in DTRs between the 2 cities; and (2) the percentage of urban built-up area in Hong Kong (23.92 %) is less than that in Shenzhen (43.21 %).

4.5. Precipitation

In studies of the impact of urbanization on precipitation in the Pearl River Delta, situated in both Shenzhen and Hong Kong, scholars arrived at different conclusions from different perspectives. Based on Landsat and monthly climate data during the period 1988–1996, Kaufmann (2007) discussed the relationship between climate and urban land use in concentric buffers around the sampling stations. He concluded that there is a causal relationship between temporal and spatial patterns of urbanization and temporal and spatial patterns of precipitation during the dry season, leading to an urban precipitation deficit in which urbanization reduces local precipitation. Meanwhile, Chen et al. (2011) analyzed the precipitation time series for a longer period of time and believed there were no significant trends for precipi-

tation in the Pearl River Delta region. Fig. 9 shows the time series of annual rainfall between the 2 cities. It shows that Hong Kong had greater overall annual rainfall than Shenzhen, which further illustrates that Hong Kong's climate is impacted by oceans more significantly than Shenzhen's climate, even though the cities are adjacent to each other and the straight-line distance between their 2 weather stations is ~30 km. A trend analysis showed that the linear regression for rainfall failed the significance test at even 0.1 CL, indicating that urbanization had no significant effect on total rainfall for both cities, which is consistent with the conclusion of Chen et al. (2011). However, another study based on data from multiple stations (Mok et al. 2006) stated that urbanization has led to spatial variation of rainfall in Hong Kong

5. ANALYSIS AND DISCUSSION

The above discussions show that although Shenzhen and Hong Kong are adjacent to each other and the straight-line distance between the CCW and HKO weather stations is only about 30 km, their recorded climate trends are very different. Since the establishment of the SEZ, Shenzhen has experienced very rapid urbanization, in which population, GDP and urban built-up area surged in a short period of time. During the same period, although Hong Kong was also growing, the rate of development was far slower than that in Shenzhen. This difference in development speed brought a very significant impact on urban climate. In Shenzhen, T_a, Tmin, Tmax and number of hot days exhibited an uptrend, while DTR, number of cold days and relative humidity decreased. The linear regression for all variables passed the significance test at 0.01 CL.

Previous findings regarding the effects of urbanization on physical climatic mechanisms can help explain climate change in Shenzhen (Carlson & Arthur 2000, Grimmond 2007), especially since these effects were particularly marked there. The artificial materials in urban built-up areas often have high heat capacity and low reflectivity, which is conducive to the increase in heat capacity and temperature of land surfaces in Shenzen. Rapid population growth and economic development compound this through anthropogenic heat emission from urban energy consumption, resulting in a faster rise in temperature. Meanwhile, at night, longwave radiation is reflected many times and absorbed by high-rise building walls and ground surfaces, creating a more significant warming effect at night. Therefore, while T_a, Tmin and Tmax increased during the period, Tmin rose more than Tmax, leading to a decrease in DTR. At the same time, because of the expansion of impermeable and impervious surfaces, more precipitation resulted in runoff into the sea, reducing the water content of the land surface, resulting in a rapid drop in relative humidity. Although Hong Kong was also growing during this period, the rate of development was far slower, and therefore the change in climate was not as dramatic as that in Shenzhen.

Correlation coefficients between urbanization statistics and climate indicators reflect the differences in the urban effect on climate between Shenzhen and Hong Kong. Because it not affected by factors such as inflation, population data are more suitable than GDP in reflecting the process of urbanization; this study therefore calculated the correlation between population and climatic indicators. Table 2 shows the correlation coefficients between populations of Shenzhen and Hong Kong with the annual averages of 5 climatic indicators over the past 46 yr. From Table 2, the absolute values of the correlation coefficients between Shenzhen's population and all 5 climatic indicators were over 0.7 and passed the significance test at 0.01 CL; T_a, Tmin and Tmax showed a positive correlation with population, while DTR and relative humidity showed a negative correlation. In contrast, although the correlation coefficients between Hong Kong's population and T_a, Tmin and Tmax passed the significance test at 0.01 CL, the values are significantly lower than those in Shenzhen. Meanwhile, the correlation coefficients between Hong Kong's population and DTR and relative humidity are near zero, which means that there are essentially no relationships between DTR/relative humidity and population growth.

Relationships between the population and climate elements were also analyzed through Granger causality, which is a statistical hypothesis test to determine whether one time series is useful in forecasting another (Granger 1969). In this test, if one time series, X, is determined to be useful in forecasting another time series, Y, X can be taken as the Granger-cause of Y. In this study, the time series of population was set as X and those of climate elements as Y. The lagged value was set as 1, which means that the response of the climate elements is assumed to be 1 yr later than the variation in population.

The results of the Granger causality test suggest that population is the Granger-cause of the variations in all climate elements in Shenzhen. However, in Hong Kong, population is only the Granger-cause of T_a, Tmin and Tmax and seems to have no impact on DTR and relative humidity, which is quite accordant with the results of the correlation analysis.

6. SUMMARY

This study compares climate change in the 2 neighboring megacities of Shenzhen and Hong Kong over the past 46 yr (1968–2013). Since these 2 cities have experienced different speeds of urbanization during this period, an intercomparison study on the climate trends in the 2 cities can further enhance our understanding of the impact of rapid urbanization on local climate. This study reached the following conclusions:

(1) Since establishment of the SEZ in 1980, Shenzhen has experienced very rapid urbanization, in which Shenzhen's population and GDP grew at a CAGR of 11.03 and 31.64%, respectively, between 1979 and 2012. In contrast, Hong Kong was already a mature metropolis since the 1960s, in which its population and GDP grew at a CAGR of only 1.12 and 9.47%, respectively, over the same period. Compared to 1980, Shenzhen's urban built-up area expanded by about 6825%, while Hong Kong's in-

Table 2. Correlation coefficients between population and different climate elements. Means not passing significance test at 0.01 level in **bold**. DTR = diurnal temperature range; RH = relative humidity; T_a = air temperature; T_m = minimum air temperature; T_M = maximum air temperature

City	T_a	T_M	T_m	DTR	RH
Shenzhen	0.82	0.71	0.84	−0.72	−0.83
Hong Kong	0.60	0.59	0.54	**−0.01**	**0.01**

creased by only 59.08%. The rate of urbanization was much faster in Shenzhen than in Hong Kong during the period.

(2) Over the last 46 yr, Shenzhen warmed by 1.63 ± 0.18°C at a rate of 0.35 ± 0.04°C per decade, while Hong Kong warmed by 0.47 ± 0.20°C at a rate of 0.10 ± 0.04°C per decade. Combined with previous studies in Hong Kong on the contribution of urbanization to warming, it is estimated that urbanization increased Shenzhen's temperature by 1.36 ± 0.18°C, with a contribution of 81.4 to 85.1%.

(3) Over the last 46 yr, Shenzhen's Tmax increased by 0.90 ± 0.19°C, Tmin increased by 2.09 ± 0.23°C and DTR decreased 1.18 ± 0.23°C, and their linear regressions passed the significance test at 0.01 CL. During the same period, Hong Kong's Tmax increased by 0.12 ± 0.28°C, but its linearized warming trend failed the significance test at 0.01 CL, while its Tmin increased by 0.55 ± 0.20°C and DTR decreased by 0.43 ± 0.27°C, and both linear regressions passed the significance test at 0.01 CL. Correspondingly, the number of hot days in a year in Shenzhen increased by 29.1 ± 2.3 d, and the number of cold days decreased by 15.5 ± 5.2 d, and both linear regressions passed the significance test at 0.01 CL. The number of hot days in a year in Hong Kong increased by only 2.1 d, with an uncertainty range of ±33.4 d, and the number of cold days decreased by 2.5 ± 2.2 d, but both linear regressions failed the significance test at 0.01 CL.

(4) Relative humidity in Shenzhen declined by 13.13 ± 1.78% over the past 46 yr, while that in Hong Kong had no significant change. Furthermore, there were no significant linear trends in rainfall for both cities during the period, and both linear regressions failed the significance test at 0.01 CL, indicating that urbanization had no significant impact on total annual rainfall in both cities.

(5) By comparing climatic indicators with urban development in Shenzhen and Hong Kong, it is shown that the pace of urbanization is well reflected on the rate of change in local climate. Rapid urbanization is at least part of the reason for significant changes in local climate in Shenzhen.

Acknowledgements. This study was funded by the National Natural Science Foundation of the People's Republic of China (Grant No. 51278308) and the R&D Foundation of Shenzhen (Basic Research Project; Grant No. JCYJ 20130325151523015). The authors also thank the Statistics Bureau of Shenzhen Municipality and the Hong Kong Census and Statistics Department for providing data on urban development.

LITERATURE CITED

➤ Bornstein RD, Johnson DS (1977) Urban-rural wind velocity differences. Atmos Environ 11:597–604

➤ Carlson TN, Arthur ST (2000) The impact of land use–land cover changes due to urbanization on surface microclimate and hydrology: a satellite perspective. Global Planet Change 25:49–65

➤ Chan HS, Kok MH, Lee TC (2012) Temperature trends in Hong Kong from a seasonal perspective. Clim Res 55: 53–63

Chandler TJ (1967) Absolute and relative humidities in towns. Bull Am Meteorol Soc 48:394–399

➤ Chen XL, Zhao HM, Li PX, Yin ZY (2006) Remote sensing image-based analysis of the relationship between urban heat island and land use/cover changes. Remote Sens Environ 104:133–146

➤ Chen J, Li QL, Niu J, Sun L (2011) Regional climate change and local urbanization effect on weather variables in southeast China. Stoch Env Res Risk A 25:555–565

➤ Cheng X, Hu F (2005) Numerical studies on flow fields around buildings in an urban street canyon and crossroad. Adv Atmos Sci 22:290–299

Chinese Academy of Social Sciences (2011) Blue book of cities in China: annual report on urban development of China No. 4. Pan J, Wei H (eds) Social Sciences Academic Press, Beijing (in Chinese)

Eliasson I (2000) The use of climate knowledge in urban planning. Landscape Urban Plan 48:31-44

➤ Ge Q, Wang F, Luterbacher J (2013) Improved estimation of average warming trend of China from 1951–2010 based on satellite observed land-use data. Clim Change 121: 365–379

➤ Goldreich Y, Surridge AD (1988) A case study of low level country breeze and inversion heights in the Johannesburg area. J Climatol 8:55–66

➤ Granger CWJ (1969) Investigating causal relations by econometric models and cross-spectral methods. Econometrica 37:424–438

➤ Grimmond S (2007) Urbanization and global environmental change: local effects of urban warming. Geogr J 173: 83–88

Hamburg M (1977) Statistical analysis for decision making, 3rd edn. Harcourt Brace Jovanovich, New York, NY

Howard L (1833) The climate of London deduced from meteorological observations, 3rd edn. Harvey and Darton, London

IPCC (2013) Summary for policymakers. In: Stocker TF, Qin D, Plattner GK, Tignor M and others (eds) Climate change 2013: the physical science basis. Contribution of Working Group I to the Fifth Assessment Report of the Intergovernmental Panel on Climate Change. Cambridge University Press, Cambridge

➤ Kaufmann RK, Seto KC, Schneider A, Liu Z, Zhou L, Wang W (2007) Climate response to rapid urban growth: evidence of a human-induced precipitation deficit. J Clim 20:2299–2306

Lane DM (2014) Introduction to statistics. Rice University. http://onlinestatbook.com/Online_Statistics_Education.pdf

➤ Lee DO (1991) Urban-rural humidity differences in London. Int J Climatol 11:577–582

➤ Lenney MP, Woodcock CE, Collins JB, Hamdi H (1996) The status of agricultural lands in Egypt: the use of multitemporal NDVI features derived from landsat TM. Remote Sens Environ 56:8–20

➤ Li Q, Zhang H, Liu X, Huang J (2004) Urban heat island effect on annual mean temperature during the last 50 years in China. Theor Appl Climatol 79:165–174

➤ Masson V, Lion Y, Peter A, Pigeon G, Buyck J, Brun E (2013) 'Grand Paris': regional landscape change to adapt city to climate warming. Clim Change 117:769–782

Mok HY, Leung YK, Lee T, Wu MC (2006) Regional rainfall characteristics of Hong Kong over the past 50 years. Conf on Changing Geography in a Diversified World, Hong Kong Baptist University, Hong Kong, PR China, 1–3 June 2006, HKO Reprint No. 646

➤ Oke TR (1981) Canyon geometry and the nocturnal urban heat island: comparison of scale model and field observations. Int J Climatol 1:237–254

Oke TR (1987) Boundary layer climates. Routledge, London

➤ Stellmes M, Roder A, Udelhoven T, Hill J (2013) Mapping syndromes of land change in Spain with remote sensing time series, demographic and climatic data. Land Use Policy 30:685–702

➤ Stewart ID, Oke TR (2012) Local climate zones for urban temperature studies. Bull Am Meteorol Soc 93: 1879–1900

➤ Wang J, Xu C, Hu M, Li Q, Yan Z, Zhao P, Jones P (2014) A new estimate of the China temperature anomaly series and uncertainty assessment in 1900–2006. J Geophys Res Atmos 119:1–9

➤ Yu M, Liu Y, Dai Y, Yang A (2013) Impact of urbanization on boundary layer structure in Beijing. Clim Change 120: 123–136

➤ Zhou L, Dickinson RE, Tian Y, Fang J and others (2004) Evidence for a significant urbanization effect on climate in China. Proc Natl Acad Sci USA 101:9540–9544

3

Our commitment to climate change is dependent on past, present and future emissions and decisions

Markku Rummukainen*

Centre for Environmental and Climate Research, Lund University, Sölvegatan 37, 223 62 Lund, Sweden

ABSTRACT: The present day global climate change is fueled by our use of fossil fuels and land use change. The already observed warming and other distinct changes in the climate system stem from these human influences and are ongoing. Due to climate system inertia, a part of the climate system's response to this historical forcing remains to manifest itself, which it will do over time. At the same time, socio-economic forces and trends imply some amount of additional emission and land use change, which compounds our commitment to even more substantial climate change. Cumulative carbon dioxide emissions are the basic determinant of the ultimate amount of anthropogenic climate change. Climate system properties, such as climate sensitivity and the carbon cycle, and also possible initiation of non-linear changes, further shape the amount and nature of the long-term change for any set amount of greenhouse gas emissions. While a changed climate is, in practice, now unavoidable, our commitment to continued climate change can be constrained by reductions of global carbon dioxide emissions, their cessation and/or negative emissions. These alternatives have different implications for the long-term unfolding of these changes, but can all considerably reduce the possibility of very large amounts of change, the need for adaptation and responses to negative impacts.

KEY WORDS: Climate change · Global warming · CO_2 emissions · Climate change commitment

1. INTRODUCTION

The ongoing climate change manifests itself as global warming, sea level rise, significant ice loss from glaciers, mass loss from ice sheets and decreasing sea ice in the Arctic region (IPCC 2013). The principal drivers of the present day climate change are human activities, primarily greenhouse gas emissions. The resulting changes in the atmospheric composition impose a radiative forcing on the climate system, the response of which is one of an overall warming trend and other subsequent changes. The observed global mean warming over the period of 1880 to 2012 was around 0.8°C. However, this temperature change was only a part of the climate system response. While some heat goes to cryospheric melting and land warming, the main part warms up the oceans (Rhein et al. 2013), which contributes to the concomitant sea level rise, but also slows down surface warming. The full response of the climate system to emissions unfolds over time. This means that historical emissions result in further change on top of what already has taken place and that, at any time, there is a commitment to additional change compared to what has already taken place. The amount and time evolution of the further change depends on climate system properties, namely climate sensitivity and ocean heat uptake, and also possible non-linear climate effects on the carbon cycle and ice sheets. Significantly, however, climate change ahead of us, at any given time, is affected by present and future emissions, as these effectively lock in additional long-term change (e.g. Friedlingstein et al. 2011). This is a form of socioeconomic commitment to further climate change and is due, for example, to fossil fuel related infrastructure investments with long depreciation times.

Thus, for the foreseeable future, we are committed to additional climate change. This means that for any

*Corresponding author: markku.rummukainen@cec.lu.se

given climate stabilization target, we can effectively end up missing it well before the target level is actually exceeded. For example, the threshold to a warming of 2°C, 4°C or 6°C is effectively crossed considerably earlier than the full warming and its effects manifest themselves. Concomitantly, in the near future, we may also commit ourselves to higher risks of non-linear changes such as massive permafrost melt, which are admittedly difficult to quantify.

This paper provides a brief overview of key issues related to our climate change commitment, beyond 2100. In addition to consideration of studies which estimate the climate change commitment for some set concentration levels or emission pathways, attention is drawn to socio-economic factors. Reflections are also made on what historical emissions alone mean for our climate change commitment.

2. COMMITTED CLIMATE CHANGE

2.1. Climate system response

The amount of long-term climate change depends on the amount of carbon dioxide emissions over time and other relevant climate forcing. The relationship between the long-term warming and sustained forcing due to doubling of atmospheric carbon dioxide concentration is coined the equilibrium climate sensitivity (ECS). IPCC (2013), based on multiple lines of evidence, assessed the ECS to *likely*[1] be in the range of 1.5 and 4.5°C. The range of uncertainty is wider, as it is *extremely unlikely* that the ECS is less than 1°C and *very unlikely* that it is larger than 6°C. Model-based estimates of global mean sea level rise due to thermal expansion range from 0.2 to 0.6 m per degree of warming (IPCC 2013). Estimates of long-term mass loss from ice sheets are less well elucidated. Contributions from the Greenland Ice Sheet and Antarctic marine ice sheets could, over millennia, add up to a few meters of sea level rise for each 1°C of warming (Ridley et al. 2010, Levermann et al. 2013, Joughin et al. 2014, Mengel & Levermann 2014), given the exceedance of specific threshold warming levels, which are estimated to be lower for West Antarctica and Greenland than for East Antarctica.

The value of ECS is of first-order importance for climate change commitment; smaller (larger) ECS leads to smaller (larger) committed change, for a set forcing. The time-dependent warming approaches

the ultimate warming level with a considerable time delay, and happens on multi-century time scales (e.g. Jarvis & Li 2011). This is mostly due to the oceans' large heat capacity and slow turnover time. The sea level rise due to past emissions will continue even longer than what it will take for the atmospheric warming to level off.

Climate sensitivity refers to atmospheric carbon dioxide concentrations. The link from emissions to concentrations involves the global carbon cycle, which is a part of the overall climate system and constitutes an active element of the climate system response. At present, about half of the overall anthropogenic emissions are taken up in terrestrial systems and in the ocean. The longer term efficacy of these uptakes, in particular beyond the 21st century, is not certain. Still, it is the global emissions that drive the anthropogenic climate change. How they evolve is the fundamental determinant of the magnitude and also the pace of climate change.

Inasmuch as climate system inertia slows down the climate system response to increasing emissions, the same applies under mitigation. The full climate effect of emission reductions unfolds over time, and efforts to curb climate change will have their full effect with a time delay.

2.2. Societal inertia

Today's greenhouse gas emissions arise from the energy supply, industry, transport, buildings, agriculture, forestry and other land use. Such emissions often involve infrastructure that has a long lifetime. Power plants, for example, can operate for 30 to 40 yr (e.g. Davis et al. 2010). This implies a 'socio-economic commitment' in the form of investments (economic commitment) to continued climate change driven by not only the historical, but also by future emissions. In the absence of alternatives of fossil fuels for energy supply, one can also speak of a societal commitment to further climate change, cf. Diffenbaugh (2013).

Davis et al. (2010) estimate the committed future emissions related to already existing infrastructure that involves burning coal, oil and natural gas, between 2010 and 2060, to around 135 gigatonnes of carbon (GtC; range 77 to 191 Gt). Together with the warming due to historical emissions, they predict a global mean warming of 1.3°C. They note that there is an additional commitment from inter alia non-energy sources such as agriculture. Continued investments in replacement of infrastructure that is decommissioned or in new production capacity in-

---

[1]Extremely unlikely: 0–5%. Likely: 66–100%. Very unlikely: 0–10%. (IPCC 2013)

crease the socio-economic commitment to emissions and thus to climate change. Guivarch & Hallegatte (2011) increased the Davis et al. (2010) estimate to around 1.7°C, by also accounting for non-emitting infrastructure such as roads and buildings, and non-CO_2 emissions. More recently, Davis & Socolow (2014) noted that the committed emissions from the power sector have further increased during 2010 to 2012, by 4 % per year, due to the continued construction of coal and natural gas power plants.

Presently, energy-related investments related to fossil fuels continue (IEA 2014). The International Energy Agency (IEA) projects a long-term global mean temperature rise of 3.6°C under the current known energy policies and measures (IEA's New Policies Scenario; IEA 2013). IEA (2012) also highlighted that continued unabated energy-related investments involving fossil fuels through 2017 imply a climate warming commitment of at least a 2°C. Socio-economic trends, thus, give rise to inertia, which drives difficult-to-avoid increases in climate change commitment and adds to the climate forcing due to past emissions.

2.3 Climate change commitment

2.3.1. Stabilized concentrations or zeroed emissions

Studies on the amount of committed climate change typically involve simple climate models or Earth System Models of Intermediate Complexity. The approach has often been to consider specific stabilized (elevated) atmospheric greenhouse gas concentrations ('constant-composition commitment'), or reductions of the global emissions to zero ('zero-emissions commitment'), with more or less idealized consideration of underlying socio-economic developments. Long scenario runs with Global Climate Models (GCMs) with more specified socio-economic framing of emissions' pathways provide further insights.

The 2 approaches are fundamentally different not only in terms of their implications for global emission pathways, but also in terms of the nature of the climate change commitment. A constant-composition implies reduced, but still ongoing non-zero emissions and continued warming beyond the stabilization of atmospheric concentrations. In contrast, zeroed emissions allow for a decrease of atmospheric concentrations over time (e.g. Zickfeld et al. 2013). In both cases, the peak warming and long-term sea level rise still closely follow the cumulative emissions of carbon dioxide. The case of constant emissions ('constant-

emissions commitment') has also been highlighted (Wigley 2005), but it would seem to agree least well with mainstream climate policy discussions and is not further considered here.

The climate response to changing aerosol and aerosol precursor emissions is fast both when they increase and when they decrease. The atmospheric residence time of methane is much shorter than that of carbon dioxide, whereas nitrous oxide has a longer lifetime. Thus, the commitment to climate change due to elevated aerosol levels and past emissions of methane is much shorter than the commitment to climate change due to past emissions for carbon dioxide, as well as other greenhouse gases with long atmospheric lifetimes (Matthews & Zickfeld 2012). Changes to aerosol amounts and gases such as methane thus have a potential for relatively immediate changes to the pace of warming, but less so for long-term change. This is due to their forcing being less than that of the more long-lived gases, in particular carbon dioxide. Thus, the zero-emissions climate change commitment would follow the carbon dioxide amount. In line with the above, the constant-composition commitments follow the carbon dioxide equivalent amount, which is the same as carbon dioxide if other forcings are not in play. Fig. 1 illustrates the longer-term temperature path for constant-composition and zero-emissions, respectively.

The constant-composition case implies a residual warming trend well beyond the stabilization of concentrations (e.g. Meehl et al. 2005). It corresponds to

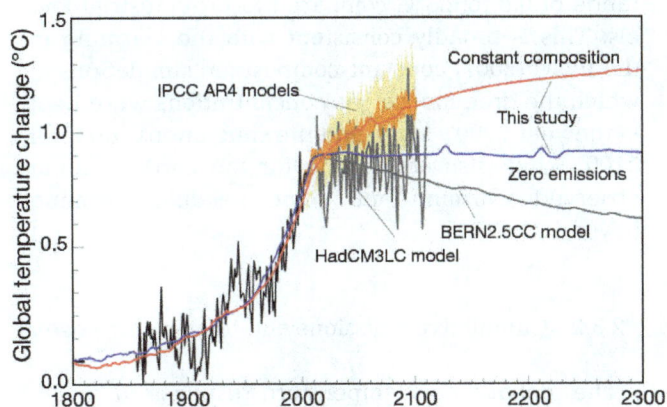

Fig. 1. Illustrative example of constant-composition (red line) and zero-emissions (blue line, grey lines; different models) global mean temperature change commitments until 2300. The orange line and the yellow shading show multi-model results as assessed in the IPCC (2007), and the black line is the observed temperature change, for reference. Figure from Matthews & Weaver (2010). Reprinted by permission from Macmillan Publishers Ltd: Nature Geoscience 3:142–143, Fig. 1

continued emissions which, although much lower than today's, are not zero-emissions. The IPCC (2007) assessed that the residual temperature increases, after a stabilization of atmospheric concentrations, to around 0.5°C over the first 100 yr, and some further increase during the subsequent centuries.

If the carbon dioxide emissions are zeroed, the atmospheric carbon dioxide concentration will begin to decline very slowly due to ocean carbon sinks. However, the global mean temperature could stabilize rather than decrease, since the ocean heat uptake would also slow down (Friedlingstein & Solomon 2005, Solomon et al. 2009). In order to achieve a reversal of the temperature trend, negative carbon dioxide emissions would be needed (Meehl et al. 2012).

Inevitability of sea level rise on multi-century time scales and longer applies for all cases above, since thermal expansion following net mixing of heat down to the deeper ocean would continue (Meehl et al. 2005, Meehl et al. 2012). Ice sheets' mass loss and other irreversible non-linear change could progress over many hundreds of years, and the regional-scale climate could continue to evolve (Gillett et al. 2011), even when the global temperature change levels off.

The equilibrium climate sensitivity (ECS) provides a way of estimating the final warming for constant-composition commitments. The *likely* range of such warming, for a number of atmospheric carbon dioxide concentration levels, is depicted in Table 1. The present atmospheric carbon dioxide level of 400 ppm[2] corresponds to a *likely* long-term warming range of 0.8 to 2.3°C compared to preindustrial levels. This is broadly consistent with the warming in the IPCC (2007) constant-composition simulations in which the greenhouse gas concentrations were kept at the year 2000's level, and the simulations run until 2100, while also accounting for the corresponding observed warming and some residual warming beyond 2100.

2.3.2. Cumulative emissions and temperature rise

The global mean temperature response in model simulations is closely related to the cumulative carbon dioxide emissions (Allen et al. 2009, Matthews et al. 2009, IPCC 2013). The global mean temperature

change per 1000 Gt C (3670 Gt CO_2) of carbon dioxide emissions is estimated to *likely* be in the range of 0.8 to 2.5°C, for cumulative emissions up to 2000 Gt C (IPCC 2013). This is coined the Transient Climate Response to cumulative carbon Emissions, TCRE. It incorporates both the climate sensitivity and the carbon cycle, which affects the amount of emissions that resides in the atmosphere over time. TCRE is a more realistic, but also a more complicated and uncertain measure compared to ECS. Fig. 2 illustrates how the global mean temperature increase is estimated to respond to cumulative carbon dioxide emissions (IPCC 2013). The black line and the grey envelope are from simulations in which only CO_2 changes. In these results, for example, a 4°C increase could ensue from cumulative carbon dioxide emissions in excess of 1800 Gt C (1500 Gt C) in the absence (in the presence of non-CO_2 emissions, here according to specific assumptions) of other anthropogenic forcings. A 2°C rise would correspond to around half of these emission amounts. The colored lines and the colored envelope are from studies which considered both carbon dioxide and other greenhouse gases, and thus exhibit a higher temperature response than for only carbon dioxide, for a given amount of carbon dioxide emissions.

The TCRE and ECS do not measure the same outcome. The former gives the peak warming due to the cumulative emissions. As a result of climate system inertia, the peak warming occurs sometime after a cessation of emissions, which assumes some lowering of the atmospheric concentration due to carbon sinks, before climate system inertia catches up. The latter is the long-term warming following the stabilization of carbon dioxide concentration, which would follow from reduced, but still non-zero, emissions.

Table 1. Probabilities of the long-term global mean temperature change (see Footnote 1 for probability ranges) since the preindustrial era for different carbon dioxide stabilization levels, following the IPCC (2013) assessment of the equilibrium climate sensitivity. Here, 280 ppm is used to represent the preindustrial atmospheric carbon dioxide concentration

Carbon dioxide (ppm)	Extremely unlikely less than (°C)	Likely range of temperature change (°C)	Very unlikely larger than (°C)
350	0.3	0.5–1.4	1.9
450	0.7	1.0–3.1	4.1
550	1.0	1.5–4.5	5.8
650	1.2	1.8–5.5	7.3
750	1.4	2.1–6.4	8.5
1000	1.8	2.8–8.3	11.0

---

[2]The carbon dioxide equivalent level is higher, but does not necessarily reflect the long-term level due to past and present emissions, as different anthropogenic greenhouse gases and aerosols have different atmospheric lifetimes

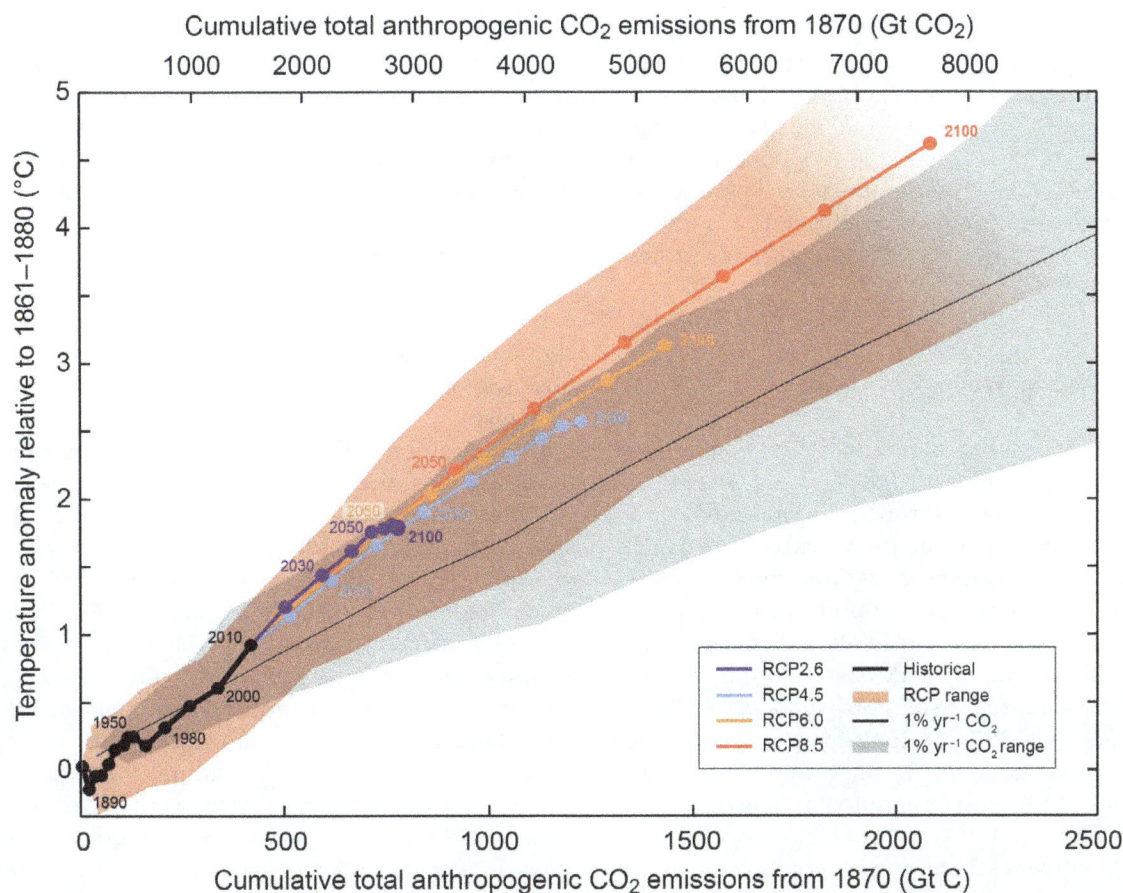

Fig. 2. Global mean surface temperature increase as a function of cumulative total global CO_2 emissions from various lines of evidence. Multi-model results from a hierarchy of climate-carbon cycle models for each representative concentration pathway (RCP) until 2100 are shown with colored lines and decadal means (dots). Some decadal means are labeled for clarity (e.g. 2050 indicating the decade 2040–2049). Model results over the historical period (1860 to 2010) are indicated in black. The colored plume illustrates the multi-model spread over the 4 RCP scenarios and fades with the decreasing number of available models in RCP8.5. The multi-model mean and range simulated by CMIP5 models, forced by a CO_2 increase of 1% per year (1% yr^{-1} CO_2 simulations), is given by the thin black line and grey area. For a specific amount of cumulative CO_2 emissions, the 1% per year CO_2 simulations exhibit lower warming than those driven by RCPs, which include additional non-CO_2 forcings. Temperature values are given relative to the 1861–1880 base period, emissions relative to 1870. Decadal averages are connected by straight lines. (Figure SPM.10 from IPCC [2013])

2.3.3. Committed temperature change for specific emission pathways

Another perspective on our climate change commitment is to consider alternative emission pathways, which correspond to alternative socio-economic developments. Here, projections based on 2 of the RCPs, including their extensions beyond 2100 (Representative Concentration Pathways, cf. van Vuuren et al. [2011]), are considered, the RCP2.6 and the RCP8.5 (Collins et al. 2013). These are the opposite extremes among the RCPs, and thus span the range of RCP-specific outcomes viz. the climate change commitment. The actual emission pathway in the future may of course follow yet another path.

Multi-model simulations under the aggressive mitigation RCP (RCP2.6), which features an imminent peaking of the global emissions followed by a steady decline and net negative emissions by ~2070, lead to the global temperature peaking around 1.6°C above preindustrial levels before 2100, and a subsequent decline on a multi-century time scale. By 2300, the atmospheric CO_2 level would be around 360 ppm, and the residual warming around 1.2°C[3] compared to preindustrial levels.

---

[3]Range: 0.6–1.8°C, which is constructed here from the observed warming contribution of 0.6°C until the CMIP5 models' reference period, and the CMIP5 multi-model projection range of 0.0–1.2°C. See IPCC (2013), their Table 12.2

At the other end of the scale is the RCP8.5 scenario with continued emission increases towards the end of the 21st century and stable high emissions for the next half-century, followed by a steady decrease until 2250, when the atmospheric CO_2 concentration balances at 2000 ppm. The global mean temperature increase is around 4°C towards the end of the 21st century and 8.4°C[4] towards 2300, compared to preindustrial levels.

2.3.4. Where we are now

The observed global mean warming since the preindustrial era is around 0.8°C, the majority of which is attributed to anthropogenic influences on the climate system, principally carbon dioxide emissions. The atmospheric carbon dioxide concentration alone stands at 400 ppm (in addition to contributions from other greenhouse gases and aerosols to net warming) and, based on the assessment of ECS, would by itself correspond to a *likely* range of 0.8 to 2.3°C for long-term warming. A global mean warming of 3.1°C would be a very unlikely, but not impossible, outcome (corresponding to ECS of 6°C).

The cumulative carbon dioxide emissions from 1750 until 2011 are estimated to be 555 Gt C (with the range of uncertainty from 470 to 640 Gt C). The central estimate suggests a global mean warming commitment between 0.4 and 1.4°C based on the climate/cumulative carbon cycle response (TCRE, 0.8 to 2.5°C per 1000 Gt C).

The ranges above are not the same. As mentioned earlier, the ECS and the TCRE are not comparable. The former relies on net non-zero future emissions, which add to the effective forcing due to the cumulative historical emissions, which the latter measure applies to.

Given the continuation of emissions, climate system inertia and effects of socioeconomic inertia, the long-term changes will exceed those that can be derived from the observed warming and past emissions. The possibility of a non-linear climate system response may further compound long-term climate change.

---

[4]Range: 3.6–13.2°C, which is constructed here from the observed warming contribution of 0.6°C until the CMIP5 models' reference period, and the CMIP5 multi-model projection range of 3.0–12.6°C. See IPCC (2013), their Table 12.2

3. SURPRISES

The considerations above do not explicitly include the effects of possible major outflow of carbon from natural sinks, such as permafrost areas or irreversible destabilization of parts of the continental ice sheets such as those in West and East Antarctica. Mobilization of permafrost carbon can be considered included in constant-composition cases, assuming that the continued anthropogenic emissions are further reduced compared to what they would be in the absence of increased carbon outflow from natural systems.

The definition of the constant-composition commitment implies that it does not as such separate between anthropogenic and induced natural emissions. The corresponding anthropogenic emissions would, however, need to be lowered, in case of weakened net natural carbon dioxide sinks. If natural carbon sinks push more carbon dioxide back into the atmosphere, the zero-emission anthropogenic commitment would provide an underestimate of longer-term climate change, as anthropogenic emissions would be compounded by natural sources. Thus, changes to the net strength of the carbon cycle would need to be compensated by lower cumulative emissions. Otherwise, the climate change commitment grows.

The long-term sea level rise due to thermal expansion is fairly well constrained. Uncertainty is greater for the possible triggering of marine ice sheet instability, which could in turn ensure a sea level rise of several meters over a long period of time, even if the global mean temperature were to stabilize or begin a slow decline, which may occur on centennial timescales following zeroed or negative emissions (see Section 2.1).

4. DISCUSSION

At any point in time, a part of climate change due to cumulative emissions remains to be realized, which will manifest over subsequent years. This constitutes a climate change commitment due to past emissions; a long-term and greater climate change than what has been observed so far.

Considerable global-scale emission reductions could lead to a stabilization of atmospheric concentrations of greenhouse gases, in particular carbon dioxide. This would constrain the global mean temperature trend considerably and, subsequently, other aspects of anthropogenic climate change; however, there would still be some residual warming, over a long period of time. Zeroed emissions would allow for a stabilization of atmospheric temperature. Negative emissions of carbon dioxide could reverse the warming trend to a cooling one.

Global emissions are still on the rise. Recent global carbon dioxide emissions from fossil fuel use, cement manufacture and land use change[5] are around 10.6 Gt C yr^{-1}. This can be coined a socio-economic commitment, which arises from the continuing investments in fossil-fuel-related energy supplies and other relevant infrastructure. During the next few years, this may effectively lock in a long-term warming of at least 2°C. Continued emissions at today's level would bring the cumulative amount up to 1000 Gt C after some 40 yr. Based on the cumulative carbon/climate relationship, it would mean a global mean warming commitment of 0.8°C to 2.5°C (excluding extreme outcomes). Assuming continuously increasing emissions at a rate of, for example, 2.5% per year, the same level would be accumulated after only 30 yr. Larger cumulative emissions would result in greater climate change.

The long-term temperature rise and sea level change estimates — our climate change commitment — can, for some set cumulative emissions or atmospheric concentration levels, be bounded by estimates of climate sensitivity, climate impacts on the carbon cycle, heat uptake in the ocean and possible ice sheet instability (e.g. Friedlingstein et al. 2011). As these estimates are subject to uncertainty, estimates of climate change commitment can also only be in probabilistic terms. For example, if the atmospheric carbon dioxide concentration reaches double the preindustrial level (i.e. to around 550 ppm), the likely assessment for long-term global mean warming is 1.5 to 4.5°C. However, there are outcomes outside this range, including much larger values, 6°C being a 'very unlikely' but still conceivable outcome. For 650, 450 and 350 ppm, the 'very unlikely' outcomes are around 7°C, 4°C and 2°C, respectively.

The constant-composition case corresponds to continued residual emissions, although on a much lower level than today's, which may or may not be realistic depending on the time horizon. If emissions are zeroed instead, climate system inertia can counteract

the level of peak warming. Negative emissions can reverse the warming trend, but not return the world to preindustrial climate conditions in any practical sense. Our commitment to continued climate change is inherently unavoidable on the multi-decadal time scale and our commitment to *a changed climate* is on a centennial-to-millennia time scale.

The societal view on which level of risk and harm is acceptable and, subsequently, on the necessary amount of mitigation and adaptation, is a value-laden decision. In terms of climate change related risks, the higher end of warming is more decisive from a risk perspective, as potential harm in general increases with the amount and pace of climate change. This highlights the importance of considering the full range of possible outcomes.

We can hardly affect the climate system inertia or historical emissions. In order to bound climate change in the medium-to-long term, managing the emission trends including the socio-economic commitment to continued emissions is crucial, as the climate change commitment due to past emissions is compounded by future emissions. Keeping the cumulative carbon dioxide emissions as small as possible is the key for reducing our climate change commitment. The present socio-economic trends, such as investment patterns and climate policy, however, continue to increase our climate change commitment, which will make its mark on the climate system on the centennial-to-millennial timescale. Conversely, management of these trends can over time curb the climate change commitment.

Acknowledgements. This study is a contribution to the Swedish strategic research area ModElling the Regional and Global Earth system, MERGE.

LITERATURE CITED

▶ Allen MR, Frame DJ, Huntingford C, Jones CD, Lowe JA, Meinshausen M, Meinshausen N (2009) Warming caused by cumulative carbon emissions towards the trillionth tonne. Nature 458:1163–1166

Collins M, Knutti R, Arblaster J Dufresne JL and others (2013) Long-term climate change: projections, commitments and irreversibility. In: Stocker TF, Qin D, Plattner GK, Tignor M and others (eds) Climate change 2013: the physical science basis. Contribution of working group I to the fifth assessment report of the Intergovernmental Panel on Climate Change. Cambridge University Press, Cambridge, p 1029–1136

▶ Davis SJ, Socolow RH (2014) Commitment accounting of CO$_2$ emissions. Environ Res Lett 9:084018

▶ Davis SJ, Caldeira K, Matthews HD (2010) Future CO$_2$ emissions and climate change from existing energy infrastructure. Science 329:1330–1333

---

[5]According to the Global Carbon Project (www.globalcarbonproject.org, last read 4 August 2014), the year 2012, CO$_2$ emissions due to fossil fuel use and cement production were 9.7 ± 0.5 GtC. Deforestation and other land use change contributed with 0.9 ± 0.5 GtC per year, on average, over 2003–2012

► Diffenbaugh N (2013) Human well-being, the global emissions debt, and climate change commitment. Sustain Sci 8:135–141

► Friedlingstein P, Solomon S (2005) Contributions of past and present human generations to committed warming caused by carbon dioxide. Proc Natl Acad Sci USA 102: 10832–10836

► Friedlingstein P, Solomon S, Plattner GK, Knutti R, Ciais P, Raupach MR (2011) Long-term climate implications of twenty-first century options for carbon dioxide emission mitigation. Nat Clim Change 1:457–461

► Gillett NP, Arora VK, Zickfeld K, Marshall SJ, Merryfield WJ (2011) Ongoing climate change following a complete cessation of carbon dioxide emissions. Nat Geosci 4:83–87

► Guivarch C, Hallegatte S (2011) Existing infrastructure and the 2°C target. Clim Change 109:801–805

IEA (International Energy Agency) (2012) World energy outlook 2012. OECD/IEA, Paris

IEA (2013) World energy outlook 2013. OECD/IEA, Paris

IEA (2014) World energy investment outlook. Special report. OECD/IEA, Paris

IPCC (2007) Climate change 2007: the physical science basis. Contribution of working group I to the fourth assessment report of the Intergovernmental Panel on Climate Change (IPCC). Cambridge University Press, Cambridge

IPCC (2013) Climate change 2013: the physical science basis. Contribution of working group I to the fifth assessment report of the Intergovernmental Panel on Climate Change, Cambridge University Press, Cambridge

► Jarvis A, Li S (2011) The contribution of timescales to the temperature response of climate models. Clim Dyn 36: 523–531

► Joughin I, Smith BE, Medley B (2014) Marine ice sheet collapse potentially under way for the Thwaites Glacier Basin, West Antarctica. Science 344:735–738

► Levermann A, Clark PU, Marzeion B, Milne GA, Pollard D, Radic V, Robinson A (2013) The multimillennial sea-level commitment of global warming. Proc Natl Acad Sci USA 110:13745–13750

► Matthews HD, Weaver AJ (2010) Committed climate warming. Nat Geosci 3:142–143

► Matthews HD, Zickfeld K (2012) Climate response to zeroed emissions of greenhouse gases and aerosols. Nat Clim Change 2:338–341

► Matthews HD, Gillett NP, Stott PA, Zickfeld K (2009) The proportionality of global warming to cumulative carbon emissions. Nature 459:829–832

Meehl GA, Washington WM, Collins WD, Arblaster JM and others (2005) How much more global warming and sea level rise? Science 307:1769–1772

► Meehl GA, Hu A, Tebaldi C, Arblaster JM and others (2012) Relative outcomes of climate change mitigation related to global temperature versus sea-level rise. Nat Clim Change 2:576–580

► Mengel M, Levermann A (2014) Ice plug prevents irreversible discharge from East Antarctica. Nat Clim Change 4:451–455

Rhein M, Rintoul SR, Aoki S Campos E and others (2013) Observations: ocean. In: Stocker TF, Qin D, Plattner GK, Tignor M and others (eds) Climate change 2013: the physical science basis. Contribution of working group I to the fifth assessment report of the Intergovernmental Panel on Climate Change. Cambridge University Press, Cambridge

► Ridley J, Gregory J, Huybrechts P, Lowe J (2010) Thresholds for irreversible decline of the Greenland ice sheet. Clim Dyn 35:1049–1057

► Solomon S, Plattner GK, Knutti R, Friedlingstein P (2009) Irreversible climate change due to carbon dioxide emissions. Proc Natl Acad Sci USA 106:1704–1709

► van Vuuren D, Edmonds J, Kainuma M, Riahi K and others (2011) The representative concentration pathways: an overview. Clim Change 109:5–31

► Wigley TML (2005) The climate change commitment. Science 307:1766–1769

► Zickfeld K, Eby M, Weaver AJ, Alexander K and others (2013) Long-term climate change commitment and reversibility: an EMIC intercomparison. J Clim 26:5782–5809

4

Effect of a high-end CO_2-emission scenario on hydrology

Ida B. Karlsson[1,3,*], Torben O. Sonnenborg[1], Lauren P. Seaby[2], Karsten H. Jensen[3], Jens Christian Refsgaard[1]

[1]Geological Survey of Denmark and Greenland (GEUS), Voldgade 10, Copenhagen 1350, Denmark

[2]Department of Environmental, Social, and Spatial Change, Roskilde University, Universitetsvej 1, Building 02, Roskilde 4000, Denmark

[3]Department of Geosciences and Natural Resource Management, Copenhagen University, Rolighedsvej 23, 1958 Frederiksberg C, Copenhagen, Denmark

ABSTRACT: In the latest IPCC report, worst case scenarios of climate change describe average global surface warming of up to 6°C from pre-industrial times by the year 2100. This study highlights the influence of a high-end 6 degree climate change on the hydrology of a catchment in central Denmark. A simulation from the global climate model, EC-Earth, is downscaled using the regional climate model HIRHAM5. A simple bias correction is applied for daily reference evapotranspiration and temperature, while distribution-based scaling is used for daily precipitation data. Both the 6 degree emission scenario and the less extreme RCP4.5 emission scenario are evaluated for the future period 2071–2099. The downscaled climate variables are applied to a fully distributed, physically based, coupled surface–subsurface hydrological model based on the MIKE SHE model code. The impacts on soil moisture dynamics and evapotranspiration show increasing drying-out tendencies for the future, most pronounced in the 6 degree scenario. Stream discharge and groundwater levels also show increased drying due to higher evapotranspiration. By comparing the 6 degree scenario with other emission scenarios, it is found that the most prominent changes in the water balance are caused by drying out of soils rather than precipitation effects.

KEY WORDS: Climate change · High-end scenarios · Hydrological modelling · Impact study

1. INTRODUCTION

During the last decade the number of hydrological impact studies has been increasing, and several studies have been published trying to quantify future hydrological changes using climate model outputs on both global (e.g. Arnell 1999), national (e.g. Bergstrom et al. 2001), and regional (e.g. van Roosmalen et al. 2007) scales. The need for ensemble approaches in climate change impact studies has been highlighted several times (e.g. Déqué et al. 2007) to encompass more of the uncertainties arising from emission scenarios as well as climate models and downscaling techniques. Impact studies have therefore increasingly focused on different combinations of scenarios, models, and techniques. Emission scenarios have traditionally been based on projections defined in Nakicenovic et al. (2000) and later by van Vuuren et al. (2011).

The most recent IPCC report (Collins et al. 2013) describes the possibility for high-end scenarios like RCP8.5 with a global climate warming of over 4°C at the end of the 21st century. Changes in this range will have profound effects on hydrology and vegetation, with a potential for increasing in large ecological and economic impacts such as water shortage, wildlife loss, crop failure, flooding, and droughts. The hydrology of a high-end scenario has however not yet been documented.

*Corresponding author: ika@geus.dk

The objective of this study is to assess the hydrological impacts for a high-end emission scenario (with a 6°C warming), from here onwards referred to as the 6 degree scenario, compared to impacts for a medium emission scenario. Due to the unusually warm climate projection used in this study, it is logical to focus primarily on consequences of drying effects within the hydrological system. Since the study area is primarily an agricultural area, we decided to emphasise factors affecting crops. This is done using the agricultural drought indices soil moisture deficit index (SMDI) and evapotranspiration deficit index (ETDI) (Narasimhan & Srinivasan 2005).

2. STUDY AREA

The study area is a sub-catchment in the Odense Fjord Basin, located on the island of Funen in central Denmark (Fig. 1). The Odense River drains the 1025 km^2 basin running from mid-Funen into the Odense Fjord in the northeast. In this study, the focus is on the upstream 486 km^2 large sub-catchment with an average river discharge of 4.6 $m^3 s^{-1}$ (1991–2010). The topography in the area varies from ~12 to 129 m from the river valley to the moraine hills. Climatologically, the area is temperate and wet with an annual mean temperature of 8.8°C and precipitation of 808 mm yr^{-1} (1991–2010). The Odense Fjord Basin is comprised of mostly agricultural lands (68%) with some urban areas (16%), woodlands (10%) and a smaller percentage of natural areas (6%) like wetlands, lakes, meadows, and grasslands. The pre-

dominant crops are spring and winter cereals, constituting 23 and 45%, respectively, of all farmland (Environment Centre Odense 2007). The geology in the catchment is dominated by end moraines to the south and southwest, and moraine hills to the southeast and northwest. Clayey moraine deposits dominate the area itself; however, aquifers constituted by meltwater sand and gravel deposits are also present (Troldborg et al. 2010).

3. METHODS

3.1. Hydrological modeling code

MIKE SHE originates from the Systeme Hydrologique Europeen (SHE) modelling system (Abbott et al. 1986) and is now a coherent modelling framework by DHI Water and Environment. The model is fully distributed and physically-based, and includes descriptions of the following processes: evapotranspiration, snow melt, overland flow, channel flow, unsaturated zone flow, drainage pipe flow, and groundwater flow. For each of these processes different numeric engines are available. The processes are calculated separately but are coupled 2-ways in every time step. All spatially distributed input data are pre-processed into grid based files of the specified grid size. Channel flow is handled by the MIKE-11 model using a kinematic routing description. The saturated flow is calculated using a 3D finite difference scheme, which is a fully distributed and physically based solver (DHI 2009b).

3.2. Emission scenarios and climate models

The basis of this study is the 6 degree high-end emission scenario; it is an idealized scenario, meaning that the run is not based on projected emissions but rather a forced CO_2-development. The atmospheric CO_2 concentration is specified to increase from a pre-industrial level with a rate of 1% yr^{-1} and serves as input to the global coupled atmosphere–ocean–sea ice model EC-Earth (Hazeleger et al. 2012). The model reaches a 6 degree warming at an atmospheric CO_2-concentration of ~1423 ppm. Hereafter, the model is run with a fixed CO_2 concentration for 29 yr. In this period the global mean surface temperature continues

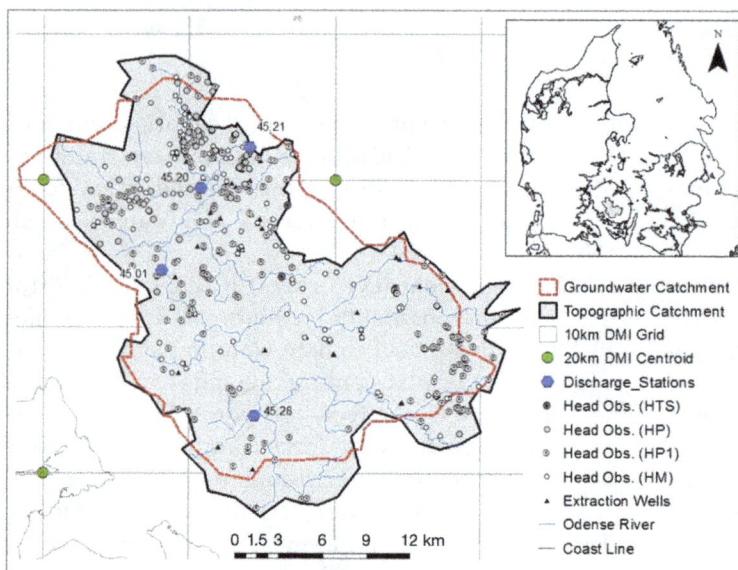

Fig. 1. Study area: Odense Fjord Basin, Denmark

to slowly increase and stabilize at a level of about 6.5 K above pre-industrial level. These 29 yr are then used to represent the years 2071–2099 as a 6 degree warming. Dynamic downscaling is carried out using the RCM HIRHAM5 (Christensen et al. 2007), see Christensen et al. (2015, this Special) for a detailed description. Results from the same general circulation model/regional climate model (GCM/RCM) pairing forced by historical data from the period 1961–2005 were used to establish the bias correction for the control period (see below).

The RCP4.5 emission scenario was also run through the same climate and hydrological model framework to provide data for comparison. The RCP4.5 scenario is based on a real emission projection founded on work by Smith & Wigley (2006), Clarke et al. (2007) and Wise et al. (2009). The RCP4.5 owes its name to the maximum radiative forcing of 4.5 W m^{-2} reached at the end of the century, followed by a stabilization of constant forcing thereafter (Thomson et al. 2011). This forcing is equivalent to a CO_2-concentration of ~540 ppm (Meinshausen et al. 2011) and results in roughly 2 degrees warming at the end of the century. Additionally, results from the study by Karlsson et al. (2014) for the same area using alternative climate models from the ENSEMBLES project (Hewitt & Griggs 2004) and the A1B emission scenario (Nakicenovic et al. 2000) are also included as an additional impact reference. The A1B scenario corresponds approximately to a radiative forcing of 6 W m^{-2} and 700 ppm CO_2-concentration at the end of the century (IPCC 2001).

Whereas results on temperature and precipitation are used directly from the RCM, daily reference evapotranspiration is estimated using the FAO Penman–Monteith equation (Allen et al. 1998) based on RCM output such as incoming and outgoing, short- and long-wave radiation, temperature, water vapor pressure, and wind speed.

3.3. Bias correction

The choice of bias correction method is based on a study by Seaby et al. (2013). They used different GCM/RCM couplings from the ENSEMBLES project (Hewitt & Griggs 2004) investigating the different responses of climate change for 6 Danish regions on a 10 km grid. Two different downscaling methods were evaluated: the delta change approach (DC) and a distribution-based scaling method (DBS; Piani et al. 2010). The DC approach was used on observed data of temperature, precipitation and reference evapotranspiration (potential evapotranspiration for a well-watered grass of uniform height) while applying a monthly change factor (RCM control to future, indirect use of climate data). The DBS method was applied for precipitation using double gamma distributions combined with a dry day correction on a seasonal basis. This method was used along with a bias removal (BR) method on reference evapotranspiration and temperature that also uses the simulated future RCM data directly. Here, the difference (bias) between the observation data and the RCM control period data is perturbed onto the simulated future data. Seaby et al. (2013) showed that the DBS and the DC approach were equally good at reproducing changes in the mean, but the DBS was superior in preserving the variance as well as capturing the precipitation extremes. In a subsequent analysis Seaby et al. (2015) found that a DBS downscaling for each individual grid resulted in reduction of spatial biases between RCM control data and observed data.

The climate model inputs in this study are thus downscaled using DBS for precipitation and BR for reference evapotranspiration and temperature. The historical control period is 1991–2005, while the future period covers 2071–2099. Both the BR and DBS methods are applied on a grid-by-grid basis for each season.

3.4. CO_2-crop factor

Increases in CO_2-concentrations have been shown to affect the stomatal conductance of the leaf surface for plants (Kimball et al. 1993), meaning that higher levels of CO_2 allow the plant to reduce the stomata opening, thus lowering water vapor loss resulting in reduced evapotranspiration (Samarakoon & Gifford 1995, Kimball et al. 1999, Conley et al. 2001, Krujit et al. 2008). This response is somewhat uncertain as other influences are still debated. In this study the approach by Rasmussen et al. (2012) is adopted, where CO_2 sensitivities of leaf conductance, relative transpiration, and transpiration share of the evapotranspiration are multiplied by the relative CO_2 increase. Wheat was chosen as a representative crop, as cereals are by far the most common in the catchment. Using the empirical and experimental values from Rasmussen et al. (2012), a plant correction was multiplied onto the 2 emission scenarios' annual CO_2 change (Meinshausen et al. 2011) and subsequently perturbed onto the RCM reference evapotranspiration. Hydrological model runs were done both with and without this correction.

3.5. Agricultural drought index

To describe agricultural drought, the Soil Moisture Deficit Index (SMDI) and Evapotranspiration Deficit Index (ETDI) were developed by Narasimhan & Srinivasan (2005); They applied the indices for each sub-basin on simulated data from the SWAT model (Arnold et al. 1998) for 6 catchments in Texas, USA. The results showed that both indices are well correlated with actual crop yields on the sites.

The SMDI uses the soil moisture deficit, which is based on weekly soil moisture data from the model and median soil moisture as well as minimum or maximum soil moisture in a reference period. The SMDI is a result of the soil moisture deficit and the SMDI of the previous week. Similarly, the ETDI is based on the weekly water stress anomaly, where the anomaly includes the median and minimum or maximum water stress ratio in a reference period. This ratio is calculated as the difference between potential and actual evapotranspiration, divided by the potential. The ETDI is a product of the water stress anomaly and the ETDI of the previous week. Both indices range between −4 and +4.

In this study, the indices ETDI and SMDI for the top 30 cm soil column are applied in a similar fashion as in Narasimhan & Srinivasan (2005); however, here the indices are applied at the grid scale and used to assess future data as well as historical data. To remove potential climate model biases, the indices are based on GCM/RCM data for both periods. The median, maximum, and minimum water contents (SMDI) and water stresses (ETDI) are based on data from the reference period (1991–2005). Hence, the same values of, for example minimum water content, are used to calculate the indices for the historical and future periods, and this means that the index range may go outside the normal ±4, as lower minimum and higher maximum values may be found in the future climate data. The choice to preserve the historical statistics as basis was chosen in order to compare future results with a known period of reference. If the complete dataset (historical and future) was chosen as basis for the statistics, the SMDI and ETDI in the historical period could not have been assumed to be correlated to the actual yield.

4. MODEL SETUP

MIKE SHE has been used extensively in the Danish National Water Resources model (DK-model) (Henriksen et al. 2003), developed at the Geological Sur-

vey of Denmark and Greenland (GEUS). The DK-model comprises 7 model domains covering the whole of Denmark (Højberg et al. 2013). The model domain covering Funen (Troldborg et al. 2010) has been used as point of departure for setting up the MIKE SHE model used in this study.

4.1. Boundary conditions and numerical setup

The groundwater catchment falls outside the topographical catchment by 12%, while 14% of the topographical catchment is outside the groundwater catchment (Fig. 1). Flow in the area is dominated by near-surface processes including overland flow, drain flow, and shallow groundwater flow. Hence, to minimize potential water balance errors the topographical catchment is used to delineate the model boundary. As horizontal boundary conditions in the model, zero flux is applied on all borders as it is assumed that the catchment boundaries represent hydrological divides. The horizontal discretisation of the model is 200 × 200 m, resulting in a total of 12 630 grids covering the whole catchment. The unsaturated flow is determined by the full Richards equation and discretized using a cell spacing of 5 cm in the top 30 cm of the soil column, with cell sizes increasing to 1 m cells at depths below 10 m. The number of layers therefore depends on the depth of the groundwater table (e.g. 12 layers if the groundwater table is at 1 m depth). The saturated zone is resolved by 7 computational layers, and the flow is calculated using a 3D finite difference scheme (DHI 2009b). Maximum time step is specified to 12 h for the overland flow and unsaturated zone, and 24 h for the saturated zone.

In connection with the farming and households within the catchment, there are 103 extraction wells. There is no irrigation in the catchment. Based on the Danish National Well database, Jupiter (GEUS 2014), reported extraction rates are used for the historical simulation period while fixed rates are used from 2012 onward.

4.2. Soil type

The soil data is based on 13 different soil types (Greve et al. 2007) distributed in 3 soil horizons: A, B, and C. This gives rise to potentially almost 2200 soil column combinations. Due to computational limitations, 50 type-soil profiles were generated based on the most common A-, B-, and C-horizon combinations (Børgesen et al. 2013). In this study, the 10 occuring type-soil profiles in the catchment with different A-, B-, C-horizon soil properties are listed here with area coverage (%) and soil profile number (in parentheses): 0.1% moraine sand soil (16); 1% diluvial sand soil (18); 15% moraine clay soil (37); 15% diluvial sand soil (38); 0.4% moraine sand soil (46); 12% moraine clay soil (47); 10% diluvial sand soil (48); 33% moraine clay soil (67); 11% diluvial sand soil (77); and 4% freshwater sand soil (998).

4.3. Land use

Land use in the area is divided into 12 categories; 3 covering natural landscapes: grassland (5%), deciduous forest (3%), and coniferous forest (2%); 7 covering farm lands: 3 types of dairy farms (18%), 2 types of pig farms (48%), and 2 types of plant production (16%); one category covering hydrology in the form of water bodies (1%); and one category covering urban areas (8%).

In reality, farm type dependent crop rotation schemes are used at each individual field. To simplify the description, the farm type's crop rotation is translated into a relative distribution of crop types within each of the 7 farm types (Table 1), which results in 12 different crop types recognized in the catchment; however, only 8 different crop inputs are used, as some crop types are very similar and are thus treated as one (Table 1). For each of the 8 crop types, information about the crop parameters including leaf area index and root depth are available from DAISY

model runs for the period 1990–2010 (Børgesen pers. comm.). DAISY (Hansen et al. 1990) is a root zone model that is physically based and simulates nitrate and carbon transport/transformation and water flow based on agricultural practices.

As the crop parameters not only depend on crop type but also on soil type, the crops are distributed such that the relative distribution is preserved within each land use–soil combination. Spatially, the grids are distributed randomly on the grid map in the cells with the appropriate land use–soil combination. The 5 additional land use categories are represented by one 'crop type' only: the forests by deciduous and coniferous trees, urban areas and water bodies by 'crops' specified to these categories, and the grassland is represented by the farm land crop type of grass. Finally, this results in a grid map where the 12 crop types and 10 soil types in combination yield 107 different crop–soil matches in total for the whole catchment.

4.4. Geology

The geology in the area is divided into units with similar hydraulic properties and is based on the work done in connection with the DK-model (Troldborg et al. 2010). The area contains 11 geological units, constituting 4 hydro-stratigraphic layers. The first layer contains the top 3 m with the geological units of top sand, top clay, top chalk, top moraine clay, peat, and other distributed on the basis of a geological map. The first layer covers the unsaturated zone and the top of the saturated zone. The second layer is a fractured

Table 1. Relative crop distribution within each farm type

Crop	Farm types							% of agricultural cells
	Plant/mixed	Pig/plant	Pig	Dairy 1	Dairy 2	Dairy 3	Unknown/mixed	
Maize	0.2	5.3	0.7	2.6	18.3	31.3	1.4	5
Grass								21
Grass	1.8	4.7	0.6	5.5	15.7	15.7	6.4	
Pasture	3.6	4.1	2.1	14.3	7.8	5.3	28.2	
Grass Seeds	9.1	5.7	5.6	6.8	2.5	0.7	5.1	
Barley								20
Spring Cereal	22.9	23.8	21	27.2	19.2	16	14.5	
Legume	0.6	1.2	0.3	0.2	1	0	0.2	
Winter Wheat	38.1	35.9	50	26.8	20.3	18.9	19.7	35
Winter Rape	4	4	7.6	2.3	2.6	3.7	1.8	5
Potatoes	1.3	1.2	0.3	0.2	0.6	0	0.5	1
Sugar beets								5
Vegetable	1	1.4	0.4	0.8	0.3	0	2.9	
Sugar Beets	5.5	4.5	4.8	4.5	3.3	3.5	2.2	
Fallow	11.9	8.1	6.7	8.7	8.4	4.9	17.3	8
% of agricultural cells	12	22	36	3	14	5	7	–

clay layer. The third layer is a low permeable formation consisting of moraine clay, post-glacial clays, and silts (geological units: quaternary clay). Below the third layer, pre-quaternary deposits containing low permeable clays and chalk from Danien (geological units: pre-quaternary clay, pre-quaternary fractured clay, pre-quaternary chalk) are situated. Within these 4 layers, 3 sand bodies consisting of outwash sands and gravel (geological units: outwash sand) are recognized and constitute the aquifers in the area. The middle sand lens represents the aquifer with the largest extent and thickness (Nyegaard et al. 2010).

4.5. Time varying input data

Daily values of precipitation, temperature, and reference evapotranspiration are specified as input to the model. As precipitation input, the 10 km DMI grid data (Scharling 1999) (Fig. 1) with dynamic gauge catch correction (Stisen et al. 2012) are used. Data for temperature and reference evapotranspiration, calculated by Makkink's formula (Makkink 1957), are specified as 20 km grid values (Fig. 1).

5. CALIBRATION

The model is run for the period 1990–2010, where 1990–1999 is used as a spin-up period, and is calibrated for the period 2004–2007. There are 2 validation periods, one before (2000–2003) and one after (2008–2010) the calibration period. The AutoCal scheme incorporated in the MIKE SHE software is used as the autocalibration tool (DHI 2009a). The most sensitive parameters were identified by a sensitivity analysis prior to the calibration.

5.1. Calibration data

The catchment includes 4 discharge stations with data in the calibration period: Stns 45.21, 45.01, 45.28, and 45.20 (see Fig.1). There are 455 wells with hydraulic head observations in the catchment. The wells are divided into 4 categories: H_{TS}, H_P, H_{P1}, and H_M. H_{TS} contains 5 wells with long time series (>3000 observations) within the calibration period. H_P contains 63 wells that only have a few observations within the calibration period. H_{P1} contains 209 wells with only 1 observation within the calibration period. The last category, H_M, holds the remaining 178 wells; these wells have no observations within

the calibration period; however, their measurements are used as average values for hydraulic head (calculated outside the calibration period).

5.2. Objective function

The objective functions are based on the performance criteria water balance (WB) and RMSE:

$$f(\theta)_{WB,X} = \sum_{k=1}^{K}\left(\frac{1}{n}\sum_{i=1}^{n}(O_i - S_i)\right) \qquad (1)$$

$$f(\theta)_{RMSE,X} = \sum_{k=1}^{K}\left(\frac{1}{n}\sum_{i=1}^{n}(O_i - S_i)^2\right)^{1/2} \qquad (2)$$

where n is the number of time steps, O_i and S_i are observed and simulated values at time $i = 1, 2, ..., n$, for $k = 1, 2, ... K$, where K is number of stations or wells. A subscript, X, after the objective function type (WB or RMSE) indicates whether the objective function is based on discharge values (Q) or on head measurements (H_{TS}, H_P, H_{P1} or H_M). For the calibration scheme, a total of 6 different objective functions are used.

The objective functions for the 4 discharge stations are aggregated such that the total objective function value is simply a summation of all stations, ensuring that small sub-catchment stations have smaller contributions to the overall objective function. Station 45.28 is situated just downstream lake Arreskov Soe, and the outflow from the lake is highly influenced by human regulations, which makes it difficult to reproduce the temporal variability of the outflow from the lake. Therefore, the station is only included in the objective function for WB. The multiple objection function, within the predefined parameter space (θ) is defined as:

$$F(\theta) = \qquad (3)$$

$$\min\{f(\theta)_{WB,Q}; f(\theta)_{RSME,Q}; f(\theta)_{RSME,H_{TS}}; f(\theta)_{RSME,H_p}; f(\theta)_{RSME,H_{ps1}}; f(\theta)_{RSME,H_M}\}$$

5.3. Optimization algorithm

The Population Simplex Evolution method (PSE; DHI 2009a) is used as the optimization algorithm in this study.

The stopping criterion is defined by either the maximum number of model evaluations or a convergence in the objective function space. This means that the objective function of the best parameter set has not changed more than a given value within a number of shuffling loops or a convergence in the parameter space, so the range of parameter values of the entire population is less than a given value (DHI 2009a).

6. RESULTS

6.1. Parameter sensitivity and parameterization

The MIKE SHE model has an extensive and complex setup and as such contains many parameters available for calibration. To decrease calibration time and to avoid non-uniqueness problems, the number of parameters has to be limited to the most sensitive set. To determine the most sensitive parameters for this model setup, 28 parameters were tested in a sensitivity analysis using the scaling procedure described by Hill (1998). The parameters investigated in the river and drainage system are the drain depth, the drainage time constant, and the stream leakage coefficient. For the unsaturated zone, the saturated hydraulic conductivity, as well as the van Genuchten parameters n and alpha for 2 soils (37 and 38) were chosen. Soil 37 is a clay type soil and tied to this value are the other dominating clay soils: 47, 67, and 68. Soil 38 is a sandy soil, and the other sandy soil, no. 48, is tied to this value. The values are tied individually for all 3 soil horizons: A, B and C. In the saturated zone the horizontal hydraulic conductivity for the geological units top sand, top moraine clay, outwash sand, quaternary clay, pre-quaternary clay with and without fractures and pre-quaternary chalk are investigated. The vertical hydraulic conductivities are tied to the horizontal values using an anisotropy factor of 10 for all but the top moraine clay where a factor of 100 is used.

This results in a total of 71 parameters where only 28 are free. The result of the aggregated sensitivity analysis is shown in Fig. 2A. The 10% limit denotes 10% of the sensitivity of the parameter with the highest sensitivity value here given by the horizontal conductivity for the quaternary clay. However, a further investigation of the parameter sensitivity for the individual objective functions shows that for the 2 least sensitive parameters above the 10% limit, the saturated hydraulic conductivity for soil 37 in horizon B is only sensitive for 1 objective function type, while the n parameter for soil 37 in horizon C is only sensitive for 2 objective function types (not shown). It was therefore chosen only to select the horizontal hydraulic conductivity for the outwash sand, the quaternary clay, and the chalk; as well as the time constant for the drains and the saturated conductivity for soil 37 of horizon C for calibration. Initial parameter values were based on the optimal parameter values from the DK-model.

6.2. Calibration and validation results

The calibration required 285 simulations to reach a satisfactory solution for the 5 parameter values (Fig. 2B). The resulting hydrograph for the main station can be seen in Fig. 2C. A relatively low WB error of –3% is found, indicating a small tendency for the model to overestimate the discharge. The Nash-Sutcliffe coefficient indicates a high degree of covariance between observed and simulated data. The plots in Fig. 2D,E show the WB and Nash-Sutcliffe coefficients for the 4 stations for both validation and calibration periods. Still looking only at the main station (Stn 45.21), the WB error is slightly elevated in the validation periods 2000–2003 and 2008–2010, while the Nash-Sutcliffe coefficient is very similar to the calibration period.

The signal is similar for Stn 45.20 as for the main station, but the performance is poorer. The poorer performance may partly be due to the smaller size of the catchment and may be affected by the non-accounted differences between the groundwater and the topographical divides in this area. It should be noted that observed data only exists for 2000–2003 and 2005. Therefore, no data are available for the second validation period. The performance for Stn 45.28 is generally low; however, this is not surprising as the hydrograph for this station is highly affected by human regulation at the outlet from the lake. Stn 45.01 has a high Nash-Sutcliffe coefficient for both validation and calibration periods. Both water balance and RMSE show elevated values in the calibration period compared to the validation periods. The reason for these discrepancies seems to originate partly from 2 peaks in the observed data in June/July and December 2007, as the WB error is reduced to 8.5% and the RMSE to 1.17 mm without these months.

The plot in Fig. 2F shows the RMSE for the 2 head objective functions H_{TS} and H_M. The H_P is not shown as few wells have measurements in the whole period, and H_{P1}-wells only have measurements in the calibration period. The H_M shows that the model performs almost equally well on the mean hydraulic head in the validation periods as in the calibration period. The H_{TS} cannot be directly compared across the 3 periods as there are a different number of measurements which the RMSE is based on, for instance, only 3 of the H_{TS}-wells have any data in the last validation period.

6.3. Downscaled climate results

The results of the downscaled climate data are presented in Fig. 3 using change factors. The 6

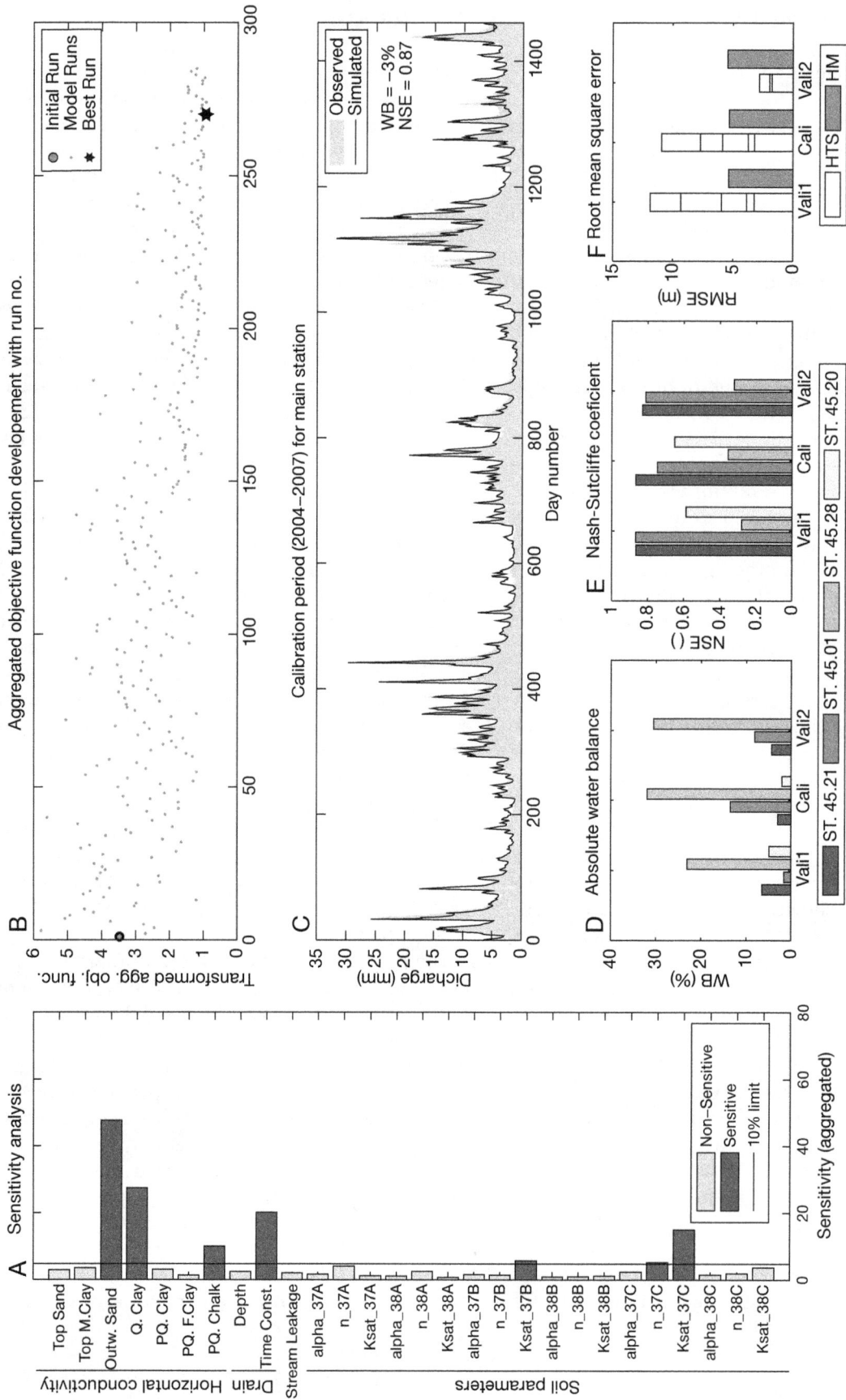

Fig. 2. (A) Sensitivity analysis of the chosen 29 parameters in the MIKE SHE model. 10 % limit: 10 % of the parameter with the highest sensitivity; Sensitivity: aggregated sensitivity across all 6 objective functions. (B) Change in the transformed and aggregated multiple objective function with model run number during the calibration. * Best run: final model run chosen. (C) Hydrograph for the final model run for the main discharge station (Stn 45.21) in the catchment; with water balance (WB) and Nash-Sutcliffe coefficient for streamflow (NSE). (D,E) WB values and Nash-Sutcliffe coefficients for each of the 4 discharge stations, and (F) RMSE for 2 of the heads objective functions (H$_{TS}$ and H$_M$) in the calibration (cali; 2004–2007) and validation periods 1 and 2 (vali; 2000–2003 and 2008–2010)

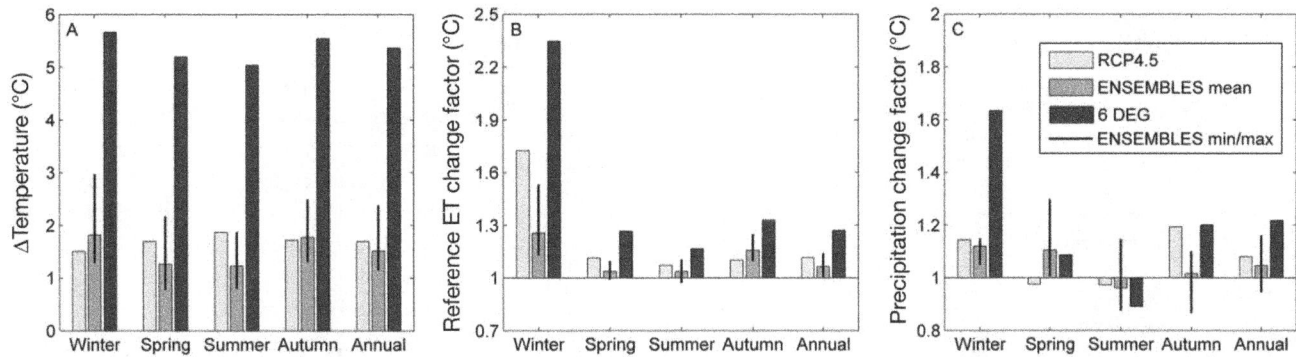

Fig. 3. Bias corrected climate variables for Funen for the 2 emission scenarios: RCP4.5 and 6 degree. Mean of the 4 ENSEM-BLES models: ECHAM5-HIRHAM5, ECHAM5-RCA3, ARPEGE-RM5.1 and HadCM3-HadRM3, and an indication of the minimum and maximum values of the 4 simulations (vertical line). (A) Relative change of mean temperature. (B) Change in reference evapotranspiration (ET) without CO_2-correction. (C) Change factors for the mean precipitation

degree emission scenario results in an annual change in temperature of 5.4°C (Fig. 3A), while the RCP4.5 induce an increase of 1.7°C. The distribution of the change is somewhat evenly spread across the seasons. For both emission scenarios the reference evapotranspiration increases (Fig. 3B), with the winter season experiencing the largest relative increase (this reference evapotranspiration is downscaled data without CO_2-correction applied). For precipitation (Fig. 3C), the signal of the 2 scenarios are quite different. Both agree on an annual increase in precipitation. For the RCP4.5, winter and autumn show an increase in precipitation, with autumn being the largest, while summer and spring precipitation are decreasing. For the 6 degree scenario, however, the largest increase is found by far for the winter precipitation, with spring and autumn having smaller increases, while only summer shows a decrease in precipitation.

Fig. 3 also shows the mean change values for 4 ENSEMBLES models from the study by Karlsson et al. (2014). The 4 models ECHA M5-HIRHAM5, ECHAM5-RCA3, ARPEGE-RM5.1, and HadCM3-HadRM3 represent a dry, wet, warm, and medium scenario for Denmark, respectively. The vertical line on the plot indicates the maximum and minimum value for the models. For some seasons and variables, the difference between the 4 models is smaller than the difference between the signals from the emission scenarios (A1B and RCP4.5 compared to the 6 degree). It is seen that the results from the RCP4.5 scenario are close to the mean of the ENSEMBLES results with respect to reference evapotranspiration and precipitation, and to some degree temperature.

6.4. Use of CO_2-crop factor

The CO_2-correction resulted in a 4 % reduction of reference evapotranspiration for the RCP4.5 emission scenario compared to the non-corrected RCP4.5 reference evapotranspiration, while the 6 degree scenario resulted in a 24 % reduction. For comparison, the A1B emission scenario results in a reduction of 7 % using the same procedure (Karlsson et al. 2014). The effect on the reference evapotranspiration compared to the baseline period can be seen in Table 2, along with other major hydrological variables. As the evapotranspiration is lower when the CO_2-correction is used, it is not surprising that the model response is overall wetter; however, for the 6 degree scenario the CO_2-correction actually leads to a complete signal reversal of the changes in actual evapotranspiration, stream discharge and groundwater level. The reduction in evapotranspiration from the 6 degree scenario seems extreme, and it raises the question of whether this correction factor type is valid when dealing with high-end scenarios with large CO_2-changes. Especially considering the still debated extent of the CO_2-effect and possible feedback, it was therefore chosen to continue further analysis of the simulation results without including the CO_2-correction on the reference evapotranspiration.

6.5. Discharge and groundwater

The mean monthly discharge of the main station can be seen on Fig. 4A. The station has high discharge in the winter and low discharge during

Table 2. Change in annual mean flux (mm) from RCM historical control period (1991–2005) to future period (2071–2099) (mm). 6 DEG: high-end 6 degree emission scenario

	Precipitation	Actual evapotranspiration	Discharge main station	Mean hydraulic head (Layer 6)
Without CO_2-correction				
RCP45	+50	+37	+25	+100
6 DEG	+58	+86	−19	−96
With CO_2-correction				
RCP45	+50	+22	+39	+171
6 DEG	+58	−12	+77	+348

summer. When imposing the climatic changes defined by the 2 climate scenarios, the result is an overall increase in discharge for the RCP4.5 and a decrease for the 6 degree scenario (Table 2). Fig. 4B shows that the increase in discharge for the RCP4.5 is primarily a result of large increases in discharge for December to March, while the response is fairly close to control period values for the other months. The 6 degree scenario results in large increases in discharge during January and February as well as noteworthy reductions in April to November. The signal from the station further upstream (Stn 45.01) is similar to the main station (not shown), while the tributary station (45.20) shows decreasing discharge for both scenarios for all months except July in the RCP4.5. The smallest station (45.28) shows decreasing discharge all year with the exception of December for the 6 degree scenario. Decreasing discharge is found in January to September (July increase excluded), and increasing discharge in October to December for the RCP4.5.

The change in hydraulic head can be evaluated in Fig. 5. The hydraulic head in the catchment is generally highest in the northwest, southwest and southeast corners of the catchment; while the lowest hydraulic head is found in the downstream end of the river valley. For the RCP4.5 scenario there is a mean groundwater level rise of 0.8 to 1.4 m in the southwestern part of the catchment, while the rest of the catchment is relatively unchanged. However, for the maximum values it is apparent that the scenario also results in a reduction in the maximum heads in the north and an increase in maximum level in south. For the 6 degree scenario there is a reduction in the mean, minimum, and maximum groundwater levels, except for a smaller rise in the southwest.

6.6. SMDI and ETDI

Fig. 6 shows the spatial distribution of the SMDI and ETDI in the study area; the top row in the figures is the driest week (from the indexing) for the historical period, the RCP4.5 and the 6 degree scenarios. The bottom row on the figures shows the percentage of weeks where the index is below −3 for each grid. Fig. 7 shows the accumulated distribution of the SMDI and ETDI index for the whole catchment. The dashed line shows the driest occurring week, corresponding to the weeks presented on Fig. 6, while the solid lines represent the distribution of the index during all summer weeks for the entire period.

The driest week in the historical period (Fig. 6A) results in scattered locations of dry areas with SMDI between −3 and −4, where 63 % of all grids are below

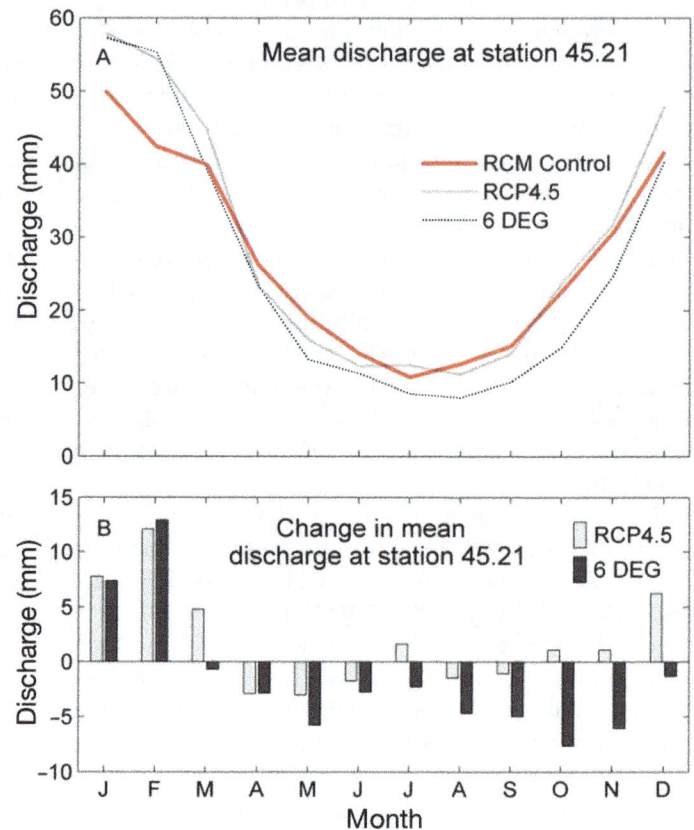

Fig. 4. (A) Mean monthly discharge at the main station (45.21) of the catchment for 2 emission scenarios (RCP4.5 and 6 degree) and the RCM control period, and (B) their respective relative change (in mm) from the RCM historical control period. Note: this is a hydrological model run without CO_2-correction on reference evapotranspiration

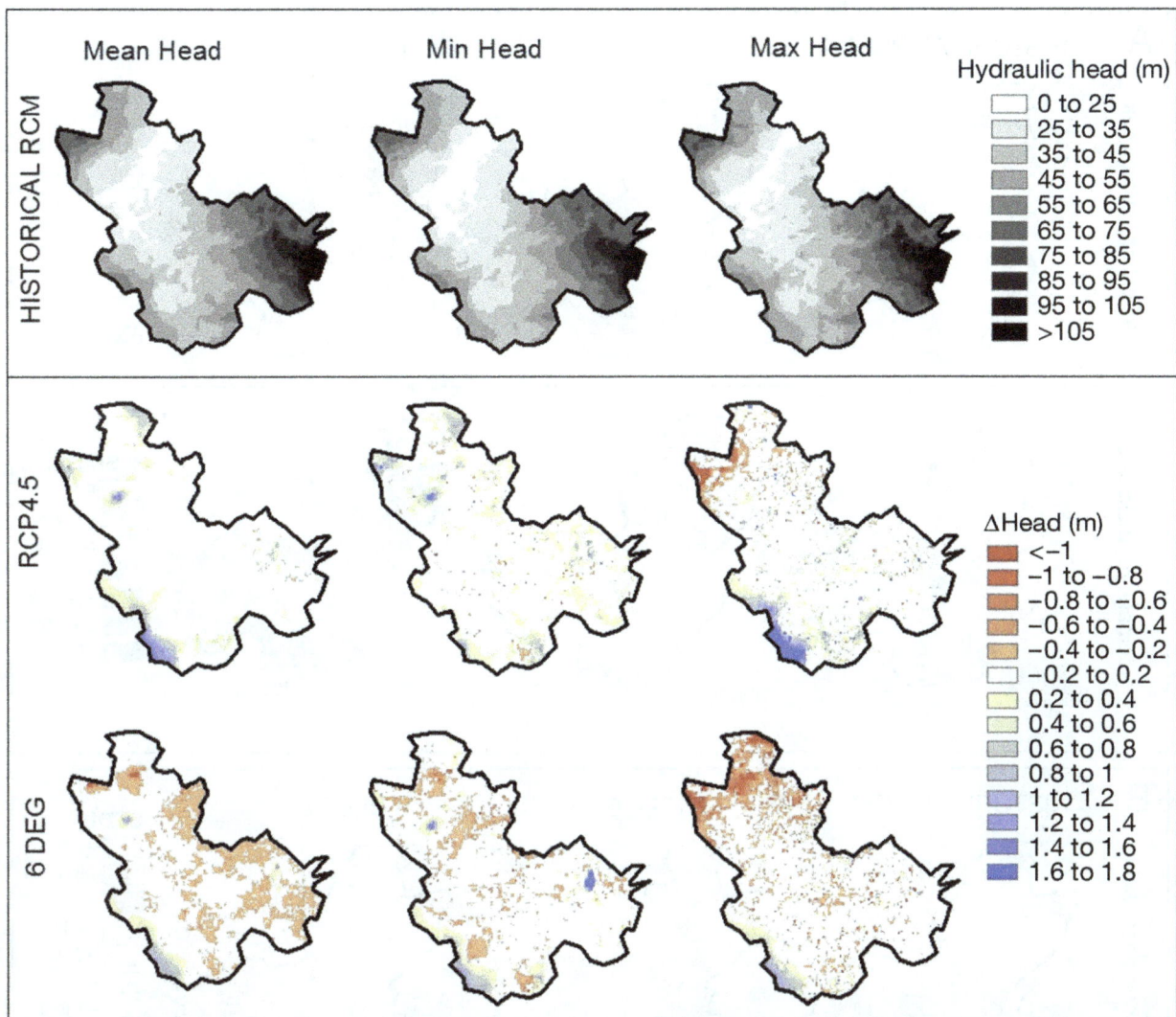

Fig. 5. Top row: mean, minimum and maximum hydraulic head for the historical RCM simulation. Bottom 2 rows: change in mean, min. and max. head from the historical RCM control period to the RCP4.5 and the 6 degree scenario future period (no CO_2-correction). Red: reduction of the head values in the future; Blue: increase in head value. Head values: top layer

−3 (Fig. 7A). For the RCP4.5 the area becomes even drier (92 % is below −3), with the driest areas present in the south. Moving on to the 6 degree scenario, a drought is realized in almost the entire catchment, with SMDI values below −4 for most of the area (96 % is below −3; 72 % is below −4). Looking at the percentage of weeks with dry conditions, a similar signal is found with increasing dryness from RCP4.5 to the 6 degree scenario, where >70 % of all grids experience SMDI below −3 more than 10 % of the time. This is also apparent on Fig. 7A for the summer weeks index distribution. Here, the graphs for the 2 future periods are both offset to the left indicating increasing dryness. The largest difference between the historical period and the 6 degree scenario is found for the driest index values.

The ETDI shows that the RCP4.5 has a less extreme driest week than the historical period (Fig. 6B), even though over the whole period the RCP4.5 has more grids with dry days. Also the total response from the summer weeks (Fig. 7B) shows a close resemblance to the historical period distribution, with a smaller tendency towards a drying offset for the driest indices. The 6 degree scenario is still the driest with 24 % of all grids below −4 in the driest week. This week, however, also shows that even though the driest indices become drier, the less dry (above −4) actually become wetter (Fig. 7B). As with the SMDI, the largest difference in index distribution for the historical and the 6 degree scenario is found in the driest indices. The overall drying signal is not as significant for the ETDI as for the SMDI, which may point out the

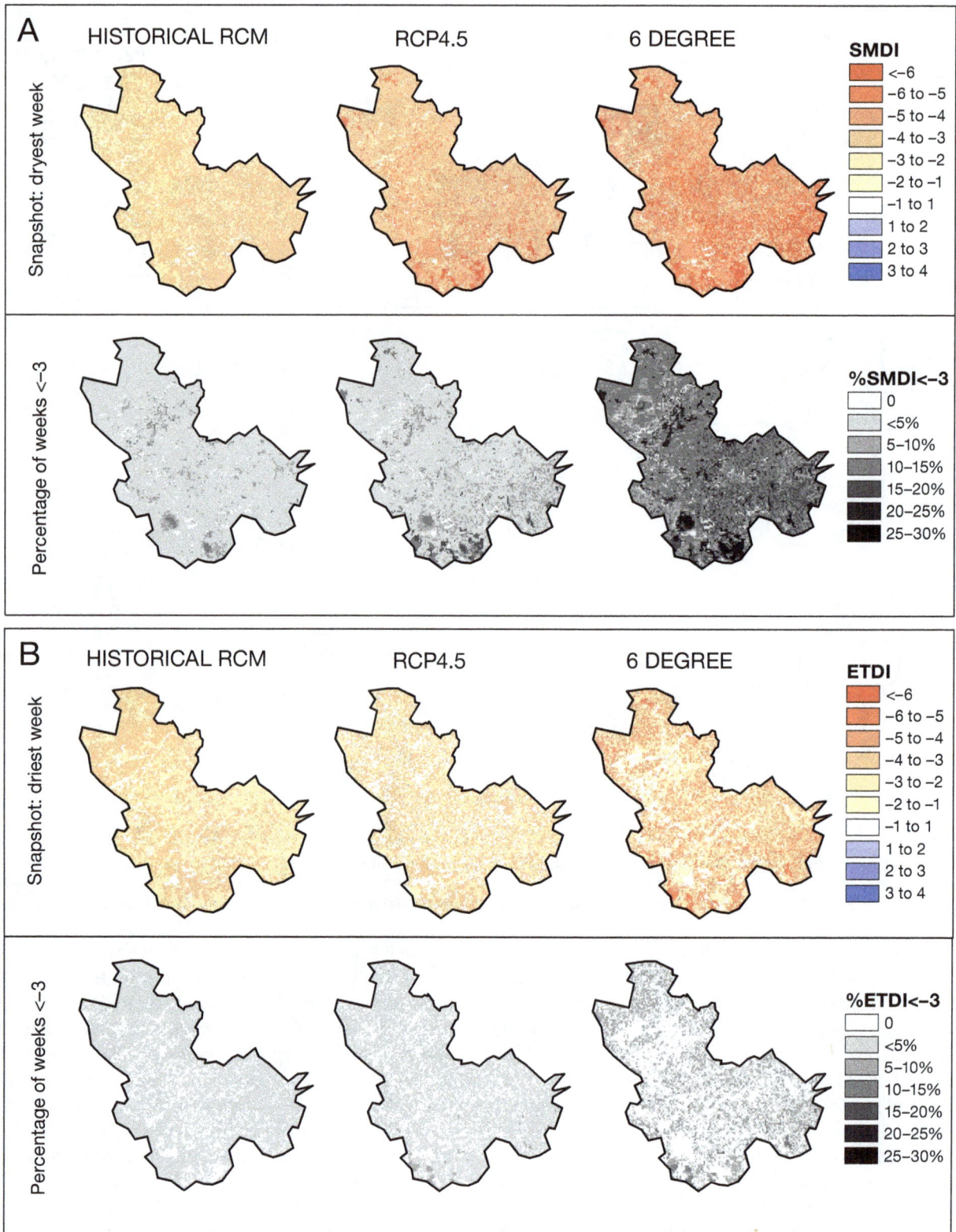

Fig. 6. (A) Soil moisture deficit index (SMDI) and (B) evapotranspiration deficit index (ETDI). Top row: driest week in the simulation period as a snapshot in time for the historical RCM simulation and the 2 future simulations from the RCP4.5 and the 6 degree scenario future period (no CO_2-correction). Bottom row: percentage of weeks during the simulation period where SMDI/EIDI were below −3 for each grid in the catchment

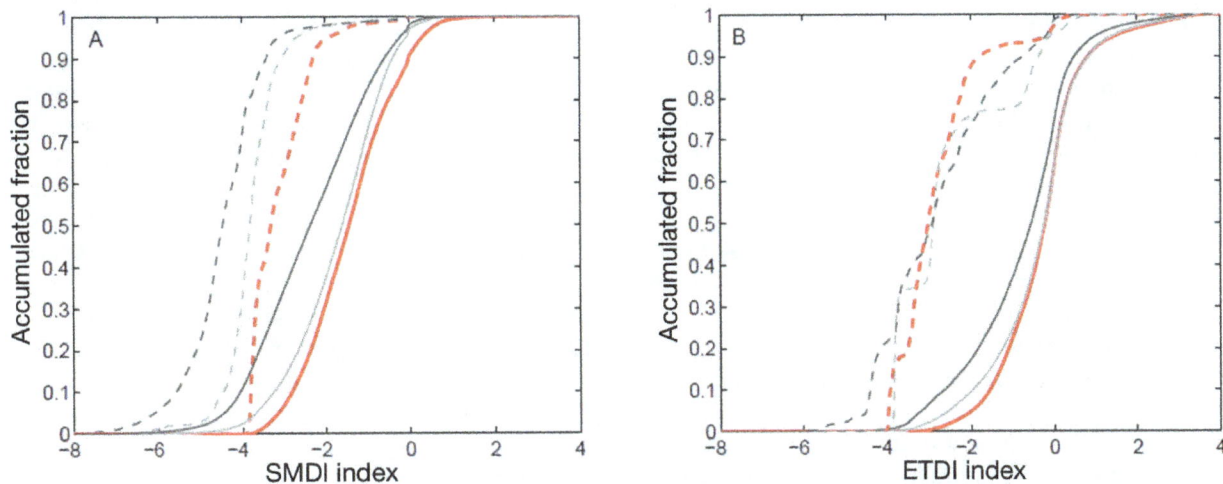

Fig. 7. Accumulated distribution of the (A) SMDI and (B) ETDI in all catchment grids in the control/future period. Red: RCM control; light grey: RCP4.5; dark grey: 6DEG. Distributions are shown both for the driest week (dashed lines) and for all the summer weeks (solid lines)

differences in the 2 indices, as SMDI only covers the top 30 cm of the root zone, while the ETDI represents the overall signal for the soil moisture available for plant transpiration in the entire root zone. On both index maps—more so for ETDI—the location of the river can be recognized (Fig. 6).

7. DISCUSSION

In this study, climate variables from a high-end 6 degree emission scenario were bias-corrected and used in a distributed hydrological model for the Odense sub-catchment in central Denmark. The impact on hydrology and agricultural drought was then evaluated. Furthermore the high-end scenario was compared with a medium emission scenario, the RCP4.5, from the same GCM/RCM model combination, and to some degree with results from A1B scenario runs with multiple GCM/RCM combinations.

7.1. Characteristic of the high-end 6 degree scenario

The 6 degree scenario is characterized by large year round temperature increases of up to 5.7°C, with increasing reference evapotranspiration as a result. Precipitation especially increases in winter and autumn, while summer experiences a decrease in precipitation. For the Odense sub-catchment, this means an annual precipitation change of +7% from the historical to future period (RCM-GCM values), compared to a +6% change in the RCP4.5. The 6 degree emission scenario results in large changes in

the hydrology of the catchment. In spite of the precipitation increase, the actual evapotranspiration increase of +17% for the 6 degree scenario causes decreasing stream discharge for most of the year. This is especially critical in the summer period where stream flow is already low, potentially leading to problematic ecological consequences as stream and wetland desiccation. Furthermore, the changes also lead to increasing difference in stream flow between the seasons as January and February have the highest discharge values and will experience further increase in the future.

The risk of drying-out is also evident from the overall lowering of the groundwater table in the catchment. The reason for the increased groundwater level in the southwest of the catchment may be due to greater depth of the groundwater table, implying that even though the groundwater level rises in response to the higher winter precipitation, it is still too deep for root zone water uptake. Conversely, the shallow groundwater table in the majority of the catchment enables the crops to extract water and thereby increase the evapotranspiration, and thus reduce the groundwater level.

7.2. CO$_2$ effect

The effect of increasing CO$_2$ is, in this study, described by a relatively simple approach where a factor is multiplied on the reference evapotranspiration to mimic the reduction in transpiration of the plants. However, the effect of CO$_2$ on plant evapotranspiration is still highly uncertain (see e.g. Zhu et al. 2012) and affected by several factors. CO_2 acts as a plant fertilizer and an increase in CO_2 could cause an increase in aboveground biomass and hence leaf area index, which will have a positive influence on evapotranspiration that may counteract the effect of stomata closure. This was shown experimentally by Qiao et al. (2010), where elevated CO_2 did not have any significant effect on evapotranspiration. These dynamic feedbacks on plant physiology are not included in this study, which could be a serious limitation. Further, it is not known at which CO_2-concentration range the approach by Krujit et al. (2008) is valid. The high-end emission scenario considered here results in extremely high CO_2-concentrations, and it is uncertain how the crops react to such a forcing. Finally, using coupled climate–carbon cycle models, Peng et al. (2014) demonstrated that the CO_2-physiological effect on transpiration also significantly affects the precipitation response of the models. Hence, it is difficult to isolate the effect of CO_2 on transpiration and not include the feedback mechanisms on cloud and precipitation formation. Because of these deficiencies in the description of the effects of increasing CO_2-concentration, it was decided not to include this effect in the final model simulations. However, this mechanism adds uncertainty to the results and more research should be done to quantify this effect.

7.3. Inter-model variability vs. emission scenario variability

Several studies have shown that the influence of the choice of climate model is more important than the choice of the emission scenario (Hawkins & Sutton 2011, IPCC 2013), as the spread in model results is large. These studies have been based on emission scenarios from the normal range of emission projections. For some variables and seasons, the present study indicates that when high-end scenarios are included in the impact studies, the choice of emission projection can be more important than the climate model choice. This finding should, however, be considered with caution, because we do not have ensemble results for the high-end scenario, but only for the moderate A1B scenario (ENSEMBLES runs).

7.4. Indices and usability

The indices SMDI and ETDI were developed by Narasimhan & Srinivasan (2005), and to our knowledge this is the first time the indices have been used

to evaluate future climate change impacts on the root zone water balance. Narasimhan & Srinivasan (2005) found a good correlation between wheat and sorghum crop yield and the index response. While further studies may be needed to document the general applicability of these indices as good indicators of crop yield, the study by Narasimhan & Srinivasan (2005) indicates that the use of indices on soil moisture data and/or evapotranspiration data from a hydrological model can be used as an initial evaluation tool to assess future climate change impact on crop yield, thereby saving substantial time on complex agricultural and hydrological model combinations.

8. CONCLUSION

This study has evaluated the impact of a high-end 6 degree emission scenario on hydrology and soil moisture in a Danish agricultural catchment. The high-end scenario showed larger temperature increases and large changes in reference evapotranspiration and precipitation. The stream discharge impact caused larger seasonal differences from winter to summer, as winter discharge increased while most of the remaining year had decreasing discharge values. Generally the climate change resulted in a lowering of the groundwater head and a substantial increase of dryness in the root zone, represented by an overall lowering of the agricultural indices. Altogether, the most prominent changes in the water balance were found due to drying out of soils rather than precipitation effects.

Acknowledgements. The present study was funded by a grant from the Danish Strategic Research Council for the Centre for Regional Change in the Earth System (CRES – www.cres-centre.dk) under contract no: DSF-EnMi 09-066868. We also acknowledge the help received from Christen Duus Børgesen, Department of Agroecology – Climate and Water at Aarhus University, Denmark in connection with the implementation of vegetation properties in the model.

LITERATURE CITED

Abbott MB, Bathurst JC, Cunge JA, O'Connell PE, Rasmussen J (1986) An introduction to the European hydrological system – Systeme Hydrologique Europeen, 'SHE'. 2. Structure of a physically-based, distributed modelling system. J Hydrol 87:61–77

Allen RG, Pereira LS, Raes D, Smith M (1998) Crop evapotranspiration: guidelines for computing crop water requirements. FAO Irrigation and Drainage Paper 56, FAO, Rome

Arnell NW (1999) Climate change and global water resources. Glob Environ Change 9(Supplement 1):S31–S49

► Arnold JG, Srinivasan R, Muttiah RS, Williams JR (1998) Large area hydrologic modelling and assessment. 1. Model development. J Am Water Resour Assoc 34:73–89

Bergström S, Carlsson B, Gardelin M, Lindström G, Pettersson A, Rummukainen M (2001) Climate change impacts on runoff in Sweden: assessments by global climate models, dynamical downscaling and hydrological modelling. Clim Res 16:101–112

Børgesen CD, Jensen PN, Blicher-Mathiesen G, Schelde K and others (2013) Udviklingen i kvælstofudvaskning of næringsstofoverskud fra dansk landbrug for perioden 2007–2011. Evaluering af implementerede virkemidler til reduktion af kvælstofudvaskning samt en fremskrivning af planlagte virkemidlers effekt frem til 2015. In: Børgesen CD, Jensen PN, Blicher-Mathiesen G, Schelde K (eds) DCA Rapport, Nr. 031, DCA, Tjele

Christensen OB, Drews M, Christensen JH, Dethloff K, Ketelsen K, Hebestadt I, Rinke A (2007) The HIRHAM Regional Climate Model Version 5 (β). DMI Tech Rep 06-17, DMI, Copenhagen

► Christensen OB, Yang S, Boberg F, Fox Maule C and others (2015) Scalability of regional climate change in Europe for high-end scenarios. Clim Res 64:25–38

Clarke L, Edmonds J, Jacoby H, Pitcher H, Reilly J, Richels R (2007) Scenarios of greenhouse gas emissions and atmospheric concentrations. Sub-report 21A of Synthesis and Assessment Product 2.1 by the US Climate Change Science Program and the Subcommittee on Global Change Research, Department of Energy, Washington, DC

Collins M, Knutti R, Arblaster J, Dufresne JL and others (2013) Long-term climate change: projections, commitments and irreversibility. In: Stocker TF, Qin D, Plattner G-K, Tignor M and others (eds) Climate Change 2013: the physical science basis. Contribution of Working Group I to the Fifth Assessment Report of the Intergovernmental Panel on Climate Change. Cambridge University Press, Cambridge

► Conley MM, Kimball BA, Brooks TJ, Pinter PJ and others (2001) CO_2 enrichment increases water-use efficiency in sorghum. New Phytol 151:407–412

► Déqué M, Rowell D, Lüthi D, Giorgi F and others (2007) An intercomparison of regional climate simulations for Europe: assessing uncertainties in model projections. Clim Change 81:53–70

DHI (2009a) AutoCal: autocalibration tool. In: DHI (ed) Mike Zero 2009. DHI, Water & Environment, Hørsholm, p 1–42

DHI (2009b) Reference guide. In: DHI (ed) MIKE SHE User Manual, Book 2. DHI, Water & Environment, Hørsholm, p 132–168 and 345–367

Environment Centre Odense (2007) Odense pilot river basin. Pilot project for river basin management planning. In: Madsen HB, Pedersen SE, Kristensen ND, Jørgensen OT (eds) Water Framework Directive Article 13. Danish Ministry of the Environment, Environment Centre Odense, Odense

GEUS (Geological Survey of Denmark and Greenland) (2014) National well database JUPITER, available at www.geus.dk/DK/data-maps/jupiter/Sider/default.aspx (accessed February 2014)

► Greve MH, Greve MB, Bøcher PK, Balstrøm T, Madsen HB, Krogh L (2007) Generating a Danish raster-based topsoil property map combining choropleth maps and point information. Geogr Tidsskr 107(2):1–12

Hansen S, Jensen HE, Nielsen NE, Svendsen H (1990) DAISY: soil–plant–atmosphere system model. NPO Res

Rep A10, The National Agency for Environmental Protection, Copenhagen

► Hawkins E, Sutton R (2011) The potential to narrow uncertainty in projections of regional precipitation change. Clim Dyn 37:407–418

Hazeleger W, Wang X, Severijns C, Ştefănescu S and others (2012) EC-Earth V2.2: description and validation of a new seamless earth system prediction model. Clim Dyn 39:2611–2629

► Henriksen HJ, Troldborg L, Nyegaard P, Sonnenborg TO, Refsgaard JC, Madsen B (2003) Methodology for construction, calibration and validation of a national hydrological model for Denmark. J Hydrol 280:52–71

► Hewitt CD, Griggs DJ (2004) Ensembles-based predictions of climate changes and their impacts. EOS Trans AGU 85:566–566

Hill MC (1998) Methods and guidelines for effective model calibration. Water Resour Invest Rep 98-4005. USGS, Denver, CO

Højberg AL, Troldborg L, Stisen S, Christensen BSB, Henriksen HJ (2013) Stakeholder driven update and improvement of a national water resources model. Environ Model Software 40:202–213

IPCC (Intergovernmental Panel on Climate Change) (2001) Climate change 2001: the scientific basis. Contribution of Working Group I to the Third Assessment Report of the Intergovernmental Panel on Climate Change. In: Houghton JT, Ding Y, Griggs DJ, Noguer M and others (eds) IPCC Third Assessment Report. Cambridge University Press, Cambridge

IPCC (2013) Climate Change 2013: the physical science basis. Contribution of Working Group I to the Fifth Assessment Report of the Intergovernmental Panel on Climate Change. In: Stocker TF, Qin D, Plattner G-K, Tignor M and others (eds) IPCC Fifth Assessment Report (AR5). Cambridge University Press, Cambridge

Karlsson IB, Refsgaard JC, Sonnenborg TO, Jensen KH (2014) Significance of hydrological model choice when doing climate change impact assessment. EGU General Assembly, 27 April– 2 May 2014, Vienna

Kimball BA, Mauney JR, Nakayama FS, Idso SB (1993) Effects of increasing atmospheric CO_2 on vegetation. Vegetatio 104-105:65–75

► Kimball BA, LaMorte RL, Pinter PJ Jr, Wall GW and others (1999) Free-air CO_2 enrichment and soil nitrogen effects on energy balance and evapotranspiration of wheat. Water Resour Res 35:1179–1190

► Krujit B, Witte JPM, Jacobs C, Kroon T (2008) Effects of rising atmospheric CO_2 on evapotranspiration and soil moisture: a practical approach for the Netherlands. J Hydrol 349:257–267

Makkink GF (1957) Testing the Penman formula by means of lysimeters. J Inst Water Eng Sci 11:277–288

► Meinshausen M, Smith SJ, Calvin K, Daniel JS and others (2011) The RCP greenhouse gas concentrations and their extensions from 1765 to 2300. Clim Change 109: 213–241

Nakicenovic N, Alcamo J, Davis G, de Vries B and others (2000) Emission scenarios. In: Nakicenovic N, Swart R (eds) Emission scenarios. Cambridge University Press, Cambridge

► Narasimhan B, Srinivasan R (2005) Development and evaluation of soil moisture deficit index (SMDI) and evapotranspiration deficit index (ETDI) for agricultural drought monitoring. Agric For Meteorol 133:69–88

Nyegaard P, Troldborg L, Højberg AL (2010) DK-model 2009: Geologisk og hydrostratigrafisk opdatering 2005–2009. GEUS Rep 2010/80, Copenhagen

▶ Peng J, Dan L, Dong W (2014) Are there interactive effects of physiological and radiative forcing produced by increased CO2 concentration on changes of land hydrological cycle? Global Planet Change 112:64–78

▶ Piani C, Haerter JO, Coppola E (2010) Statistical bias correction for daily precipitation in regional climate models over Europe. Theor Appl Climatol 99:187–192

▶ Qiao Y, Zhang H, Dong B, Shi C, Li Y, Zai H, Lui M (2010) Effects of elevated CO2 concentration on growth and water use efficiency of winter wheat under two soil water regimes. Agric Water Manag 97:1742–1748

▶ Rasmussen J, Sonnenborg TO, Stisen S, Seaby LP, Christensen BSB, Hinsby K (2012) Climate change effects on irrigation demands and minimum stream discharge: impact of bias-correction method. Hydrol Earth Syst Sci 16:4675–4691

▶ Samarakoon AB, Gifford RM (1995) Soil water content under plants at high CO_2 concentration and interactions with the direct CO_2 effects: a species comparison. J Biogeogr 22:193–202

Scharling M (1999) Klimagrid Danmark: nedbør 10 × 10 km. Metodebeskrivelse. DMI Tech Rep 99-15. DMI, Copenhagen

▶ Seaby LP, Refsgaard JC, Sonnenborg TO, Stisen S, Christensen JH, Jensen KH (2013) Assessment of robustness and significance of climate change signals for an ensemble of distribution-based scaled climate projections. J Hydrol 486:479–493

▶ Seaby LP, Refsgaard JC, Sonnenborg TO, Højberg AL (2015) Spatial uncertainty in bias corrected climate change projections and hydrological impacts. Hydrol Processes (in press) doi:10.1002/hyp.10501

Smith SJ, Wigley T (2006) Multi-gas forcing stabilization with MiniCAM. Energy J Spec Issue 3:373–392

▶ Stisen S, Højberg AL, Troldborg L, Refsgaard JC, Christensen BSB, Olsen M, Henriksen HJ (2012) On the importance of appropriate precipitation gauge catch correction for hydrological modelling at mid to high latitudes. Hydrol Earth Syst Sci 16:4157–4176

▶ Thomson AM, Calvin KV, Smith SJ, Kyle GP and others (2011) RCP4.5: a pathway for stabilization of radiative forcing by 2100. Clim Change 109:77–94

Troldborg L, Højberg AL, Nyegaard P, Stisen S, Christensen BSB, Ondracek M (2010) DK-model2009: Modelopstilling og kalibrering for Fyn. GEUS Rep 2010/76, Copenhagen

▶ van Roosmalen L, Christensen BSB, Sonnenborg TO (2007) Regional differences in climate change impacts on groundwater and stream discharge in Denmark. Vadose Zone J 6:554–571

▶ van Vuuren D, Edmonds J, Kainuma M, Riahi K and others (2011) The representative concentration pathways: an overview. Clim Change 109:5–31

▶ Wise M, Calvin K, Thomson A, Clarke L and others (2009) Implications of limiting CO_2 concentrations for land use and energy. Science 324:1183–1186

▶ Zhu Q, Jiang H, Peng C, Liu J and others (2012) Effects of future climate change, CO_2 enrichment, and vegetation structure variation on hydrological processes in China. Global Planet Change 80–81:123–135

5

Projecting the future ecological state of lakes in Denmark in a 6 degree warming scenario

Dennis Trolle[1,2,*], **Anders Nielsen**[1,3], **Jonas Rolighed**[1], **Hans Thodsen**[1],
Hans E. Andersen[1], **Ida B. Karlsson**[4], **Jens C. Refsgaard**[4], **Jørgen E. Olesen**[2,3],
Karsten Bolding[1], **Brian Kronvang**[1], **Martin Søndergaard**[1], **Erik Jeppesen**[1,2]

[1]Department of Bioscience, Aarhus University, Vejlsøvej 25, 8600 Silkeborg, Denmark
[2]Sino-Danish Centre for Education and Research (SDC), Beijing, PR China
[3]Department of Agroecology, Aarhus University, Blichers Allé 20, 8830 Tjele, Denmark
[4]Geological Survey of Denmark and Greenland (GEUS), Øster Voldgade 10, 1350 Copenhagen, Denmark

ABSTRACT: Lakes are highly sensitive to climate change, and climate warming is known to induce eutrophication symptoms in temperate lakes. In Denmark, climate is projected to cause increased precipitation in winter and increased air temperatures throughout the year by the end of the 21st century. Looking further into the future, the warming trend is projected to continue and likely reach a 6°C increase around the 22nd century (relative to a baseline period of 1986–2005). In the present study, we evaluate the consequences of such extreme changes for temperate Danish lakes. We use a multifaceted modelling approach by combining an eco-hydrological model to estimate future water runoff and catchment nutrient exports with both mechanistic and empirical lake models, describing key biogeochemical indicators in lakes, in order to quantify the effects of future nutrient loads and air temperature on lake ecosystems. Our model projections for the future scenario suggest that annual water runoff will increase (46%), driving also increases in exports of nitrogen and phosphorus (13 and 64%, respectively). Both the mechanistic and empirical modelling approaches suggest that phytoplankton biomass will increase and that potentially toxin-producing cyanobacteria may become a dominant feature of the phytoplankton community from spring. Warming and increased nutrient loads also affect the food webs within the lakes in the direction of higher fish control of algae-grazing water fleas, further reinforcing eutrophication. To be able to mitigate these eutrophication effects, external nutrient loading to the lakes must be reduced considerably.

KEY WORDS: Extreme warming · Impacts · Lakes

1. INTRODUCTION

Lakes are highly sensitive to climate change (Williamson et al. 2008). In Denmark, the projected climate change in this century suggests, on balance, increased precipitation in winter months, and consequently a higher export of nutrients from catchments to lakes (Andersen et al. 2006, Jeppesen et al. 2009, 2011). This may enhance eutrophication and the risk of dominance by potentially toxin-producing cyanobacteria (Trolle et al. 2014). These algae are further stimulated by higher temperatures (i.e. the kind of temperature expected in the future) (Kosten et al. 2012), and summer heat waves are known to favor blooms of potentially harmful cyanobacteria (Jöhnk et al. 2008).

Climate impact studies have typically been carried out using moderate emission scenarios such as A1B, yielding a projected warming of 2° to 3°C by 2100 relative to present time (van der Linden & Mitchell 2009). In its 5th assessment report, the Intergovern-

*Corresponding author: trolle@bios.au.dk

mental Panel for Climate Change (IPCC) noted that the present emissions follow a trajectory that is on or above the RCP8.5 scenario, leading to a likely global warming between 2.6° and 4.8°C for the period 2081–2100 compared to 1986–2005 (Collins et al. 2013). However, the warming will continue beyond 2100, and the expected (global mean) temperature increase under the Representative Concentration Pathway scenario RCP8.5 by 2181–2200 is 6.5°C, with a likely range between 3.3° and 9.8°C, rising even further by 2300. Hence, as long as the current emissions continue to increase, there is a global warming potential equivalent to 6°C, either before or, most likely, shortly after 2100 (Christensen et al. 2015, this Special). In the present study we evaluate the consequences of such extreme changes for temperate Danish lakes. In this context, Christensen et al. (2015) analyzed to what extent the actual projected RCP emission scenarios (van Vuuren et al. 2011) reach the same temperature level as the 6°C scenario. They found that 8 out of 9 model runs for the RCP8.5 scenario reached a 6°C warming before 2300. This indicates that even though a 6°C warming seems unlikely within the present century (Collins et al. 2013), it is plausible and very likely to occur at some time if the present level of emissions continues.

It is evident from empirical space-for-time studies (Meerhoff et al. 2012, Jeppesen et al. 2014) and studies of effects of extreme events, such as heat waves and severe precipitation (Jöhnk et al. 2008, Elliott 2010), that a 6°C warming scenario could have drastic effects on ecosystem functioning, ecological state and water quality of lakes. However, no studies relating to the impacts on lakes of such high-end climate change scenarios have yet, to the best of our knowledge, been reported in the literature.

In the present study, we elucidate the potential consequences of an extreme climate warming scenario on lake ecosystems by applying 2 different mechanistic ecosystem models to 4 different Danish temperate lakes, and by the use of existing and new empirical regression models describing relations between temperature and nutrient loads, and key biological indicators based on a dataset comprising 909 Danish lakes (and up to 1270 lake years).

2. METHODS

2.1. Climate scenarios

The climate change scenario used is a 6°C warming scenario generated by perturbing a 1% CO_2

increase every year onto a climate model until the response reaches 6°C. It is therefore not a truly projected CO_2-emission scenario as those reported in the IPCC reports (e.g. Nakicenovic et al. 2000), but rather an artificially created climatic state. The emission scenario was run in the global and regional climate model (GCM-RCM) coupling of ECEARTH-HIRHAM (Christensen et al. 2007, Hazeleger et al. 2012). After reaching a 6°C increase in air temperature relative to a baseline period comprising the years 1986–2005, the CO_2 concentration was held constant, while the model was run for an additional 29 yr. This period was then used as a representation of the future period, for convenience referred to as 2071–2099, although it may occur at another point in time. Further information on the climate model can be found in Christensen et al. (2015). Due to scale differences and climate model biases, the GCM-RCM output needs to be downscaled and bias corrected before it can be used to force hydrological and lake models. In this study, data were downscaled using the Distribution Based Scaling method (Piani et al. 2010) for precipitation data, and the bias removal approach (Seaby et al. 2013) for temperature data. The historical control period was set to 1991–2005 (for which observation data were readily available), and the downscaling was performed on a grid basis for all variables. Further information about the applied downscaling/bias correction methods is provided by Seaby et al. (2013) and Karlsson et al. (2015, this Special). For simplicity, and also as not all the lake models used in this study are able to account for climate forcing other than temperature, the future climate scenarios where generated through changes to air temperatures and precipitation alone (the latter for quantifying water and nutrient runoff), which we assume are the main drivers of changes in hydrology and nutrient dynamics, thus disregarding any changes that may occur in other meteorological variables.

For projections of the potential future state of Danish lake ecosystems, we take advantage of a range of mechanistic aquatic ecosystem model applications, which have previously been applied and undergone adaptation and rigorous testing with 4 Danish lakes, and also well-established empirical regression models supplemented with a few models of recent origin between biological indicators and temperature and nutrient loads (Table 1). These models have been set up to reflect certain, and somewhat differing, time periods, but all within the climate baseline period. To utilize the entire but differing time periods that these models represent (including preservation of the nat-

Table 1. Basic description of lakes used in the analysis: mechanistic modelling (4 lakes) and an empirical modelling approach

Lake	Model application period	Mean depth (m)	Surface area (km^2)	References
Lake Ravn	1989–2002	15	1.8	Trolle et al. (2008a,b)
Lake Arreskov	1990–2010	1.9	3.2	Nielsen (2013), Nielsen et al. (2014)
Lake Søbygaard	1990–2010	1.1	0.38	Rolighed (2013), Rolighed et al. (unpubl. data)
Lake Engelsholm	1999–2001	2.4	0.44	Trolle et al. (2014)
Empirical data (n = 909 lakes)	1989–2010	0.5–16.2	0.01–40	Jeppesen et al. (2009) (unpubl. data)

Table 2. Scenarios (SC) simulated for the 4 lake case studies (see Table 1) using mechanistic ecosystem models. Air temperature was used as the primary determinant for scenarios simulated by the lake ecosystem models (i.e. median, 10th and 90th percentiles [pctl] of daily air temperature time series), whereas downscaled precipitation data were only used as a baseline mean (750 mm yr^{-1}) and future mean (931 mm yr^{-1}) by the SWAT model to generate estimates of future nutrient exports (SC7–9)

Scenario	Climate data	Annual mean air temperature (°C)	External nutrient load data
SC1	Base (median)	8.6	Base
SC2	Base (10 pctl)	5.3	Base
SC3	Base (90 pctl)	11.4	Base
SC4	Future (median)	13.9	Base
SC5	Future (10 pctl)	11.2	Base
SC6	Future (90 pctl)	16.4	Base
SC7	Future (median)	13.9	Monthly changes (SWAT-estimated)
SC8	Future (10 pctl)	11.2	Monthly changes (SWAT-estimated)
SC9	Future (90 pctl)	16.4	Monthly changes (SWAT-estimated)

ural year to year variation in external nutrient loads), we derived annual time series of a national average daily air temperature for both the baseline and future climate periods, based on the climate model simulation. In total, a series including 6 annual time series with daily data was generated, representing the median, 10th and 90th percentile of daily air temperatures for the baseline and future periods (Table 2). In creation of baseline and future scenarios, these yearly time series of daily air temperatures were applied and repeated for the entire duration of the original lake model applications.

2.2. Eco-hydrological modelling for nutrient export estimates

In addition to direct climate effects on lakes represented by changes in air temperature, we derived an additional scenario where both temperature forcing and future nutrient loads are changed. The semi-distributed SWAT model (Soil and Water Assessment Tool) by Arnold et al. (1998) has recently been applied to the entire island of Funen, Denmark (Thodsen et al. unpubl. data). This set-up was utilized to obtain

an indication of potential future changes in nutrient exports from Danish catchments, which may eventually contribute to the loading of downstream lakes. We generated a single future water runoff and nutrient export scenario to be combined with the multiple scenarios of future climate warming. The changes in hydrology and nutrient exports in the future scenario are here influenced only by changes in air temperature and precipitation from the climate change scenario. We do acknowledge that other factors, and in particular land use management, will likely change considerably with climate (Olesen et al. 2011), and thus this will ultimately also shape the response of catchment hydrology and nutrient exports to climatic changes. In our study, however, we are interested in an overall indication of the magnitudes by which seasonal nutrient exports could change. An eco-hydrological model such as SWAT, set up to represent the current Danish landscape and land uses, can provide just that. Also, given the relatively small geographical extent of Denmark, and the fact that the island of Funen is to some extent representative of the median climate in Denmark in terms of net precipitation (precipitation minus evapotranspiration; e.g. Henriksen et al. 2003), we assume that model simulations of Funen provide a reasonable indicator for the behavior of other regions in Denmark. The catchment of the entire island of Funen (3528 km^2) was delineated by SWAT into 115 sub-basins and 7685 Hydrological Response Units (HRUs). Followed by a warm-up period (model spin-up) for 5 yr (1995–1999), SWAT was calibrated for discharge and total nitrogen (TN) and total phosphorus (TP) exports from 11 individual sub-basins scattered throughout Funen for the 6 yr period 2000–2005, and subsequently validated for the 4 yr period 2006–2009. Further details on the SWAT application are provided by Thodsen et al. (unpubl. data). For simplicity, the climate change scenario simulated by SWAT was implemented by a monthly mean delta change factor for air temperature and precipitation derived from the mean future climate change scenario period relative to the baseline period (where data for the future period was downscaled as described above). The monthly delta change factors were applied to the period on which all required input data for SWAT was available (1990–2009) using a 10 yr warm-up period followed by a 10 yr period for which the results were extracted and compared between the baseline and future climate scenario. Monthly average changes in the export of TN and TP were extracted and subsequently applied in additional lake ecosystem scenario simulations (Table 1).

2.3. Mechanistic ecosystem modelling

Mechanistic aquatic ecosystem models have previously been applied and have undergone rigorous testing on 4 Danish lakes (Table 1). This includes the PCLake model (Janse 1997) adapted to 3 shallow lakes (Lakes Arreskov, Søbygaard, and Engelsholm) and the DYRESM-CAEDYM model (Hamilton & Schladow 1997) adapted to a deep lake (Lake Ravn). Both model types have built in the effect of temperature (and nutrient loads) on a wide range of ecosystem processes, including, for example, mineralization rates of organic matter, growth rates specific for individual phytoplankton groups, sediment oxygen demand, etc. While differing somewhat in the conceptual basis (see Mooij et al. 2010 and Trolle et al. 2014 for details), including how temperature shapes particular processes, these models encompass a range of key components of aquatic ecosystems, including a closed nutrient cycle, and phytoplankton and zooplankton dynamics. In the original model applications for the Danish lakes (Table 1), water col-

umn temperature (and vertical mixing in case of the deep lake model application) was determined using a hydrodynamic model driven by observed climate forcing data for air temperature (°C), wind speed (m s^{-1}), shortwave radiation (W m^{-2}), relative humidity (%), and cloud cover (%). In this study, we adapted the climate forcing data of the original model applications to reflect that of the baseline and future scenario periods (Table 2) based on changes to the air temperature input to the models, whereas other meteorological variables remained the same as in the original model application. Thus, the yearly time series (represented by the median, 10th and 90th percentiles) of the climate model simulated daily air temperatures for the baseline and future scenarios were looped for the entire duration of the lake model applications, thereby allowing each application to retain its individual time series length, and consequently also the natural year to year variations in hydrology and external nutrient loads. By contrast, air temperatures had the same seasonality from year to year for each individual percentile. For the scenarios where future climates were combined with changed future nutrient loadings (estimated by the SWAT simulations), the future nutrient load was implemented by monthly change factors for phosphorus and nitrogen loads, respectively (allowing increasing loads in some months and decreasing in others, depending on the estimates of the eco-hydrological model simulations). When extracting the scenario results, the first year of simulation was discarded for each lake in all scenarios to reduce the effects of water temperature and biogeochemical state variable initializations (which were not changed between scenarios) on simulated outcomes. To enable scenario comparisons between lakes, we derived a yearly time series from daily data for each individual model application, representing the average of key state variables across the entire duration of the individual model applications, TP, TN, total chlorophyll a (chl a), the contribution of cyanobacteria biomass to total phytoplankton biomass (%), and the zooplankton: phytoplankton biomass ratio.

2.4. Empirical relations

A series of empirical regressions between lake water quality attributes (e.g. biological indicators), water temperature, hydraulic load (inferred by retention time) and nutrient load were established to enable extrapolation to the future scenarios. To calculate the annual mean lake TP concentration (TP$_{lake}$) from the discharge-weighted annual mean inlet TP concentration (TP_{in}), we used the simple empirical Vollenweider model (OECD 1982) model:

$$TP_{lake} = TP_{in}(1 + \sqrt{TW})^{-1} \qquad (1)$$

where TW is the hydraulic retention time (in yr) in the lake. This model has given reasonably good results for Danish lakes (Jeppesen et al. 1991). We also used established relationships (Jeppesen et al. 1997) between annual means of TP_{lake} and lake mean depth (Z, m) and chl a ($\mu g\ l^{-1}$) in the surface water, as well as water transparency (Secchi depth, m) from a dataset based on Danish lakes (TP_{lake} range: 0.017–1.91 mg P l^{-1}, mean depth range: 0.7–16.5 m):

$$chl\ a = \exp[5.78 + 0.85\log(TP_{lake}) - 0.26\log(Z)] \qquad (2)$$

Secchi depth =
$$\exp[-1.23 + 0.45\log(TP_{lake}) + 0.42\log(Z)] \qquad (3)$$

with $R^2 = 0.72$, n = 60 for Eq. (2) and $R^2 = 0.75$, n = 59 for Eq. (3). We further developed new empirical regression models for August TP concentrations, cyanobacteria biomass and the size of cladocerans (a zooplankton-based indicator of fish predation, Jeppesen et al. 2010) in the present study, based on a general linear modelling (GLM) approach using forward selection of variables (input variables included TP_{lake}, Z, and lake temperature). Using these regression models, we analyzed the effects of changes to temperature and nutrient load.

3. RESULTS

3.1. Future nutrient exports

The SWAT-simulated response of catchment hydrology and nutrient exports to the future change in mean air temperature and precipitation showed marked seasonality. The flow-weighted monthly discharge rates averaged across 11 stations on Funen suggest that discharge may decrease in late summer to early autumn (September to October) and increase in the remaining months (Fig. 1). The discharge will increase most during the winter months, up to 86 % in January, and decrease most in October (8 %). The changes in discharge patterns influence nutrient exports, which tend to decrease for a longer period ranging from July to October, while increasing in the remaining months. The changes in nutrient export dynamics are most dramatic for TP, increasing up to 119 % in January relative to the baseline, and decreasing 26 % in August, while the corresponding figures for TN is a 41 % increase and a 42 % decrease,

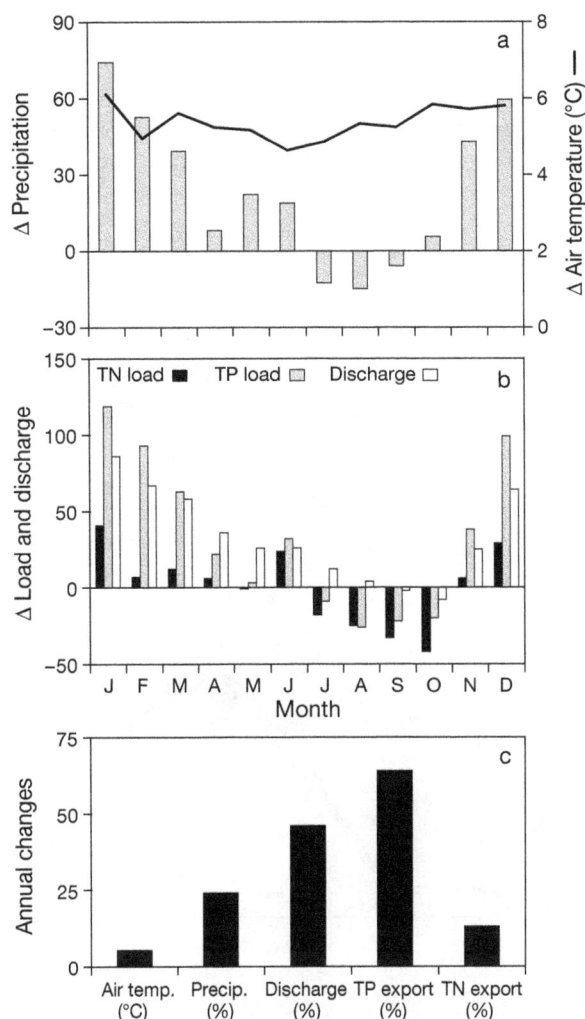

Fig. 1. (a) Monthly changes in precipitation and air temperature between the baseline and future climate periods, and (b) simulated monthly and (c) annual changes in discharge and nutrient exports (air temperature and precipitation also included in [c]). Variables are in %, except temperature (°C). TP: total phosphorus; TN: total nitrogen

respectively. On an annual basis, precipitation increases by 24 %, discharge rates by 46 %, TP exports by 64 %, and TN exports by 13 % (Fig. 1), illustrating also that much of the increased precipitation falls during winter months with low actual evaporation (thus causing relative increases of runoff to change more than that of precipitation on an annual basis).

3.2. Changes and variability in key properties of the aquatic ecosystems projected by the mechanistic models

The natural year to year variability in temperature and phytoplankton biomass of the baseline scenario relative to the future scenario for the individual lakes reveals that the future will bring about changes that go beyond what we experience from the year to year variations of the present (Fig. 2). For example, the future minimum summer phytoplankton biomass in

Lake Arreskov is generally greater than the highest biomass achieved under the present climate (Fig. 2d). There is also a general tendency for the spring bloom to occur earlier in the future in all the lakes. Phytoplankton biomass generally increases in the future

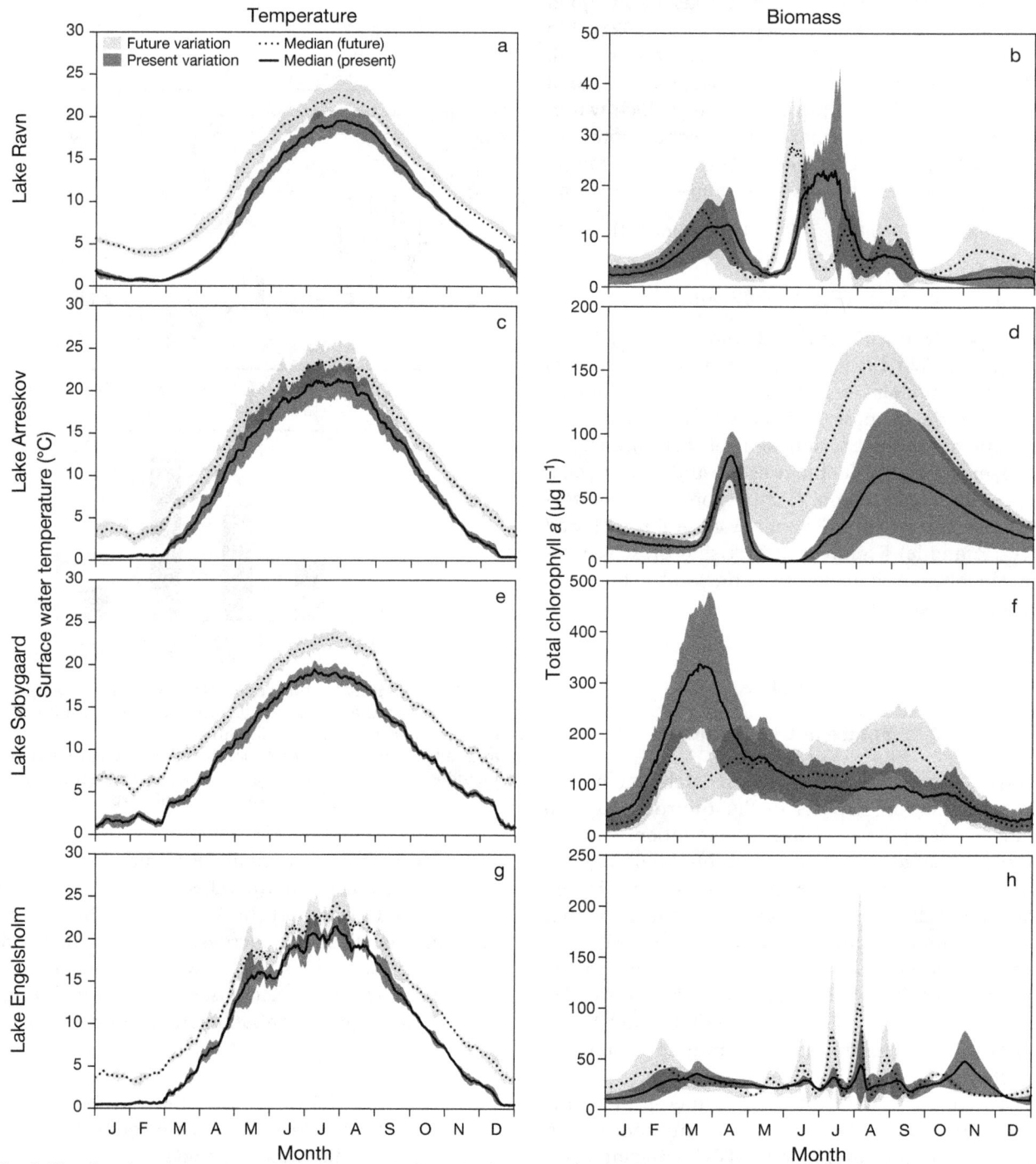

Fig. 2. Simulated surface water temperature and phytoplankton biomass (total chl *a*) for the 4 individual lakes for the baseline (SC1) and the future period with both increased air temperature and changed nutrient load (SC7). Variation bands are derived from the SD of year to year variations in surface water temperature and phytoplankton biomass (e.g. for Lake Arreskov, represented by 21 yr of simulation). See Table 3 for details on all scenarios

scenario, in particular in late summer months, infer-
ring that cyanobacteria become more dominant.

To disentangle the causes of the diversity in the
simulated responses to climate change of the 4 case
studies, we took a closer look at the simulated time
series of key water quality attributes for Lake Ravn,
representing a mesotrophic deep lake, and Lake
Arreskov, representing a eutrophic shallow lake
(Fig. 3). Deep Lake Ravn generally exhibited surface
water temperatures 1° to 1.5°C below those simu-

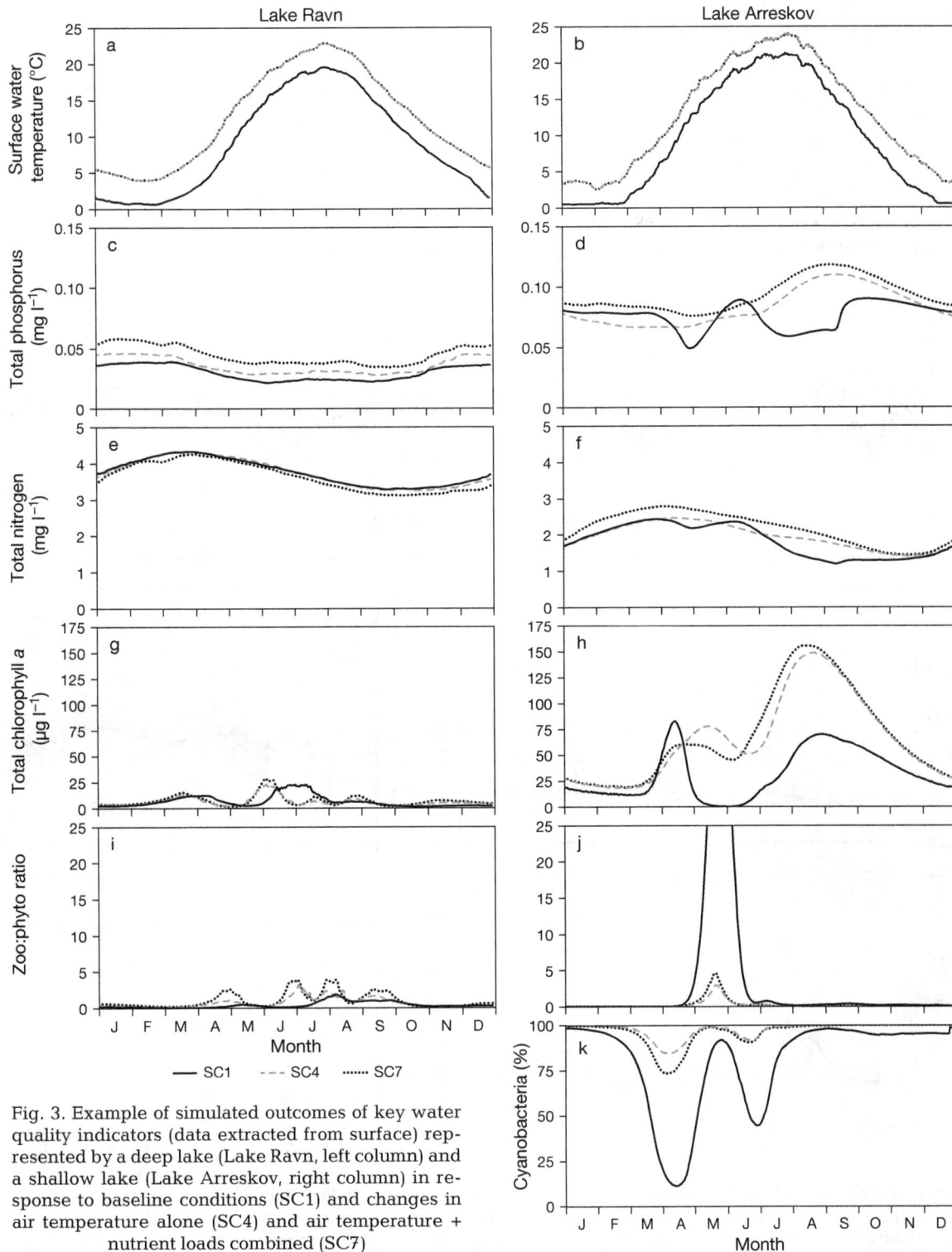

Fig. 3. Example of simulated outcomes of key water
quality indicators (data extracted from surface) represented by a deep lake (Lake Ravn, left column) and
a shallow lake (Lake Arreskov, right column) in response to baseline conditions (SC1) and changes in
air temperature alone (SC4) and air temperature +
nutrient loads combined (SC7)

lated for Lake Arreskov for both the baseline and the future scenarios. The responses to climate change and increased nutrient loads appeared to be more uniformly distributed for Lake Ravn throughout the year, in particular for lake nutrient concentrations, while the changes simulated for Lake Arreskov were considerably higher in the summer months, likely reflecting internal phosphorus loading in summer, which is typical for this lake (Nielsen et al. 2014) and other shallow eutrophic temperate lakes (Søndergaard et al. 2013). The seasonality of phytoplankton biomass in Lake Ravn generally shows an earlier onset of the summer peak in biomass in the future scenario, reflecting earlier onset of stratification, while there were no apparent changes in the timing of the summer peak in Lake Arreskov, which, how-

ever, was markedly higher relative to the baseline level. The occurrence of the clear-water phase between spring and summer blooms was reduced in the future scenarios for Lake Arreskov, and the simulated cyanobacteria biomass suggests that blooms may become a dominant feature of the phytoplankton community during spring, while diatoms dominate in spring in the baseline scenario.

When looking at the responses, and specifically the absolute changes between the median baseline and future scenarios, the effects of increased air temperature alone (SC4) suggest that key water quality attributes including TP, TN, and chl a concentrations may increase during summer, while remaining close to, or below, baseline levels in the other seasons (Fig. 4). The simulations also suggest that cyano-

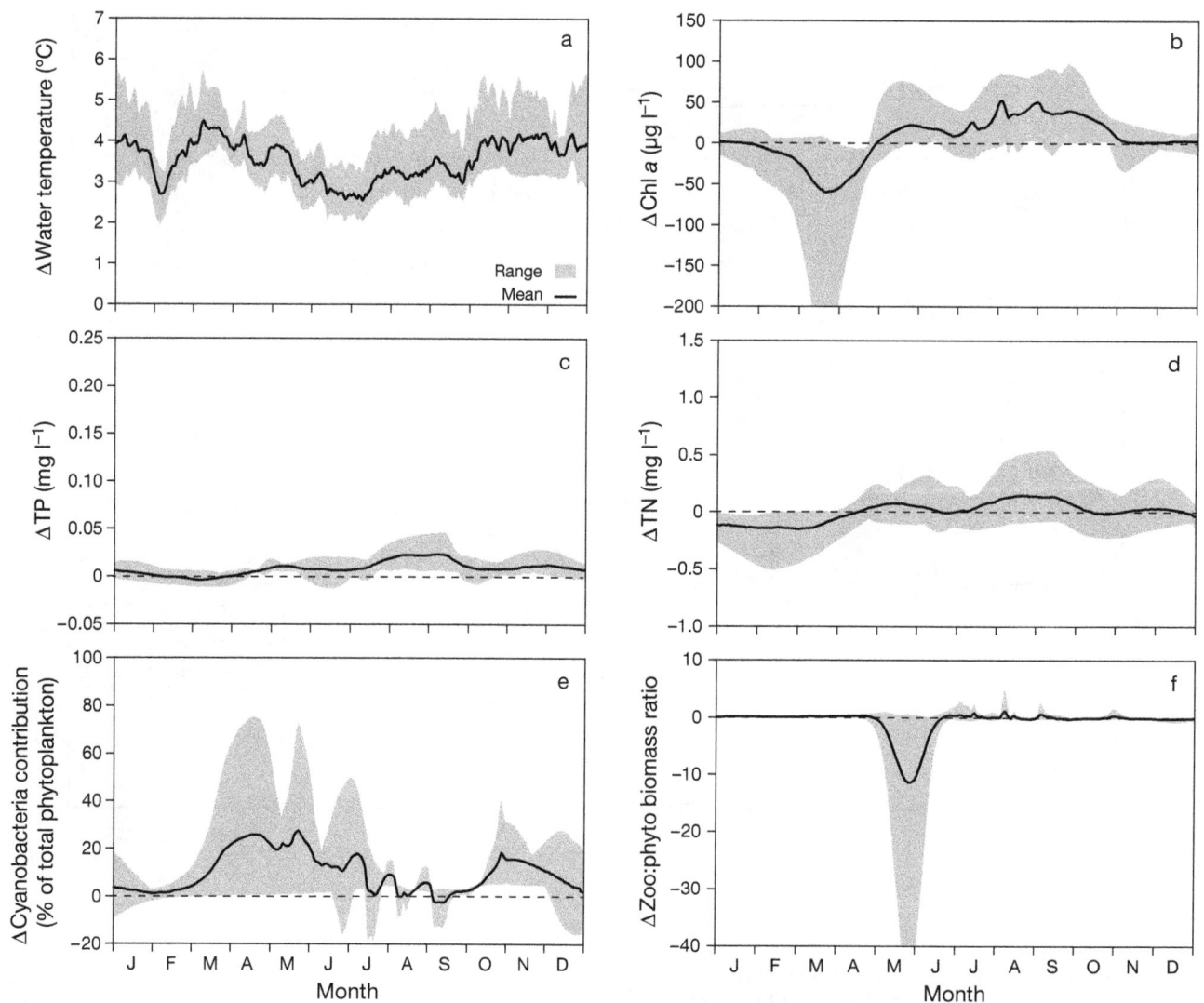

Fig. 4. Simulated seasonal changes in key water quality indicators across the 4 lakes (data extracted from surface) represented by mechanistic ecosystem models in response to median changes in air temperature (SC4 minus SC1). Gray band: range in simulated changes; black line: mean change across all 4 case study lakes; TP: total phosphorus; TN: total nitrogen

bacteria will become a more dominant feature of the phytoplankton community from spring, while there is no marked change during the summer months where they are already a dominant feature of the phytoplankton community in the baseline scenario (i.e. for the case studies that included cyanobacteria, including lakes Arreskov, Søbygaard, and Engelsholm). There are, however, also considerable discrepancies between the simulated responses for the 4 different lakes, as represented by the wide band of model outcomes around the mean response for cyanobacteria (Fig. 4). When comparing simulations of the median future scenario with that of the 90th percentile of air temperatures in the baseline scenario (Table 3), it is apparent that the future scenario is different from even the warmest days of the present time (i.e. the baseline scenario). For example, the simulated water temperature of the median year of the future is ~1.5°C warmer than the 90th percentile of the baseline scenario on both an annual basis and during the summer months. Further, when comparing simulated water temperatures, the 10th percentile of the future scenario is within the range of the 90th percentile of the baseline scenario.

As future air temperatures are combined with the estimates of future nutrients loads, the simulated outcomes change considerably (Fig. 5), and at the same time the discrepancy between simulated responses of the 4 case studies increase (comparing the gray bands in Fig. 5 with those in Fig. 4). For example, while TP only increased in the summer months in the scenarios where only air temperature was changed, it increased throughout the year in scenarios involving changes in both temperature and nutrient loads. In the temperature+nutrient load change scenario, lake TN concentrations increased in the winter months (Fig. 5d) but were generally below baseline levels in winter in the temperature-only scenario (Fig. 4d). For the remaining water quality attributes, the patterns in simulated responses were similar between scenarios where only air temperatures were changed to those where both air temperature and nutrient loads were changed.

3.3. Changes in the key properties of the aquatic ecosystems projected by the empirical relations

Based on data from Danish lakes, we established a relationship between August TP_{lake} (TP_{lake_aug}) and annual mean TP_{lake} and Z:

$$TP_{lake\_aug} = \exp[0.745 \pm 0.033 + (1.166 \pm 0.014)\log (TP_{lake}) - (0.175 \pm 0.014)\log(Z)] \quad (4)$$

where $R^2 = 0.85$, n = 1270 (number of lake years and not number of individual lakes); the range of variables being: 0.006 – 1.96 mg l^{-1} for TP_{lake_aug}, 0.2–16.5 m for Z.

We further developed a relationship between biomass of cyanobacteria (mm^3 l^{-1}) in August (TP_{lake_aug}), August mean water temperature ($Temp_{W\_aug}$,°C), and Z:

$$Cyano_{aug} = \exp[-6.75 \pm 0.26 + (1.120 \pm 0.087)\log (TP_{lake\_aug}) + (2.84 \pm 0.87)\log (Temp_{W\_aug}) + (0.319 \pm 0.12)\log (Z)] \quad (5)$$

where $R^2 = 0.21$, n = 1029; the range of input variables being: 0.006–4.10 mg l^{-1} for TP_{lake_aug}, 0.2–15 m for Z, and 11.1–26.1°C for $Temp_{W\_aug}$.

Finally, we derived a regression between the size (inferred by biomass) of cladocerans ($ClaSize_{aug}$; µg ind.$^{-1}$) and TP_{lake_aug}, $Temp_{W\_aug}$ and Z:

$$ClaSize_{aug} = \exp[-(0.02 \pm 0.01)\log(TP_{lake\_aug}) - (0.76 \pm 0.22)\log(Temp_{W\_aug}) + (0.42 \pm 0.04)\log(Z)] \quad (6)$$

where $R^2 = 0.16$, n = 953; the range of input variables being: 0.006–2.45 mg l^{-1} for TP_{lake_aug}, 0.3–15 m for Z, and 11.1–26.6°C for $Temp_{W\_aug}$.

When forced with the future nutrient load estimates (TP load increases by 64 %), the empirical relationships suggest an increase in annual mean TP_{lake}, ranging from 43 % in lakes with a retention time of 0.01 yr to ~28 % with a retention time of 20 yr, leading to a Secchi depth decline to ~89 and 83 % of the baseline levels, respectively (Fig. 6). The concentrations of annual mean chl a and Secchi depth in lakes with different TP_{lake} concentrations and different depths are given in Fig. 7 for the baseline annual mean TP loading and temperature and for the predicted future loading, showing the highest chl a and lowest Secchi depth in shallow lakes. The cyanobacteria biomass in August is predicted to increase substantially depending on the TP_{lake} concentration, depth, temperature and retention time, and the size of cladocerans is projected to decrease. In Fig. 8 this is illustrated for 2 examples: (1) a shallow lake with a mean depth of 3 m and a 1 yr retention time, and (2) a deep lake with a mean depth of 15 m and a 10 yr retention time. The differences between the 3 baseline scenarios (SC1–3) and the future scenario with increased temperatures+nutrient loads (SC7–9) are overall larger for the shallow lake type, and the difference increases with increasing TP_{lake}. The maximum level is also higher for the shallow lake, reflecting the effect of depth on both TP_{lake_aug} and on cyanobacteria biomass (see equations in 'Methods' section). The body size of cladocerans in August declines with increas-

Table 3. Annual and summer (Jun–Aug) averages of modeled state variables for all 4 mechanistic models and all scenarios (SC1–9). TP: total phosphorus (mg l^{-1}); TN: total nitrogen (mg l^{-1}); Zoo(Phyto): zoo-(phyto-)plankton; Cyano: cyanobacteria (%). Water temperature is in °C; chl a is in mg l^{-1}

Scenario	State variable	Lake Ravn		Lake Arreskov		Lake Engelsholm		Lake Søbygaard	
		Annual	Summer	Annual	Summer	Annual	Summer	Annual	Summer
SC1	Water temperature	9.07	17.81	9.73	19.84	9.60	19.08	9.54	18.02
	TP	0.03	0.02	0.08	0.07	0.05	0.05	0.29	0.44
	TN	3.77	3.68	1.81	1.84	1.21	0.69	2.37	1.69
	Chl a	6.60	13.00	31.52	30.07	24.94	26.17	120.44	97.36
	Zoo:Phyto	0.41	0.68	3.50	4.66	0.38	0.83	0.27	0.47
	Cyano			79.89	77.61	34.67	72.25	4.50	7.27
SC2	Water temperature	7.75	16.77	8.45	18.92	8.26	18.14	7.98	16.62
	TP	0.03	0.02	0.07	0.05	0.04	0.05	0.29	0.43
	TN	3.84	3.73	1.70	1.65	1.29	0.98	2.40	1.73
	Chl a	6.46	12.74	25.15	19.29	24.68	25.66	131.81	98.93
	Zoo:Phyto	0.38	0.59	8.85	22.22	0.43	0.81	0.27	0.50
	Cyano			64.41	74.07	29.21	68.86	2.82	5.40
SC3	Water temperature	10.97	19.55	11.30	21.11	11.23	20.38	11.89	20.01
	TP	0.03	0.03	0.08	0.09	0.05	0.05	0.30	0.46
	TN	3.77	3.69	2.04	2.23	1.15	0.68	2.30	1.64
	Chl a	6.60	11.54	55.34	77.99	24.71	29.61	108.43	109.11
	Zoo:Phyto	0.53	1.04	0.96	0.87	0.43	0.91	0.26	0.38
	Cyano			91.06	89.13	40.72	77.13	7.21	9.90
SC4	Water temperature	12.55	20.96	12.84	22.29	12.73	21.58	14.02	21.91
	TP	0.04	0.03	0.08	0.09	0.06	0.07	0.30	0.45
	TN	3.75	3.69	1.94	2.00	1.21	0.90	2.26	1.57
	Chl a	6.48	9.26	65.97	94.94	27.20	37.97	108.74	131.09
	Zoo:Phyto	0.73	1.43	0.20	0.19	0.63	1.16	0.24	0.30
	Cyano			97.55	97.35	40.87	76.68	10.66	10.23
SC5	Water temperature	10.79	19.90	11.17	21.50	11.07	20.77	11.70	20.65
	TP	0.03	0.03	0.08	0.09	0.05	0.05	0.30	0.47
	TN	3.78	3.69	2.04	2.25	1.20	0.69	2.29	1.64
	Chl a	6.79	12.68	51.40	65.70	24.30	30.18	116.99	112.48
	Zoo:Phyto	0.52	1.00	1.67	1.11	0.42	0.93	0.24	0.37
	Cyano			87.61	81.05	40.43	76.38	7.15	10.16
SC6	Water temperature	14.06	22.50	14.30	23.55	14.15	22.89	16.14	23.98
	TP	0.04	0.03	0.08	0.09	0.05	0.05	0.30	0.44
	TN	3.79	3.73	1.86	1.88	1.07	0.65	2.23	1.52
	Chl a	6.71	8.98	72.08	110.89	26.65	34.69	125.48	192.08
	Zoo:Phyto	0.75	1.43	0.05	0.08	0.53	0.99	0.21	0.16
	Cyano			99.44	99.88	42.71	79.15	12.20	7.65
SC7	Water temperature	12.56	20.99	12.84	22.29	12.73	21.58	14.02	21.91
	TP	0.04	0.04	0.09	0.10	0.07	0.08	0.36	0.48
	TN	3.64	3.57	2.17	2.29	1.15	0.79	2.39	1.53
	Chl a	7.35	10.59	68.29	105.76	28.09	37.17	108.15	134.67
	Zoo:Phyto	1.05	1.88	0.27	0.18	0.62	1.26	0.24	0.28
	Cyano			96.18	96.81	44.67	78.13	11.55	11.03
SC8	Water temperature	10.80	19.92	11.17	21.50	11.07	20.78	11.70	20.65
	TP	0.04	0.04	0.10	0.11	0.06	0.06	0.36	0.50
	TN	3.64	3.57	2.28	2.53	1.17	0.70	2.41	1.59
	Chl a	7.15	10.80	53.30	65.82	26.56	32.60	116.48	114.68
	Zoo:Phyto	0.85	1.61	2.04	1.24	0.58	1.11	0.23	0.36
	Cyano			86.16	77.78	38.10	78.19	7.79	10.78
SC9	Water temperature	14.06	22.51	14.30	23.55	14.15	22.90	16.14	23.98
	TP	0.05	0.04	0.09	0.10	0.06	0.06	0.36	0.47
	TN	3.67	3.60	2.08	2.13	1.09	0.75	2.37	1.48
	Chl a	7.56	10.33	75.32	119.77	28.49	39.87	118.07	181.01
	Zoo:Phyto	1.09	2.01	0.06	0.07	0.73	1.20	0.21	0.18
	Cyano			99.00	99.86	41.58	78.04	15.02	9.26

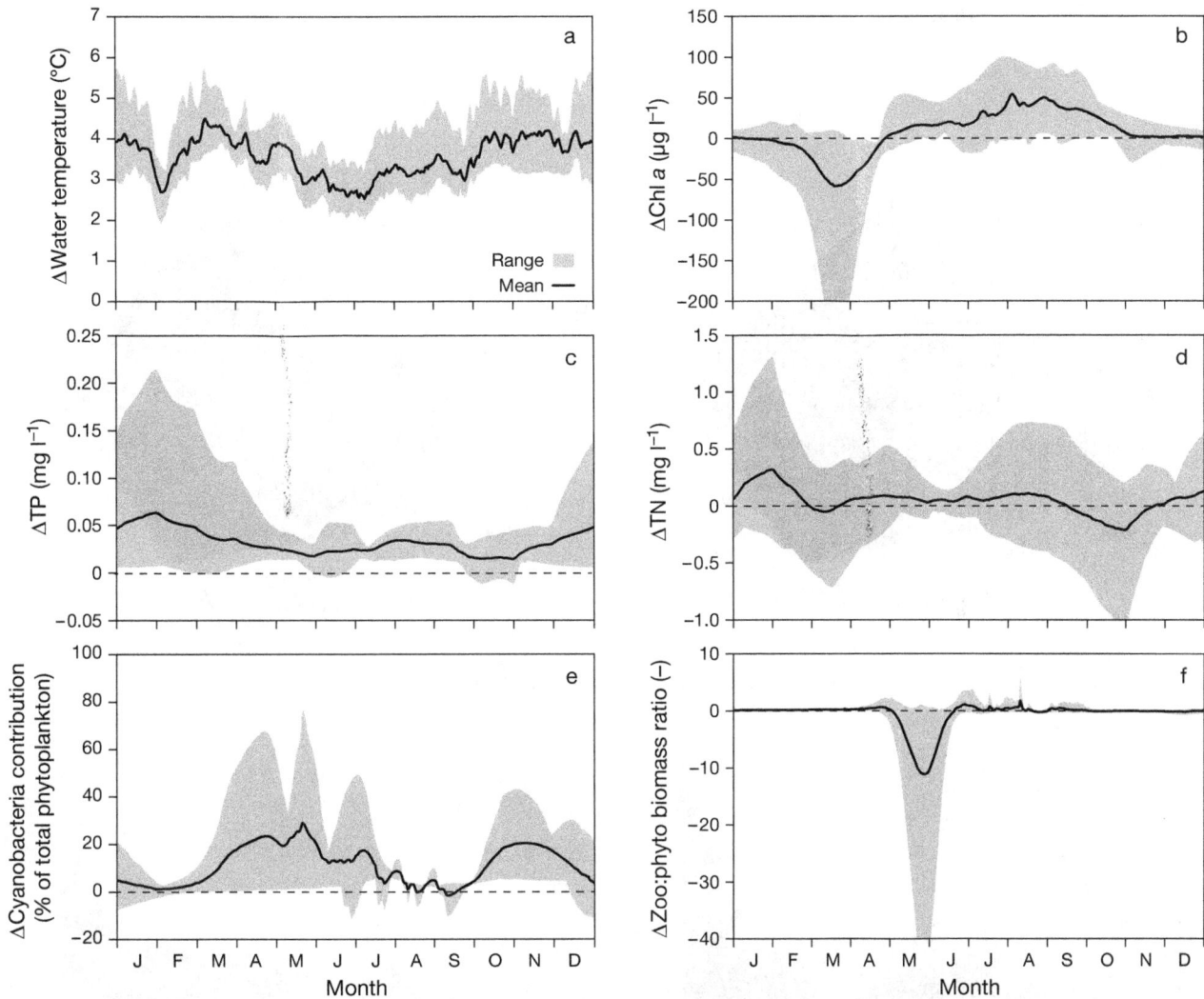

Fig. 5. Simulated seasonal changes in key water quality indicators across the 4 lakes (data extracted from surface) represented by mechanistic ecosystem models in response to changes in air temperature + changes in nutrient load (SC7 minus SC1). See Fig. 4 for definitions

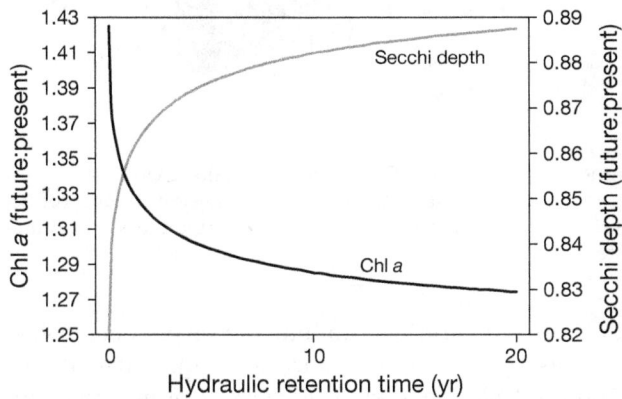

Fig. 6. Effect of increased loading and shorter retention time in the future scenario (data from Fig. 1) on the annual mean lake concentration of chl *a* and Secchi depth in Danish lakes with contrasting water retention time. Effects are shown as ratios between future:present-day situation

ing temperature, most markedly in the deep lakes, as the size is already small in the present-day situations (SC1–3) in shallow lakes (Fig. 8) due to expected high fish predation (Jeppesen et al. 2000). Also, the differences among SC1–3 and SC7–9, respectively, are larger in the deep lakes when evaluating the change in size of cladocerans.

4. DISCUSSION

4.1. Projected effects of climate change vs. empirical evidence

Both the mechanistic and empirical models projected considerably increased warming-induced eutrophication as well as increased nutrient loadings in

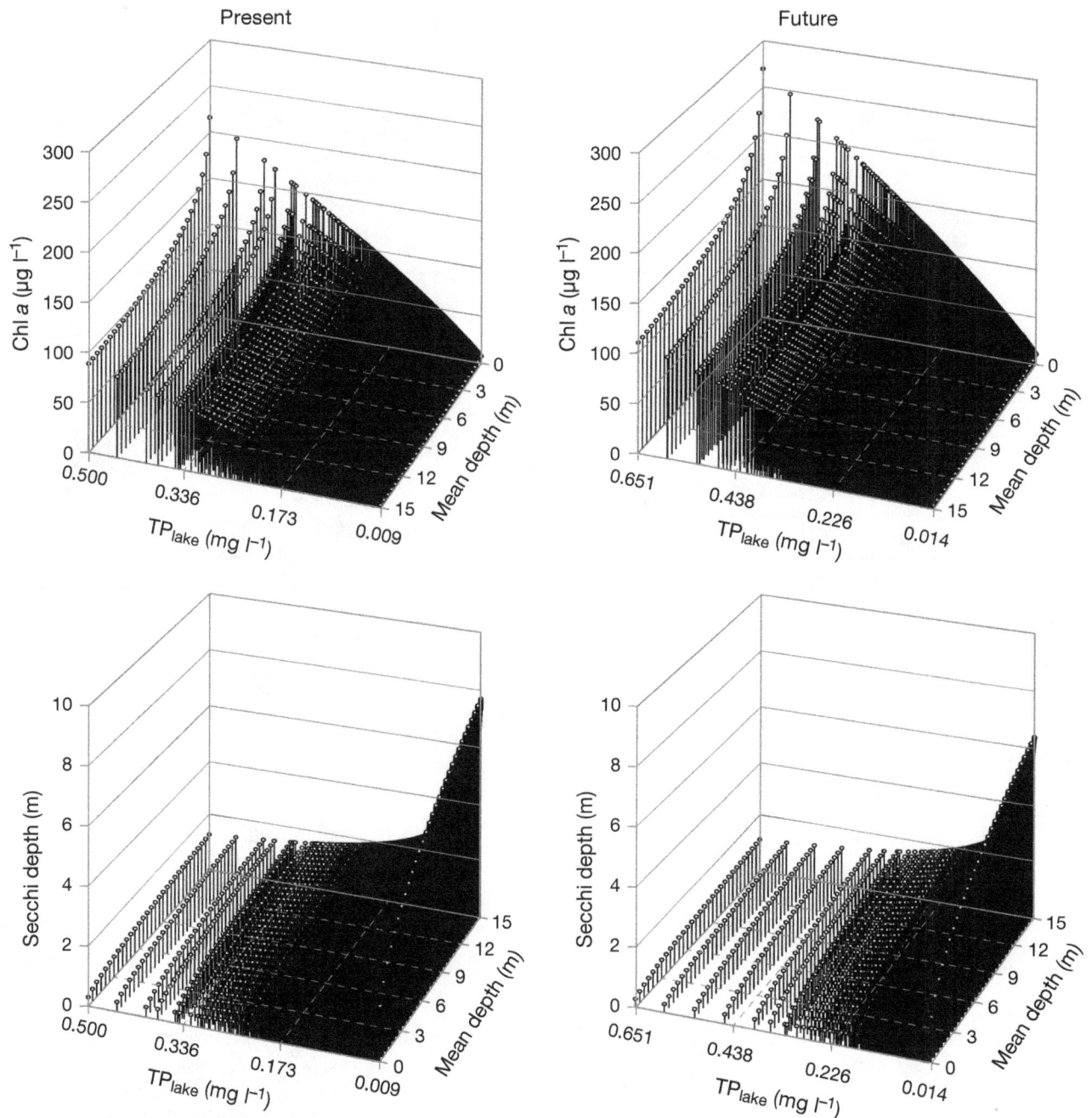

Fig. 7. Present-day (left) and predicted (right) concentration of annual mean chl *a* and Secchi depth in lakes with contrasting total phosphorus concentrations (TP_{lake}) and depths. Note differing TP_{lake} axes, as the concentrations in the right panels reflect the higher future concentrations. Note also the reversed depth axes from the upper to the lower panels to enhance visibility of the data

the future for Danish lakes. While the additional nutrient load in the future scenario (SC7) had significant impact in the lakes' nutrient levels, minor effects were seen on biological variables relative to the effects of increased temperatures alone (SC4). This suggests that as the lakes get warmer, and nutrient cycling processes intensify, productivity may become decreasingly nutrient limited, and instead controlled increasingly by temperature and light availability. The total phytoplankton biomass is projected to increase the largest amount during the summer months, and the presence of potentially toxin-producing cyanobacteria is projected to be amplified and likely to become a dominant feature of the phytoplankton

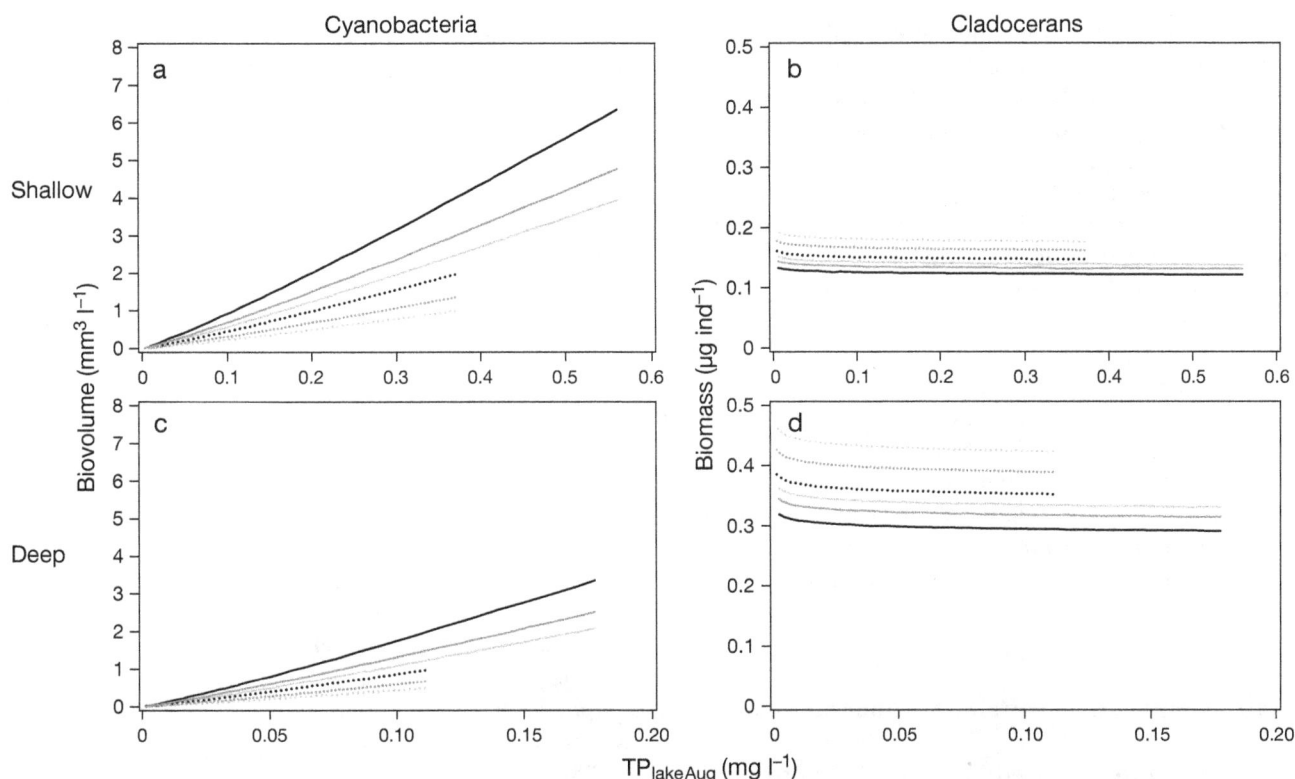

Fig. 8. Outcome of the empirical modelling of (a,c) cyanobacteria biomass and (b,d) average biomass of cladocerans (µg dry wt ind.$^{-1}$) in August in both (a,b) shallow and (c,d) deep lakes with mean depths of 3 and 15 m and 1 and 10 yr retention times, respectively, run at contrasting lake water phosphorus concentrations (TP_{lake}). Dotted lines: present-day scenario (SC1–3); solid lines: future scenario (SC7–9). Increasing shade of gray: 10th percentile, median, and 90th percentile

community from spring. According to the model projections, this shift towards cyanobacteria dominance in spring is accompanied by a shift in the balance of the ecosystem, with a generally declining grazing pressure by zooplankton on phytoplankton; this is evidenced by the lower zooplankton to phytoplankton biomass ratio, particularly during spring. Such shifts in ecosystem structure and balance have also been reported in literature based on empirical observations. For example, Jeppesen et al. (2010) showed that a warming-triggered shift in fish community structure and abundance is to be expected, potentially leading to dominance by small-sized fish, lower zooplankton grazing, stronger cyanobacterial dominance, and higher internal nutrient loading. Moss et al. (2011) suggested that the future warming and increased nutrient loads will work synergistically and amplify such eutrophication effects, and that lake ecosystems therefore face 'allied attacks'. Monitoring data from Danish lakes collected in late summer (15 August to 15 September) from 1989 to 2006 also revealed a significant increase in the proportion of small perch *Perca fluviatilis* and bream *Abramis brama* with increasing temperature (Jeppesen et al.

2010) despite an overall major reduction in nutrient levels (reduction of the external nutrient loading), the latter of which (seen in isolation) should have resulted in a lower proportion of small fish (Jeppesen et al. 2000). Moreover, a long-term study conducted in shallow Lake Søbygaard showed a major decline in the average size of key fish species (Jeppesen et al. 2012). The decline coincided well with the change in average summer air temperature, indicating that the changes might reflect the enhanced ambient temperature. In accordance with this, the empirical data analyses (Fig. 8) predict a future decrease in the size of cladocerans, which concurs with higher fish predation, as the fish generally have a preference for larger-bodied individuals (Gliwicz 2003). Fish abundance and size may also be influenced by lower ice cover in winter, allowing higher survival of small fish with cascading effects on the food web and, with this, stronger symptoms of eutrophication (Jackson et al. 2007, Balayla et al. 2010, Ruuhijärvi et al. 2010). Cross-latitude gradient studies of shallow lakes have shown that the ratio of fish biomass (expressed as catches per net in multi-mesh sized gillnets) to zooplankton biomass increased from northern to southern Europe, while the zooplankton:phytoplankton biomass ratio decreased, both substantially (Gyllström et al. 2005). Thus, the projections by the models used in the present study are generally supported by empirical evidence. However, the strong changes in the food web structure observed in the empirical data are not yet well-implemented in the predictive models, emphasizing that the projected eutrophication effects of climate warming by the models may be conservative (i.e. the effects may be greater than our projections).

Both the mechanistic and empirical models predict an overall higher concentration of TP in the lakes, and thus higher concentrations in the lake outlets. Combined with higher discharge rates, the nutrient loading eventually reaching the coastal area will also increase quite substantially and contribute to the deterioration of the ecological state of these environments.

4.2. Pitfalls and uncertainties of our approaches

The 2 approaches used in this study to project the future state of lake ecosystems, based on mechanistic and empirical lake models, respectively, are associated with uncertainty. Projections by mechanistic models are only valid as long as the forcing data (e.g. temperature) are within the range for which the process descriptions in the models are still valid (e.g. the influence of temperature on the growth rates of phytoplankton). Empirical models are only applicable within the range of forcing data for which they have originally been developed (typically a more narrow range relative to the validity range of mechanistic models), and great care must be taken when extrapolating beyond this range. Based on hydrological modelling studies, Refsgaard et al. (2014) also demonstrated that model performance continues to deteriorate when the models are driven by forcing data differing increasingly from the forcing data for which the models had originally been applied and calibrated. Given the rather extreme warming scenarios simulated in this study, both the mechanistic and empirical models are applied at the periphery of their validity (and in reality extrapolated beyond the range of validity in the case of the empirical models); thus our projections should be treated with caution.

In our study, the future climate is based only on a single CO_2-emission scenario in combination with one GCM-RCM model, and is downscaled with one bias-correction approach. Recent studies have highlighted the need for using an ensemble of climate models (Déqué et al. 2007) as the climate change signal has been shown to differ substantially between climate models (Seaby et al. 2013). Therefore, care should also be taken when analyzing impacts from only one GCM-RCM combination, and a larger ensemble of climate models would help to evaluate the findings of the present study. Karlsson et al. (2015) compared the ECEARTH-HIRHAM model projections for RCP4.5 to 4 other models from the ENSEMBLES project based on the A1B scenario. They found that the ECEARTH-HIRHAM for RCP4.5 had a temperature signal comparable to the median of the 4 ENSEMBLES A1B projections. As RCP4.5 is a scenario with generally less changes in forcing relative to A1B (Collins et al. 2013), with the median of the A1B projections corresponding to the high end of the RCP 4.5 range, it appears that the ECEARTH-HIRHAM may provide projections that are warmer and show higher temperatures than the average of an ensemble of models. However, this is not critical for the conclusions of our study, since we focus on the impacts occurring when, at some future point in time, warming has increased by 6°C.

The results of comparing scenarios SC4 and SC7 clearly show the large changes in nutrient exports, and the subsequent consequences for the concentrations of TP and TN, in the Danish lakes (Table 3). In our scenarios we assumed no major adaptation to climate change in the agricultural practices that contribute a considerable portion of nutrient exports to waterways. In reality, farmers would shift to growing crop species and varieties with higher productivity under climate change, which in northern Europe in particular means shifting from the currently grown cereals to grain maize (Elsgaard et al. 2012) and cereals with a longer growth duration (Olesen et al. 2012). Enhancing crop growth may reduce some of the large projected losses of nutrients (e.g. Patil et al. 2012). However, such effects of adaptation are likely to be small, as are those of other options to mitigate nutrient losses from agriculture, for instance growing catch crops (Doltra et al. 2014), since the major effects leading to enhanced loss of N and P under climate change are related to the land use practices of growing cereal crops. The major uncertainties associated with the estimated enhanced loading of N and P to the lakes are therefore associated with future land use practices, in particular whether the current high proportion of arable land in Denmark is maintained or whether some of that land use is converted to perennial cropping systems such as grassland for producing biomass for feed, energy, and other purposes (Smith & Olesen 2010).

4.3. Ways forward for research into climate change effects on lake ecosystems

There are many approaches available to examine the influence of climate forcing on ecosystem dynamics (Jeppesen et al. 2014). These all have their individual strengths and weaknesses, including "space-for-time substitution" (a challenge to separate climate effects from other anthropogenic, natural, and biogeographical effects); time-series and monitoring (a challenge to make realistic projections of the long-term changes related to ecosystem structures that may not be present under current climate conditions and also, for example, evolution and invasions); experiments (a challenge to undertake experiments at a realistic scale and sufficiently long-term duration); paleoecology (a challenge to disentangle the climate effect signals from eutrophication and other stressors); and modelling (a challenge to make models flexible and able to accurately reflect the real world ecosystems and their functions). As has already been seen for weather forecasts and global circulation models, an ensemble approach using different models (or different data series to drive a model) may also provide more robust projections of the future state of aquatic ecosystems than when using only a single model (e.g. Nielsen et al. 2014, Trolle et al. 2014). It is evident, however, that the existing models, based mainly on experiences from temperate lakes, are currently unable to fully tackle the major shifts in trophic structure and dynamics of the lake ecosystems to be expected when lakes get considerably warmer. Improvements are therefore needed in order to build more reliable models, a task that is challenged by difficulties in obtaining good quality data to parameterize them. There are many ways in which the reliability of mechanistic models may be improved, as reviewed in detail by Mooij et al. (2010) and Robson (2014). The use of bioclimatic models, which are mostly applied to terrestrial environments, to predict changes in the geographical distribution of species as well as the likelihood of biological invasions under the projected climate change scenarios (Jeschke & Strayer 2008) should ideally also incorporate biotic interactions (Fernandes et al. 2013). However, in many cases, basic niche information for several species is completely lacking, currently preventing this task. Besides, projecting distribution changes by present niche characteristics might not always result in a reliable prediction, as even in the short term the flora and fauna may adapt to the new conditions, in part by microevolution.

4.4. Ways to mitigate the effects of climate change

As many of the symptoms of warming are similar to those following from enhanced nutrient loading (e.g. Trolle et al. 2011), measures should be aimed at increasing the natural resilience of ecosystems against external perturbations (Moss et al. 2011). Measures taken to reduce the nutrient input to freshwaters (Doody et al. 2012), beyond those already implemented or planned, are relatively straightforward and may, among other measures, include: control of fertilizer application rates based on soil retention capacity and crop needs in order to reduce diffuse nutrient losses from cultivated land; establishment of riparian buffer strips to reduce nutrient transfers to water bodies (e.g. Hoffmann et al. 2009, Stutter et al. 2012); re-meandering of channelized streams and (re)establishment of wetlands to increase retention of organic matter and nutrients (e.g. Hoffmann et al. 2011, 2014); improved design of sewage works and rainwater basins to cope with the consequences of flood events and low flows in receiving waters. Special attention should clearly be targeted to modulate the risk of enhanced nutrient loading from catchments during periods of extreme weather events (notably focussing on precipitation and drought-flood effects), which are generally projected to occur more frequently in the future.

However, the needed reduction in external nutrient loadings in the 6°C scenario is likely so substantial that it will probably not be possible to obtain a 'good' ecological status according to the European Water Framework Directive (e.g. Nielsen et al. 2014) without substantially interfering with land use and crop production, and this will in particular involve reducing the area allocated to agriculture. Since projections of the global food demand for 2050 and beyond point towards a doubling of the current demand, and given that this is not likely to be satisfied by cultivating new lands (Tilman et al. 2011), there will be increasing pressure on current lands to enhance agricultural outputs. This pressure will be particularly large in the areas of the world which are less impacted by climate change, such as the regions of northern Europe, leading to enhanced nutrient use and losses to the environment (Olesen et al. 2011). The conflicts between the needs to maintain food supply and to maintain good ecological status of lakes (and other natural environments) will thus grow markedly in a much warmer world. Solving these issues would need radical new technologies and changes in human life styles.

5. CONCLUSIONS

We evaluated the impacts of an extreme 6°C warming scenario on temperate Danish lakes using a multifaceted modelling approach. We combined an ecohydrological model to estimate future water runoff and catchment nutrient exports with both mechanistic and empirical models, describing key biogeochemical indicators in lakes, which enabled us to quantify the effects of future nutrient loads and air temperature on lake ecosystems. Our model projections suggest that annual water runoff will increase (46%), driving also increases in catchment exports of nitrogen and phosphorus (13 and 64%, respectively). The mechanistic and empirical lake modelling approaches both suggest that phytoplankton biomass will increase and that potentially toxin-producing cyanobacteria may become a dominant feature of the phytoplankton community from spring. The food webs within the lakes are also likely to be affected in the direction of higher fish control of algae-grazing water fleas, further reinforcing eutrophication. To be able to mitigate these eutrophication effects, external nutrient loading to the lakes must be reduced considerably. Consequently, compliance with existing legislation, such as the European Union Water Framework Directive, will become increasingly challenging in the future.

Acknowledgements. The present study was funded by a grant from the Danish Council for Strategic Research for the project Centre for Regional Change in the Earth System (CRES — www.cres-centre.dk) under contract no.: DSF-EnMi 09-066868, the MARS project (Managing Aquatic ecosystems and water Resources under multiple Stress) funded under the 7th EU Framework Programme, Theme 6 (Environment including Climate Change), contract no.: 603378 (www.mars-project.eu), the Danish projects CIRCE and CLEAR2 (a Villum Kann Rasmussen Centre of Excellence project). We thank Anne Mette Poulsen for valuable editorial comments and Tinna Christensen for figure layout.

LITERATURE CITED

Andersen HE, Kronvang B, Larsen SE, Hoffmann CC, Jensen TS, Rasmussen EK (2006) Climate-change impacts on hydrology and nutrients in a Danish lowland river basin. Sci Total Environ 365:223–237

Arnold JG, Srinivasan R, Muttiah RS, Williams JR (1998) Large area hydrologic modeling and assessment. I. Model development. J Am Water Resour Assoc 34:73–89

Balayla DJ, Lauridsen TL, Søndergaard M, Jeppesen E (2010) Larger zooplankton in Danish lakes after cold winters: Are fish kills of importance? Hydrobiologia 646:159–172

Christensen OB, Drews M, Christensen JH, Dethloff K, Ketelsen K, Hebestadt I, Rinke A (2007) The HIRHAM Regional Climate Model Version 5 (β). In: DMI Tech Rep, Book 06-17. DMI, Copenhagen

Christensen OB, Yang S, Boberg F, Maule CF and others (2015) Scalability of regional climate change in Europe for high-end scenarios. Clim Res 64:25–38

Collins M, Knutti R, Arblaster J, Dufresne JL and others (2013) Long-term climate change: projections, commitments and irreversibility. In: Stocker TF, Qin D, Plattner G-K, Tignor M and others (eds) Climate change 2013: the physical science basis. Contribution of Working Group I to the Fifth Assessment Report of the Intergovernmental Panel on Climate Change. Cambridge University Press, Cambridge

Déqué M, Rowell D, Lüthi D, Giorgi F and others (2007) An intercomparison of regional climate simulations for Europe: assessing uncertainties in model projections. Clim Change 81:53–70

Doltra J, Lægdsmand M, Olesen JE (2014) Impacts of projected climate change on productivity and nitrogen leaching of crop rotations in arable and pig farming systems in Denmark. J Agric Sci 152:75–92

Doody DG, Archbold M, Foy B, Flynn R (2012) Approaches to the implementation of the Water Framework Directive: targeting measures at critical source areas of diffuse pollution in Irish catchments. J Environ Manag 93:225–234

Elliott JA (2010) The seasonal sensitivity of cyanobacteria and other phytoplankton to changes in flushing rate and water temperature. Glob Change Biol 16:864–876

Elsgaard L, Børgesen CD, Olesen JE, Siebert S and others (2012) Shifts in comparative advantages for maize, oat, and wheat cropping under climate change in Europe. Food Addit Contam A 29:1514–1526

Fernandes JA, Cheung WWL, Jennings S, Butenschön M and others (2013) Modelling the effects of climate change on the distribution and production of marine fishes: accounting for trophic interactions in a dynamic bioclimate envelope model. Glob Chang Biol 19:2596–2607

Gliwicz ZM (2003) Between hazards of starvation and risk of predation: the ecology of offshore animals. In: Kinne O (ed) Excellence of ecology, Book 12. International Ecology Institute, Oldendorf/Luhe

Gyllström M, Hansson LA, Jeppesen E, Garcia-Criado F and others (2005) The role of climate in shaping zooplankton communities of shallow lakes. Limnol Oceanogr 50:2008–2021

Hamilton DP, Schladow SG (1997) Prediction of water quality in lakes and reservoirs. 1. Model description. Ecol Modell 96:91–110

Hazeleger W, Wang X, Severijns C, Ştefănescu S and others (2012) EC-Earth V2.2: description and validation of a new seamless earth system prediction model. Clim Dyn 39:2611–2629

Henriksen HJ, Troldborg L, Nyegaard P, Sonnenborg TO, Refsgaard JC, Madsen B (2003) Methodology for construction, calibration and validation of a national hydrological model for Denmark. J Hydrol (Amst) 280:52–71

Hoffmann CC, Kjaergaard C, Uusi-Kämppä J, Hansen HCB, Kronvang B (2009) Phosphorus retention in riparian buffers: review of their efficiency. J Environ Qual 38:1942–1955

Hoffmann CC, Kronvang B, Audet J (2011) Restoration and monitoring of nutrient buffering capacities in Danish riparian wetlands. Hydrobiologia 674:5–24

Hoffmann CC, Brix H, Kronvang B (2014) SWS European

Chapter Meeting on wetland restoration—challenges and opportunities. Ecol Eng 66:1–5

➤ Jackson LJ, Søndergaard M, Lauridsen TL, Jeppesen E (2007) A comparison of shallow Danish and Canadian lakes and implications of climate change. Freshw Biol 52: 1782–1792

➤ Janse JH (1997) A model of nutrient dynamics in shallow lakes in relation to multiple stable states. Hydrobiologia 342–343:1–8

Jeppesen E, Kristensen P, Jensen JP, Søndergaard M, Mortensen E, Lauridsen TL (1991) Recovery resilience following a reduction in external phosphorus loading of shallow, eutrophic Danish lakes: duration, regulating factors and methods for overcoming resilience. Mem Ist Ital Idrobiol 48:127–148

➤ Jeppesen E, Jensen JP, Søndergaard M, Lauridsen TL, Pedersen LJ, Jensen L (1997) Top-down control in freshwater lakes: the role of nutrient state, submerged macrophytes and water depth. Hydrobiologia 342/343:151–164

➤ Jeppesen E, Jensen JP, Søndergaard M, Lauridsen T, Landkildehus F (2000) Trophic structure, species richness and biodiversity in Danish lakes: changes along a phosphorus gradient. Freshw Biol 45:201–213

➤ Jeppesen E, Kronvang B, Meerhoff M, Søndergaard M and others (2009) Climate change effects on runoff, catchment phosphorus loading and lake ecological state, and potential adaptations. J Environ Qual 38:1930–1941

➤ Jeppesen E, Meerhoff M, Holmgren K, González-Bergonzoni I and others (2010) Impacts of climate warming on lake fish community structure and potential effects on ecosystem function. Hydrobiologia 646:73–90

➤ Jeppesen E, Kronvang B, Olesen JE, Audet J and others (2011) Climate change effects on nitrogen loading from cultivated catchments in Europe: implications for nitrogen retention, ecological state of lakes and adaptation. Hydrobiologia 663:1–21

➤ Jeppesen E, Søndergaard M, Lauridsen TL, Davidson TA and others (2012) Biomanipulation as a restoration tool to combat eutrophication: recent advances and future challenges. Adv Ecol Res 47:411–487

➤ Jeppesen E, Meerhoff M, Davidson TA, Søndergaard M and others (2014) Climate change impacts on lakes: an integrated ecological perspective based on a multi-faceted approach, with special focus on shallow lakes. J Limnol 73:84–107

➤ Jeschke JM, Strayer DL (2008) Usefulness of bioclimatic models to study climate change and invasive species. Ann NY Acad Sci 1134:1–24

➤ Jöhnk KD, Huisman J, Sharples J, Sommeijer B, Visser PM, Stroom JM (2008) Summer heatwaves promote blooms of harmful cyanobacteria. Glob Change Biol 14: 495–512

➤ Karlsson IB, Sonnenborg TO, Seaby LP, Jensen KH, Refsgaard JC (2015) Effect of a high-end CO_2-emission scenario on hydrology. Clim Res 64:39–54

➤ Kosten S, Huszar VLM, Bécares E, Costa LS and others (2012) Warmer climate boosts cyanobacterial dominance in lakes. Glob Change Biol 18:118–126

➤ Meerhoff M, Teixeira-de Mello F, Kruk C, Alonso C and others (2012) Environmental warming in shallow lakes: a review of effects on community structure as evidenced from space-for-time substitution approaches. Adv Ecol Res 46:259–350

➤ Mooij WM, Trolle D, Jeppesen E, Arhonditsis G and others (2010) Challenges and opportunities for integrating lake ecosystem modelling approaches. Aquat Ecol 44: 633–667

➤ Moss B, Kosten S, Meerhoff M, Battarbee RW and others (2011) Allied attack: climate change and eutrophication. Inland Waters 1:101–105

Nakicenovic N, Alcamo J, Davis G, de Vries B and others (2000) Emission scenarios. In: Nakicenovic N, Swart R (eds) Emission scenarios. Cambridge University Press, Cambridge

Nielsen A (2013) Predicting the future state of freshwater lake ecosystems influenced by climate and land-use changes. PhD thesis, Aarhus University

Nielsen A, Trolle D, Bjerring R, Søndergaard M, Olesen JE, Jeppesen E (2014) Effects of climate and nutrient load on the water quality of shallow lakes assessed through ensemble runs by PCLake. Ecol Appl 24:1926–1944

OECD (Organisation for Economic Co-operation and Development) (1982) Eutrophication of waters. Monitoring, Assessment and Control. Rep OECD, Paris

➤ Olesen JE, Trnka M, Kersebaum KC, Skjelvåg AO and others (2011) Impacts and adaptation of European crop production systems to climate change. Eur J Agron 34: 96–112

➤ Olesen JE, Børgesen CD, Elsgaard L, Palosuo T and others (2012) Changes in flowing and maturity time of cereals in Northern Europe under climate change. Food Addit Contam 29:1527–1542

Patil R, Lægdsmand M, Olesen JE, Porter JR (2012) Sensitivity of crop yield and N losses in winter wheat to changes in mean and variability of temperature and precipitation in Denmark using the FASSET model. Acta Agric Scand B 62:335–351

➤ Piani C, Haerter JO, Coppola E (2010) Statistical bias correction for daily precipitation in regional climate models over Europe. Theor Appl Climatol 99:187–192

➤ Refsgaard JC, Madsen H, Andréassian V, Arnbjerg-Nielsen K and others (2014) A framework for testing the ability of models to project climate change and its impacts. Clim Change 122:271–282

Robson B (2014) State of the art in modelling of phosphorus in aquatic systems: review, criticisms and commentary. Environ Model Softw 61:339–359

Rolighed J (2013) Climate change makes re-oligotrophication more difficult: a PCLake model study of shallow, Danish Lake Søbygaard. MSc thesis, Aarhus University

➤ Ruuhijärvi J, Rask M, Vesala S, Westermark A, Olin M, Keskitalo J, Lehtovaara A (2010) Recovery of the fish community and changes in the lower trophic levels in a eutrophic lake after a winter kill of fish. Hydrobiologia 646:145–158

➤ Seaby LP, Refsgaard JC, Sonnenborg TO, Stisen S, Christensen JH, Jensen KH (2013) Assessment of robustness and significance of climate change signals for an ensemble of distribution-based scaled climate projections. J Hydrol (Amst) 486:479–493

➤ Smith P, Olesen JE (2010) Synergies between mitigation of, and adaptation to, climate change in agriculture. J Agric Sci 148:543–552

➤ Søndergaard M, Bjerring R, Jeppesen E (2013) Persistent internal phosphorus loading during summer in shallow eutrophic lakes. Hydrobiologia 710:95–107

➤ Stutter MI, Chardon W, Kronvang B (2012) Riparian buffer strips as a multifunctional management tool in agricultural landscapes: introduction. J Environ Qual 41:297–303

➤ Tilman D, Balzer C, Hill J, Befort BL (2011) Global food

demand and the sustainable intensification of agriculture. Proc Natl Acad Sci 108:20260–20264

➤ Trolle D, Hamilton DP, Pilditch CA, Duggan IC, Jeppesen E (2011) Predicting the effects of climate change on trophic status of three morphologically varying lakes: implications for lake restoration and management. Environ Model Softw 26:354–370

➤ Trolle D, Elliott A, Mooij WM, Janse JH, Bolding K, Hamilton DP, Jeppesen E (2014) Advancing projections of phytoplankton responses to climate change through ensemble modelling. Environ Modell Softw 61:371–379

van der Linden P, Mitchell JFB (2009) ENSEMBLES: climate change and its impacts. Summary of research and results from the ENSEMBLES project. Met Office Hadley Centre, Exeter

➤ van Vuuren D, Edmonds J, Kainuma M, Riahi K and others (2011) The representative concentration pathways: an overview. Clim Change 109:5–31

➤ Williamson CE, Dodds W, Kratz TK, Palmer MA (2008) Lakes and streams as sentinels of environmental change in terrestrial and atmospheric processes. Front Ecol Environ 6:247–254

6

Evaluating adaptation options for urban flooding based on new high-end emission scenario regional climate model simulations

K. Arnbjerg-Nielsen[1,*], L. Leonardsen[2], H. Madsen[3]

[1]Department of Environmental Engineering, Technical University of Denmark, Miljovej, Building 113, 2800 Lyngby, Denmark
[2]Copenhagen Municipality, Njalsgade 13, 1505 Copenhagen V, Denmark
[3]DHI, Agern Alle 5, 2970 Hørsholm, Denmark

ABSTRACT: Climate change adaptation studies on urban flooding are often based on a model chain approach from climate forcing scenarios to analysis of adaptation measures. Previous analyses of climate change impacts in Copenhagen, Denmark, were supplemented by 2 high-end scenario simulations. These include a regional climate model projection forced to a global temperature increase of 6°C in 2100 as well as a projection based on a high radiative forcing scenario (RCP8.5). With these scenarios, projected impacts of extreme precipitation increase significantly. For extreme sea surges, the impacts do not seem to change substantially compared to currently applied projections. The flood risk (in terms of expected annual damage, EAD) from sea surge is likely to increase by more than 2 orders of magnitude in 2100 compared to the present cost. The risk from pluvial flooding in 2000 is likely to increase by almost 4 and 8 times the current EAD for the RCP8.5 and 6°C scenario, respectively. For both hazards, business-as-usual is not a possible scenario, since even in the absence of policy-driven changes, significant autonomous adaptation is likely to occur. Copenhagen has developed an adaptation plan to pluvial flooding that makes the urban areas more robust and reduces the risk of flooding under the current climate to a very low level. The reduction in flood risk for the A1B scenario is substantial (corresponding to 0.2–0.3 times the current EAD in 2100), and even in the high-end scenarios, the risk is significantly reduced (corresponding to 0.6–1.0 and 1.2–2.1 times the current EAD for the RCP8.5 and 6°C scenario, respectively).

KEY WORDS: Climate change adaptation · Pluvial flooding · High-end scenarios · Copenhagen · Precipitation · Climatic extremes

1. INTRODUCTION

Studies focusing on how to adapt urban areas to increased flood risk from climate change are often based on a model chain approach consisting of the following elements: (1) climate forcing scenarios, (2) global and regional climate model projections, (3) statistical downscaling, (4) hydrological modelling and (5) flood hazard analysis (see e.g. Dankers & Feyen 2009, Wilby & Dessai 2010, Madsen et al. 2014). The use of different scenarios, climate projections, downscaling methods and hydrological mod-

els may lead to very different results. Thus, it is generally recommended to use several options at the different stages of the model chain, in order to quantify the uncertainty in the impact assessment (Wilby & Harris 2006, Arnbjerg-Nielsen et al. 2013). Several studies have used an ensemble of projections from global and regional climate models for evaluating changes in flood frequency (see e.g. Veijalainen et al. 2010 or Rojas et al. 2012). Some studies have also included alternative statistical downscaling methods (see e.g. Willems & Vrac 2011, Sunyer et al. 2012, 2015) or the inclusion of hydro-

*Corresponding author: karn@env.dtu.dk

logical modelling uncertainty (Lawrence & Haddeland 2011).

Climate change adaptation in response to increased risk of flooding has been high on the political agenda in a number of countries during recent years, and hence climate change adaptation planning has occurred in a number of cities worldwide, including New York (Rosenzweig et al. 2011) and London (Penning-Rowsell et al. 2013, Ranger et al. 2013). In Denmark, the need for rapid reaction has been further exacerbated by a large increase in the number of extreme precipitation events (Gregersen et al. 2013). These events have resulted in pluvial flooding in most major cities in Denmark during the last 10 yr. One of the largest of these storms hit the capital city of Copenhagen on 2 July 2011. During this event, precipitation intensities higher than 2000 yr return periods were recorded at the most critical locations. The damage to the city was substantial; insurance companies reported that >30% of building owners in the municipality of Copenhagen filed insurance claims after the event, and the total claims exceeded 800 million euros. The municipality has responded by developing master plans for climate change adaptation and is currently developing more detailed planning for all major urban catchments in the Copenhagen region.

Development of such adaptation studies is based on recommendations of projected changes of climate extremes. Copenhagen Municipality (2011) made an assessment of future impacts on which their climate change adaptation plan is based. Based on this assessment, pluvial flooding and sea surges were selected as key areas where strategic initiatives were needed. An initial screening indicated that from an economic point of view, measures against sea surges should be implemented 30 to 40 yr from now, whereas it was highly advantageous to start implementing adaptation plans immediately, in order to mitigate the impacts of increased pluvial flooding risks. An assessment of all climate change impacts on the water cycle in Denmark confirmed that pluvial flooding is probably the most important factor to consider (Refsgaard et al. 2013).

In this study, we analyse anticipated changes of precipitation extremes and sea surges and their impacts on Copenhagen in a business-as-usual scenario, as well as in a scenario where the Copenhagen adaptation plan against pluvial flooding (the 'cloudburst' plan) is fully implemented. The focus will be on assessing changes in impacts that can be attributed to differences between a moderate climate change scenario such as an A1B scenario (van der Linden & Mitchell

2009) and high-end scenarios, including a 6°C global warming scenario (Christensen et al. 2015, this Special) and the most intense Representative Concentration Pathway (RCP8.5) scenario (Mayer et al. 2015).

2. DATA AND METHODS

2.1. Climate change impact assessment

Projections of extreme sea surges and extreme precipitation using high-end scenarios were developed for Copenhagen for the year 2100. They were then compared to the projections that the Copenhagen Municipality has developed and used in its planning for adapting to climate change. The aim of this planning was to estimate changes over a 100 yr horizon from 2010 to 2110 according to an A2 scenario (Copenhagen Municipality 2011). Projections of climate change impacts for extreme precipitation were in accordance with national guidelines at the time (Arnbjerg-Nielsen 2012). For sea surges, no national guidelines existed, and hence the municipality derived its own projection.

2.1.1. Extreme sea surges

Mean sea level rise and changes in storminess will have an impact on extreme sea surges. Grinsted et al. (2015, this Special) developed probabilistic projections of expected mean sea level rise in northern Europe in the year 2100 based on the RCP8.5 scenario. They estimated a median sea level rise in Copenhagen of 68 cm and a 5–95% uncertainty range of 29–162 cm. To estimate the expected changes in sea surges, model simulations have been carried out with a hydrodynamic model covering the North Sea, Baltic Sea and inner Danish waters (M. Rugbjerg & P. Jensen unpubl.). The model was forced with simulated wind and atmospheric pressure fields for current (using the control period 1961–1990) and future conditions (2071–2100) from 3 regional climate models from the ENSEMBLES data archive (van der Linden & Mitchell 2009). The results showed an unchanged extreme sea surge signal in 2100 (relative to mean sea level) in the waters around Copenhagen, i.e. only changes in mean sea level contribute to changes in extreme sea surges. The results of the studies are summarised in Table 1 and compared to the expected increases in extreme sea surges used in the climate change adaptation plan developed by the Copenhagen Municipality.

Table 1. Summary of increases in extreme sea surges (cm) in Copenhagen in 2100 according to the projection used in the Copenhagen climate change adaptation plan (Copenhagen Municipality 2011) and the projection for the RCP8.5 scenario in Grinsted et al. (2015)

Projection	Return period	
	10 yr	100 yr
Current planning	94	103
RCP8.5 scenario, median	68	68
RCP8.5 scenario, 95% uncertainty quantile	162	162

Table 2. Summary of climate factors (dimensionless) for extreme precipitation. Current planning used by the Copenhagen Municipality is based on Arnbjerg-Nielsen (2012), RCP8.5 according to Sørup et al. (2015) and the estimated climate factors for the 6°C scenario

	Return period	
	10 yr	100 yr
Hourly precipitation		
Current planning	1.3	1.4
RCP8.5	1.8	2.2
Global 6°C scenario	2.2	2.7
Daily precipitation		
Current planning	1.3	1.4
RCP8.5	1.6	1.8
Global 6°C scenario	1.5	1.7

2.1.2. Extreme precipitation

The recommendation regarding the most likely climate change impacts on extreme precipitation in Denmark used by the Copenhagen Municipality is based on Arnbjerg-Nielsen (2012). The recommendation has recently been revised based on an assessment including different climate forcing scenarios, climate model projections and statistical downscaling methods (Gregersen et al. 2014). A key input to the revised recommendation is the use of a multi-model ensemble of regional climate projections in Europe from the ENSEMBLES project (van der Linden & Mitchell 2009) and application of 3 different downscaling methods (Sunyer et al. 2015). In addition, results based on a high-end scenario are included using a projection of the RCP8.5 scenario (Mayer et al. 2015) downscaled by means of a stochastic weather generator (Sørup et al. 2015). Based on this ensemble of information, 'mean' and 'high' changes are suggested as a function of the projection horizon and return period. Sunyer et al. (2015) showed that, in general, the recommendation given until now is conservative in comparison to the ensemble results using the A1B scenario, but that the RCP8.5 scenario leads to substantially larger changes in extreme precipitation. The new guideline recommends use of 'mean' changes that correspond to the recommendations given so far, but recommends additional analysis to be carried out with a higher change. The changes in extreme precipitation used by the Copenhagen Municipality corresponding to current 'mean' recommendations are summarised in Table 2.

To estimate changes in extreme precipitation based on the 6°C scenario simulation, 2 different statistical downscaling methods were applied, viz. a direct change factor methodology and a space-for-time approach, representing 2 very different methods based on extreme and average properties of climate change, respectively. Both methods were also used in the

analysis of the ENSEMBLES data by Sunyer et al. (2015). The methods are briefly outlined below.

The direct change factor approach is based on estimating the changes in extreme precipitation statistics from the regional climate model projections. In this case, climate model simulations are available for the periods 1976–2005 and 2071–2100 (Christensen et al. 2015). For each period, a regional extreme value analysis is applied that includes all land points covering Denmark. The regional model uses a partial duration series (PDS) approach, similar to the method applied in the analysis of Danish rainfall extremes by Madsen et al. (2002).

The PDS model includes precipitation events above a threshold, which in this case is defined as the level that includes on average 3 events yr^{-1} (i.e. for a 30 yr period, the 90 most extreme events are included in the analysis). A generalized Pareto distribution is fitted to the data using a regional estimated shape parameter, following the regional L-moment approach by Hosking & Wallis (1997). For each land grid point covering Denmark, extreme precipitation statistics can then be estimated for current (1976–2005) and future (2071–2100) conditions. A climate factor is finally obtained as:

$$CF_{T,i} = \frac{\hat{x}_{T,i,p}}{\hat{x}_{T,i,c}} \qquad (1)$$

where $\hat{x}_{T,i,c}$ and $\hat{x}_{T,i,p}$ are the estimated extreme precipitation for a return period of T years in grid point i for current (c) and projected (p), respectively. Climate factors are estimated for both hourly and daily precipitation extremes.

The direct change factor approach focusses specifically on estimating extreme precipitation properties. An alternative method is to search for areas where the current climate resembles the projected future climate at the location of interest (often referred to as space-for-time or climate analogue method). The first application of such an approach within climate change impacts was by Hallegatte et al. (2007). Here we used the method applied by Arnbjerg-Nielsen et al. (2015) for identifying climate analogue regions for Denmark based on the ENSEMBLES A1B scenario simulations. In their study, the following 3 indices were given the highest priority when defining climate analogues: (1) monthly distribution of mean temperature, (2) monthly distribution of mean precipitation and (3) extreme value statistics of daily precipitation.

For the 2 indices based on monthly values, a metric S_i is calculated that measures the similarity between the projected future climate at a specified location with the current climate at other locations:

$$S_i = \frac{\sqrt{\sum_{k=1}^{12}\left(x_{c,k,i} - x_{p,k,i}\right)^2}}{\sqrt{\sum_{k=1}^{12}x_{p,k,i}^2}} \qquad (2)$$

where $x_{c,k,i}$ is the monthly value (k) of index i of the current climate (c), and $x_{p,k,i}$ is the projected (p) monthly value of index i at the specified location. A low value of the metric corresponds to high climatic similarity. Similarly, for the index based on extreme value statistics, the metric is calculated as:

$$S_i = \frac{\sqrt{\sum_{k=1}^{N}\left(x_{c,k,i} - x_{p,k,i}\right)^2}}{\sqrt{\sum_{k=1}^{N}x_{p,k,i}^2}} \qquad (3)$$

where N is the number of estimated extreme precipitation events for given return periods, $x_{c,k,i}$ is the estimated event of the current climate, and $x_{p,k,i}$ is the projected event at the specified location. In the current setup, N is set to 2, using 1 and 10 yr return periods. For all 3 metrics, the E-OBS data of daily precipitation and temperature (Haylock et al. 2008) are used to define the current climate throughout Europe.

2.2. Vulnerability and adaptation assessment

Vulnerability is measured in monetary terms given that precipitation and sea surge hazards occur in the present climate and in the projected future climate (2100). The costs of flooding are based on assessment of flooded areas

for a range of return periods and for each return period registering a unit cost of flooding for the following types of assets: repair costs of flooded basements, houses, businesses (including inventory replacement) and infrastructure, as well as costs due to disruption of traffic and electricity. These calculations are summarised by COWI (2010). The costs for different return periods are then converted into an annual risk for the area in question by calculating the expected annual damage (EAD), i.e. the damage that on average will occur due to flooding. Here, p denotes the inverse of the return period, i.e. the probability of such an event occurring in any given year. The EAD is then calculated as (Olsen et al. 2015):

$$EAD = \int_0^1 D(p)dp \qquad (4)$$

where $D(p)$ is the cost at probability p. It is assumed that no damage occurs for return periods below 1 yr. Olsen et al. (2015) provided details of various ways of solving this integral.

The analysis of present flood risk indicated that flood damage already starts to occur at return periods below the design criteria (typically 10 yr). This is in accordance with other case studies and most likely because city development sometimes occurs without taking flood risk into account (see e.g. Ward et al. 2011, Zhou et al. 2012a, Olsen et al. 2015). It is therefore assumed that future adaptation scenarios also will entail small flood damages at return periods below the design level.

The Copenhagen Municipality has developed its adaptation plan under the assumption that the urban

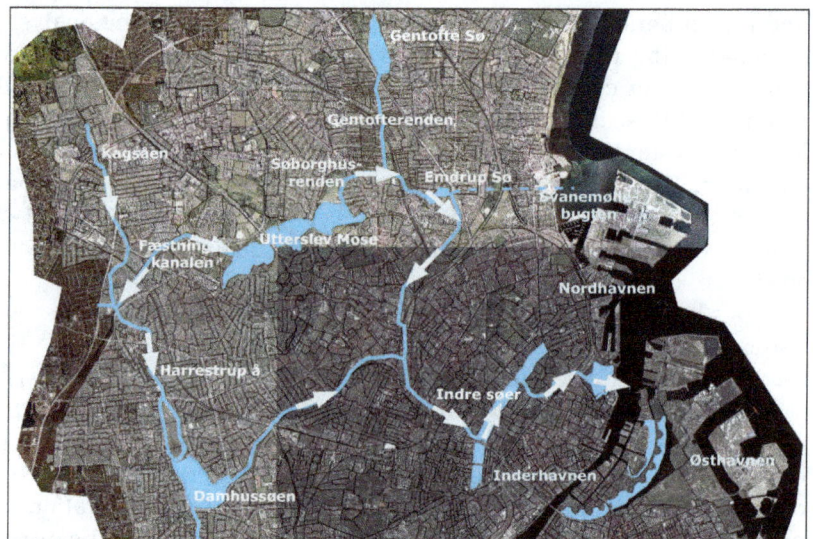

Fig. 1. Overview of surface waters and conveyance routes in Copenhagen, Denmark. The map covers an area of 10 × 12 km

fabric should change as little as possible. This implies that none of the affected assets (houses, businesses etc.) are moved per se, but instead the adaptation measures increase the return period at which damage will occur. This is in accordance with similar studies (see e.g. Zhou et al. 2012b). In cases where the adaptation measures are based on retaining water in depressions within the catchment, this may be an optimistic assessment, because once the depression is filled, the efficiency of the measure is reduced. In order to compensate for this process, it is assumed that the cost of exceeding the volumetric capacity of the depression varies between 0 and half the cost of not implementing the measure at all.

Using these assumptions, relative EAD was calculated for both precipitation and sea surge hazards under a business-as-usual scenario and for precipitation hazards assuming that the cloudburst adaptation plan is fully implemented. For the purposes of this study, the EAD calculations of the adaptation plan were adjusted to cover the period 2000–2100 to align the projection periods.

3. CURRENT COPENHAGEN CLIMATE CHANGE ADAPTATION PLAN FOR PROJECTED INCREASES IN PRECIPITATION EXTREMES

Copenhagen does not have large natural rivers or lakes. An overview of the main surface waters and conveyance routes is given in Fig. 1. The 5 surface waters (Damhussøen and Indre søer) in the region were all constructed during the 18th century as part of a system of defence structures against military attacks on the city, and the natural rivers have very low flow due to very low groundwater levels. Because of the low natural water flow, the rivers are piped, and from a legal point of view they are considered an integral part of the urban drainage system.

Based on an initial screening on the city-level of anticipated current (2010) and future (2110) costs of damage caused by precipitation extremes, the following were concluded in Copenhagen Municipality (2011):
• Currently, the expected annual cost of flooding due to precipitation extremes exceeds €1000 person^{-1} yr^{-1}.
• In the old parts of the city, buildings with basements are predominant. The value-at-risk in relation to these basements is high, and suitable measures to protect these are difficult to identify if executed by the municipality and/or the utility.
• If an optimal combination of surface and subsurface adaptation measures were implemented,

even a very substantial reduction in risk would have a positive net present value. Under some very crude assumptions, the highest net present value was found to be around a 90 % reduction of current costs of flooding. A major assumption is that urban storm water (and potentially sewage) running on city surfaces have no cost, enabling a substantial saving because some pipes may not need to be re-sized even though they cannot comply with current requirements of hydraulic capacity.

The main objectives to be met with the climate change adaptation plan towards precipitation extremes were then set as follows:
• All basements should be disconnected from the public sewer system and/or private non-return valves installed to ensure that basements cannot be flooded by sub-surface flows.
• City planning or local initiatives will ensure that up to 10 cm of water can run on streets and other surfaces without damaging the urban fabric.
• In general, a level of 10 cm of water on streets must not occur more frequently than 1 in 100 yr in 2110 given current anticipated climate change impacts (corresponding to a current 420 yr return period, see Fig. 6). Roads where this objective cannot be met will be managed by means of contingency measures as well as compensation schemes to affected citizens. These measures are not yet defined, but will be specified as the adaptation plan is developed further in collaboration with regional and national emergency management agencies.

Based on these criteria, 'cloudburst adaptation plans' have been developed for each urban hydrological catchment. An example is shown in Fig. 2. Great care has been taken to ensure that the existing urban fabric can be retained, both in terms of buildings and existing main corridors of transportation. A combination of retaining storm water in most parts of the city has been combined with storage by means of lowering the water level in a lake as well as conveyance corridors (i.e. roads) to a central tunnel that downstream will have a capacity of 27.1 m^3 s^{-1}.

4. RESULTS

4.1. Climate change impact assessment

Considering the projected changes in extreme sea surges used by the Copenhagen Municipality and the changes projected by the A2 scenario (Table 1), it is more than likely that the adaptation plan also accounts for the impact of the RCP8.5 scenario.

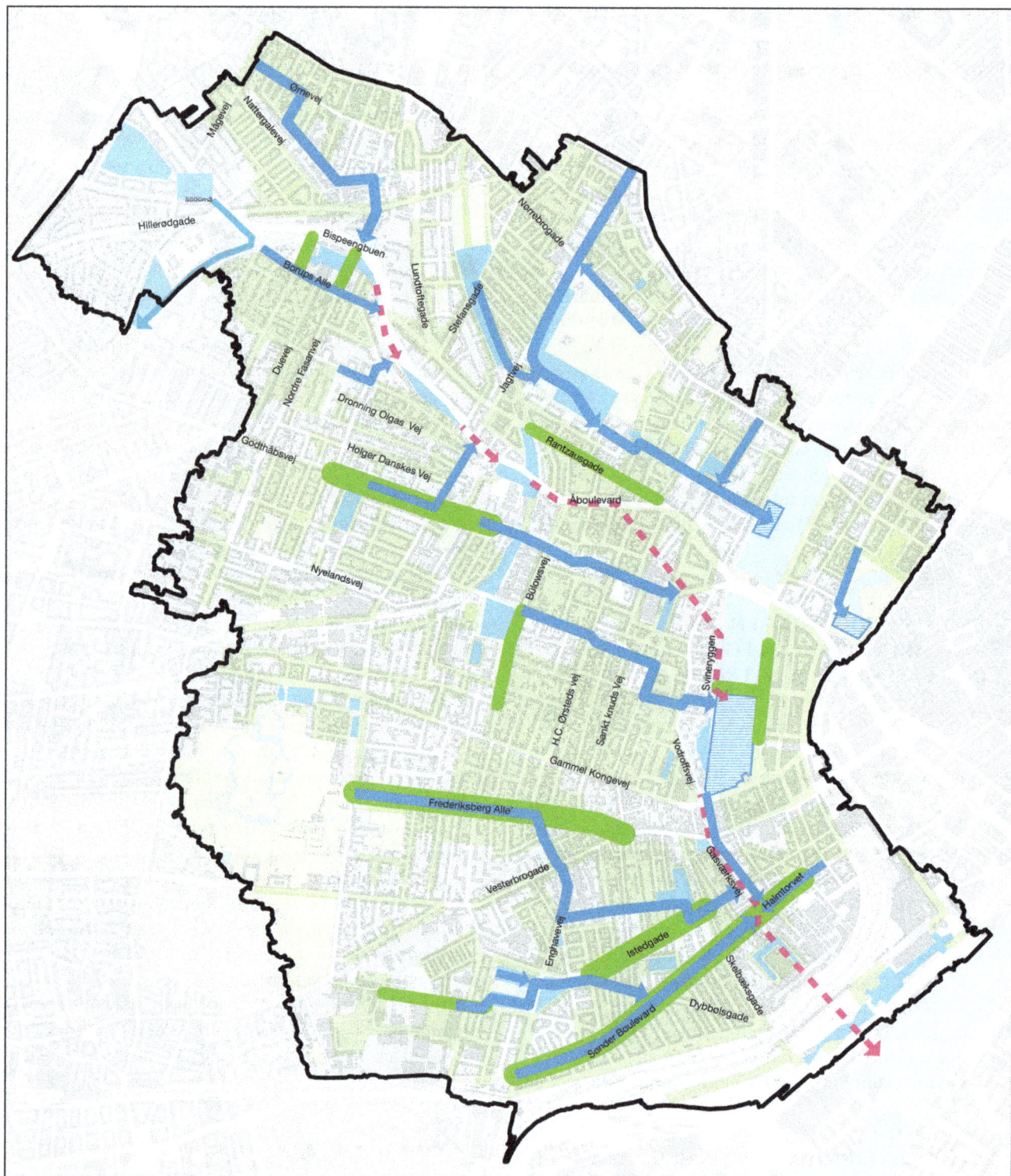

Fig. 2. Overview of adaptation measures planned for 1 urban catchment in Copenhagen (10.8 km², corresponding to ~20% of the city's area). Local retention measures are implemented in all areas marked in light green, all roads marked with dark blue indicate that the service level of 10 cm of water must be managed by local measures (e.g. channels), and mid-blue areas show locations where water will be stored temporarily during major storms. Dark green roads indicate where traffic will be affected because of retention measures implemented on the road itself. Blue hatched areas indicate where water levels will be lowered permanently to allow storage of storm water during heavy storms. A tunnel (dashed red line) with a downstream capacity of 27.1 m³ s⁻¹ must still be constructed to ensure that the design criterion is achieved

Therefore, the measures to be implemented for this hazard seem to be sufficient in the currently developed planning, and no further costs are foreseen under the high emission scenarios when considering this hazard alone.

For precipitation, the estimated climate factors for the 6°C scenario for different return periods based on the direct change factor method are shown in Fig. 3. The figure shows the average climate factor and the corresponding 16–84% variability (68% coverage interval) over the land points covering Denmark.

The seasonal distribution of extreme precipitation events for current and future conditions in the 6°C scenario is shown in Fig. 4. The frequency of precipitation extremes shows a shift towards more events in autumn and winter. The average magnitude of the extreme events increases in all seasons. These changes could likely have an impact on the occurrence of concurrent events of extreme precipitation and sea surges.

The results of the climate analogue metrics for the 6°C scenario are shown in Fig. 5. When calculating

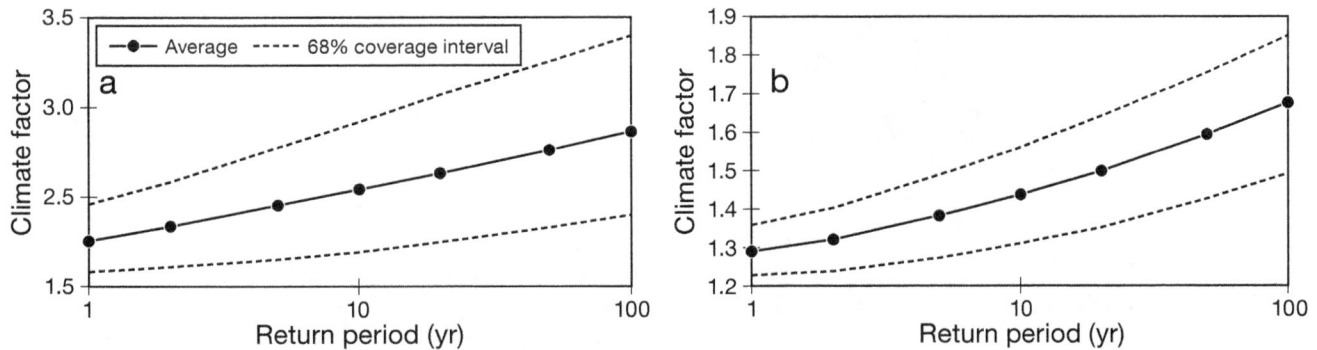

Fig. 3. Calculated climate factors for Denmark for the global 6°C scenario as a function of return period for (a) hourly and (b) daily precipitation extremes

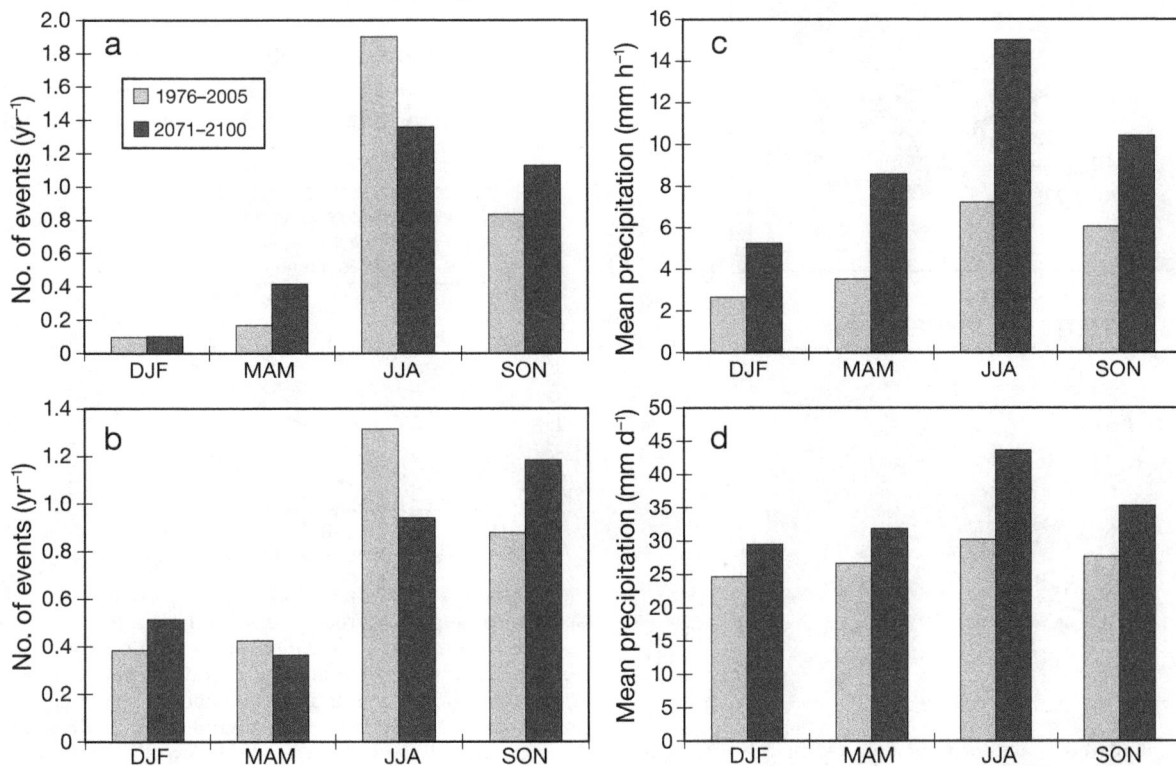

Fig. 4. Seasonal distribution in the (a,b) number of extreme events and (c,d) mean event magnitude for (a,c) hourly and (b,d) daily precipitation extremes for the 6°C scenario

these metrics for the ENSEMBLES A1B multi-model simulations, analogue locations with a current climate that correspond to a future climate in Denmark are centred around the coastal area of northern France (Arnbjerg-Nielsen et al. 2015). However, in the 6°C scenario simulation, each metric has a unique pattern, and combining them to define analogue locations seems difficult (see Fig. 5). The mean

temperature metric points towards locations in coastal southern Europe, the mean precipitation metric points towards locations in the coastal region around the British Channel, the North Sea and the Baltic Sea, and the metric on precipitation extremes varies widely, but mainly identifies locations away from the coastal regions, i.e. in central Europe, Sweden and the parts of southern Europe where the other 2 metrics indicate low climatic similarity. In conclusion, it is difficult to identify a location within Europe where the current climate resembles the future climate in Denmark based on the 6°C scenario simulation. Therefore, the results of the direct change factor method were used in the vulnerability and adaptation assessment.

The estimated climate factors are summarised in Table 2, and the corresponding changes in hazard for hourly precipitation extremes are shown in Fig. 6. For the 6°C scenario, the change in hazard is more than an order of magnitude. Severe storms with intensities corresponding to what was observed during the event in Copenhagen on 2 July 2011 could occur approximately every 40 yr in 2100 according to this scenario.

4.2. Vulnerability and adaptation assessment

The projected vulnerabilities for the business-as-usual scenario are shown in Fig. 7. These vulnerabilities are used to calculate relative EADs for precipita-

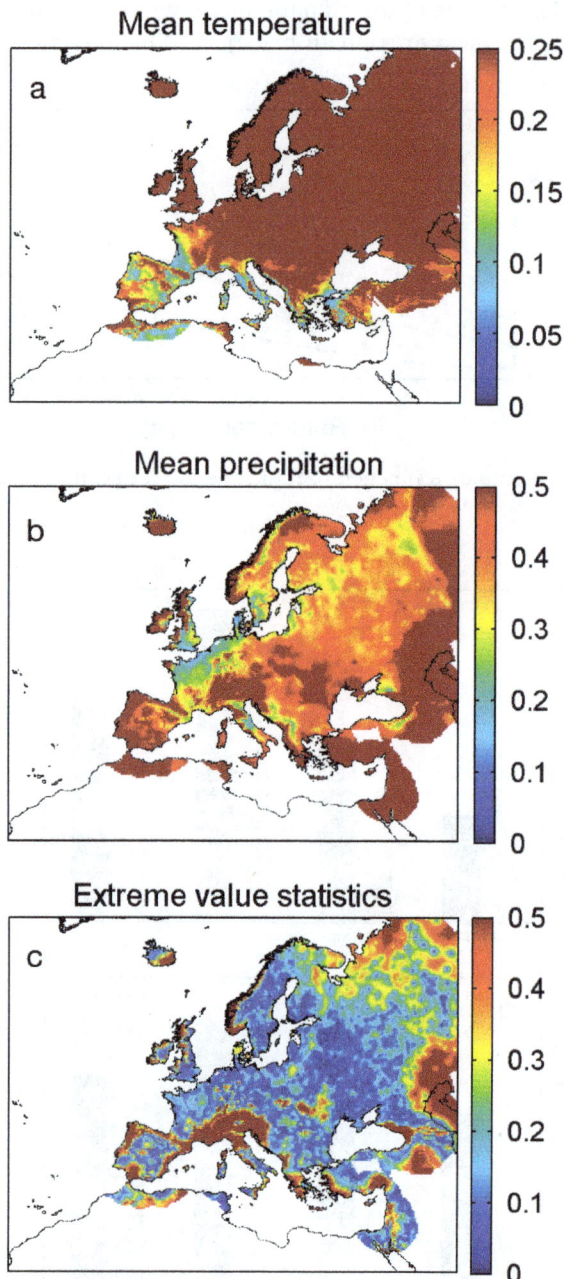

Fig. 5. Calculated space-for-time metrics for the 6°C scenario for locations within Europe for (a) mean temperature, (b) mean precipitation and (c) extreme precipitation. Low values of the metrics correspond to high climatic similarity

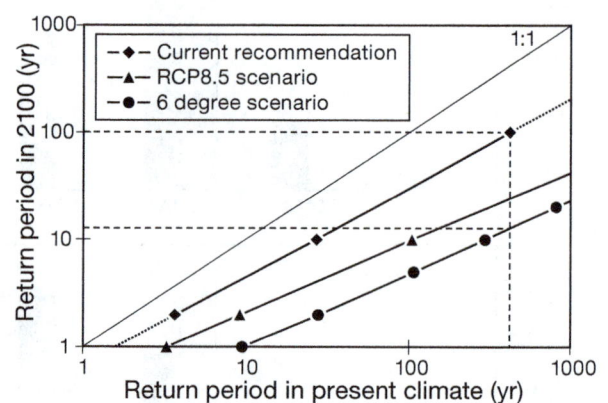

Fig. 6. Change in return periods in hourly extreme precipitation for Copenhagen from present to future for the current planning scenario and the 2 high-end scenarios. Current recommendation only covers return periods between 2 and 100 yr. Extrapolation of this recommendation is indicated by dotted lines. Dashed lines indicate the design criteria of the Copenhagen climate change adaptation plan. For the 6°C scenario, this level corresponds to a return period close to 14 yr as opposed to a 100 yr return period anticipated in the adaptation plan

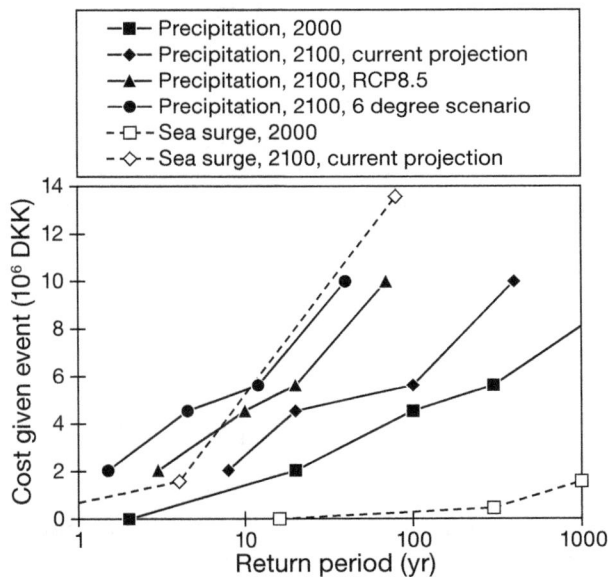

Fig. 7. Vulnerability assessment for precipitation and sea surge hazards. Costs of flooding in present and future (2100) climate under the A2 scenario used during development of the Copenhagen adaptation plan are derived from COWI (2010). The curves for precipitation hazards for 2100 for the RCP8.5 and 6°C scenario are calculated based on the assumptions described in Section 2.2. 1 DKK (Danish krone) is approximately 0.13 euro at time of writing

tion and sea surge hazards under a business-as-usual scenario and for a scenario of full implementation of the cloudburst adaptation plan (Table 3).

Based on the A2 scenario projection used by the Copenhagen Municipality, the largest change in risk is expected to be due to sea surges. The EAD increases by more than 2 orders of magnitude, from

Table 3. Calculated expected annual damage (EAD) based on the cost curves shown in Fig. 7. The uncertainty of these estimates is high, and hence relative values are reported, setting current costs of precipitation to an index value of 1

	EAD
Precipitation, business-as-usual scenario	
Present	1.0
2100, current projection	1.7
2100, RCP8.5	3.7
2100, 6°C scenario	7.7
Precipitation, cloudburst adaptation plan fully implemented	
Present	0.1
2100, current projection	0.2–0.3
2100, RCP8.5	0.6–1.0
2100, 6°C scenario	1.2–2.1
Sea surge, business-as-usual scenario	
Present	≤0.1
2100, current projection	6.6

very low values to almost 7 times current annual costs of precipitation extremes. However, for the 6°C scenario, the increase in risk for precipitation is very high, and although the relative change in risk is smaller than for sea surges, the EAD for extreme precipitation exceeds the EAD anticipated for sea surges. It is also noteworthy that current adaptation planning is quite efficient even under a 6°C scenario, because a high design criterion has been chosen (100 yr return period for flooding) compared to current design standards (10 yr return period). The EAD will be higher than today, and, most importantly, it will be based on fewer and larger events.

5. DISCUSSION

The purpose of the discussions below is not aimed at constructing a climate change adaptation plan that complies with the design criteria in a 6°C scenario, but rather to discuss what assumptions in the current planning scheme are challenged if the 6°C scenario should be taken into account.

5.1. Limits to making marginal analyses

The vulnerabilities given the changes in hazards due to climate change impacts are shown in Fig. 7. They are calculated under the assumption that no adaptation will occur and that the city fabric will not change in the future. The main costs are repair costs of buildings and replacement costs of stock and inventory in businesses and stores. As indicated by Figs. 6 & 7, the frequency of flooding in the 6°C scenario is likely to increase by more than an order of magnitude, implying that some basements could be flooded several times per year, and even the ground floor in some buildings could be flooded with a frequency of 1 yr. This implicitly means that business-as-usual is not an option, as the use of these buildings will change as flooding becomes too frequent to allow insuring the assets or even allowing physical repairs to occur between flood events. Hence, a marginal analysis assuming a strategy of business-as-usual is not feasible, because substantial autonomous adaptation will take place. In a business-as-usual scenario, the main subway line connecting northern parts of Copenhagen with the central and southern parts of the city is also projected to be flooded at several train stations with a return period below 10 yr in the 6°C scenario. It is unlikely that the citizens will allow such frequent disruptions of critical infrastructure.

5.2. Water infrastructure design considerations recognizing the impacts corresponding to a 6°C scenario

The tradition within design of large infrastructure has been to ensure that the performance of the structure is satisfactory until the end of the technical lifetime of key components. This has led to very long design lifetimes of such systems, typically with mean expected lifetimes exceeding 100 yr (e.g. Larsen et al. 2003, Zhou et al. 2012b). Similar considerations have also been the basis for the current recommendations regarding climate change adaptation for urban drainage systems in Denmark, as briefly outlined in Section 2.1.2.

This paradigm is challenged by some researchers, arguing that the large uncertainties in relation to climate change impacts call for smaller, incremental investments that are cheaper because they are more flexible and lock-in situations are avoided (Gersonius et al. 2013). On the other hand, this practice sometimes leads to sub-optimal solutions, known as the 'tyranny of incremental decisions' (Kahn 1966). Still, the massive changes needed in the urban fabric to accommodate a city that is robust in a 6°C scenario probably would lead to some sort of incremental investments, where changes over a shorter time span are considered.

The adaptation plan developed by Copenhagen is more ambitious than needed according to the current design guidelines. Large technical problems prevail, notably on how to technically allow 10 cm of water to run on most streets without damaging properties or increasing human exposure to pathogens. The choice between increasing storage and increasing conveyance of extreme water flows is also critical, and that choice is particularly sensitive to the actual climate change impact. The choice has been to retain as much water as can be achieved by using all available spaces on which there are no buildings, i.e. parks, parking lots, squares and even heavy modifications of a lake. The needed conveyance in the form of tunnels and channels has been designed to be able to hydraulically convey the excess water. The capacity of all of these planned facilities could, however, be exceeded on average every 14 yr under the 6°C scenario (see Fig. 6), leading to uncontrolled flooding of

the city. This appears to be a very high frequency, and the current adaptation plan also aims at reducing the frequency of such events (implementing a 100 yr return period for flooding). In any case, it would be advantageous to consider how to manage the residual risk by means of contingency measures, informing and educating the public, restricting spatial planning and considering measures aimed at increasing resilience in the region.

5.3. Concurrent events

The cost of an extreme precipitation event will be exacerbated if the sea level is high at the same time. Pedersen et al. (2012) showed that an important climate change impact may be a higher probability of such concurrent extreme events because of a shift in the seasonal pattern of precipitation extremes. The 6°C scenario simulation shows that the seasonal pattern of occurrence of precipitation extremes shifts towards more extreme events in autumn and winter and the mean magnitude of the extremes increases substantially (Fig. 4). Hence, since extreme storm surges primarily occur in autumn and winter, the probability of concurrent events is likely to increase, although it is difficult to quantify how much. However, this underpins the importance of developing adaptation plans that are robust to concurrent events, since the costs of climate change will otherwise be underestimated.

5.4. Practical aspects of implementing climate change adaptation

The current adaptation plan will change the urban layout in more than half of the total area of the city by changing the city layout to either store higher water volumes or convey water faster. Many of the adaptation measures are believed to improve the livability of the local neighbourhoods by increasing the visibility of water in the city. Still many of the measures are novel and remain to be tested in a setting where water scarcity is not an issue. Also, the concept of allowing 10 cm of water to flow on most streets without causing damage is challenging because of other critical functions of the city, notably ease-of-access for disabled people and flow of traffic.

The amount of disturbance to the everyday functions of the city when executing the climate change adaptation plan is also a challenge. Currently, a new metro is being built, and the disturbance to residential areas and traffic leads to much debate. The tunnel shown in Fig. 2 does not need to be as large as a full metro-tunnel; however, the downstream capacity exceeds by far what can be conveyed in a traditional sewer pipe, and thus tunnelling is likely needed. It remains to be seen how the public will react to disturbances from construction solely for the purpose of management of precipitation extremes. A concrete measure could be to identify and monitor indicators of progress in terms of the implementation of the climate change adaptation plan and the corresponding benefits to ensure that the objectives of the adaptation plan are retained during implementation.

6. CONCLUSION

Using an ensemble projection of the effects of climate change on extreme precipitation based mainly on the A1B scenario points towards moderate impacts on pluvial flooding in urban areas in Denmark, whereas using the results from 2 high-end scenarios, the impacts are projected to be much larger. In the 6°C scenario, business-as-usual is not an option, since even in the absence of policy-driven adaptation, large-scale autonomous adaptation will occur. Essentially, the most vulnerable urban areas could be abandoned under such a scenario because of damage to assets in these areas that is too frequent and too extensive. In this sense, adaptation implicitly has a 'tipping point' (Kwadijk et al. 2010) between a moderate and a high-end scenario.

We found that for sea surges, the corresponding 'adaptation tipping point' is below current projections, and hence even in a moderate scenario, major adaptation based on policy-driven responses is needed. After a major storm in July 2011, a very ambitious climate change adaptation plan was developed for urban pluvial flooding. If the plan is fully implemented, the Copenhagen region will be well protected even in a high-end scenario. It will take several decades to implement the plan even under the most optimistic circumstances, and it remains to be seen whether the political awareness will remain high if the frequency of extreme precipitation becomes relatively low at some future point in time.

Acknowledgements. This work was carried out with the support of the Danish Council for Strategic Research as part of the projects 'Centre for Regional Change in the Earth System', contract no. 09-066868 and 'Risk Based Design in a Changing Climate (RiskChange)', contract no. 0603-00390B.

LITERATURE CITED

ä Arnbjerg-Nielsen K (2012) Quantification of climate change effects on extreme precipitation used for high resolution hydrologic design. Urban Water J 9:57–65

▶ Arnbjerg-Nielsen K, Willems P, Olsson J, Beecham S and others (2013) Impacts of climate change on rainfall extremes and urban drainage systems: a review. Water Sci Technol 68:16–28

▶ Arnbjerg-Nielsen K, Funder SG, Madsen H (2015) Identifying climate analogues for precipitation extremes for Denmark based on RCM simulations from the ENSEMBLES database. Water Sci Technol 71:418–425

Christensen OB, Yang S, Boberg F, Fox Maule C and others (2015) Scalability of regional climate change in Europe for high-end scenarios. Clim Res 64:25–38

Copenhagen Municipality (2011) Københavns Klimatilpasningsplan (Copenhagen climate change adaptation plan). Copenhagen Municipality, Copenhagen. Available at http://kk.sites.itera.dk/apps/kk_pub2/pdf/1270_UEsoeIrcLC.pdf (accessed 29 May 2015)

COWI (2010) Muligheder og konsekvenser af klimasikring af København mod oversvømmelser (Possibilities and consequences of climate change adaptation of Copenhagen in relation to flooding). Report to the Copenhagen Municipality, October 2010. COWI, Lyngby

▶ Dankers R, Feyen L (2009) Flood hazard in Europe in an ensemble of regional climate scenarios. J Geophys Res 114:D16108, doi:10.1029/2008JD011523

▶ Gersonius B, Ashley R, Pathirana A, Zevenbergen C (2013) Climate change uncertainty: building flexibility into water and flood risk infrastructure. Clim Change 116:411–423

▶ Gregersen IB, Madsen H, Rosbjerg D, Arnbjerg-Nielsen K (2013) A spatial and nonstationary model for the frequency of extreme rainfall events. Water Resour Res 49:127–136

Gregersen IB, Madsen H, Linde JJ, Arnbjerg-Nielsen K (2014) Opdaterede klimafaktorer og dimensionsgivende regnintensiteter (Updated climate change factors and extreme precipitation used for design intensities). Paper no. 30. The Water Pollution Committee of The Society of Danish Engineers. Available at https://ida.dk/sites/prod.ida.dk/files/svk_skrift30_0.pdf (accessed January 2015)

Grinsted A, Jevrejeva S, Riva REM, Dahl-Jensen D (2015) Sea level rise projections for northern Europe under RCP8.5. Clim Res 64:15–23

▶ Hallegatte S, Hourcade JC, Ambrosi P (2007) Using climate analogues for assessing climate change economic impacts in urban areas. Clim Change 82:47–60

▶ Haylock MR, Hofstra N, Klein Tank MG, Klok EJ, Jones PD, New M (2008) A European daily high-resolution gridded data set of surface temperature and precipitation for 1950–2006. J Geophys Res 113:D20119, doi:10.1029/2008JD010201

Hosking JRM, Wallis JR (1997) Regional frequency analysis: an approach based on L-moments. Cambridge University Press, New York, NY

▶ Kahn AE (1966) The tyranny of small decisions: market failures, imperfections, and the limits of economics. Kyklos 19:23–47

▶ Kwadijk JCJ, Haasnoot M, Mulder JPM, Hoogvliet MMC and others (2010) Using adaptation tipping points to prepare for climate change and sea level rise: a case study in the Netherlands. WIREs Clim Change 1:729–740

Larsen OD, Moelgaard C, Kampmann J (2003) Flooding of the Copenhagen Metro, Denmark. Struct Eng Int 13: 231–234

Lawrence D, Haddeland I (2011) Uncertainty in hydrological modelling of climate change impacts in four Norwegian catchments. Hydrol Res 42:457–471

Madsen H, Mikkelsen PS, Rosbjerg D, Harremoës P (2002) Regional estimation of rainfall intensity-duration-frequency curves using generalized least squares regression of partial duration series statistics. Water Resour Res 38:1239, doi: 10.1029/2001WR001125

Madsen H, Lawrence D, Lang M, Martinkova M, Kjeldsen TR (2014) Review of trend analysis and climate change projections of extreme precipitations and floods in Europe. J Hydrol (Amst) 519:3634–3650

Mayer S, Fox Maule C, Sobolowski S, Christensen OB and others (2015) Identifying added value in high-resolution climate simulations over Scandinavia. Tellus Ser A Dyn Meterol Oceanogr 67:24941

Olsen AS, Zhou Q, Linde JL, Arnbjerg-Nielsen K (2015) Comparing methods of calculating expected annual damage in urban pluvial flood risk assessments. Water 7: 255–270

Pedersen AN, Mikkelsen PS, Arnbjerg-Nielsen K (2012) Climate change-induced impacts on urban flood risk influenced by concurrent hazards. J Flood Risk Manag 5: 203–214

Penning-Rowsell EC, Haigh N, Lavery S, McFadden L (2013) A threatened world city: the benefits of protecting London from the sea. Nat Hazards 66:1383–1404

Ranger N, Reeder T, Lowe J (2013) Addressing 'deep' uncertainty over long-term climate in major infrastructure projects: four innovations of the Thames Estuary 2100 Project. EURO J Decis Process 1:233–262

Refsgaard JC, Arnbjerg-Nielsen K, Drews M, Halsnæs K and others (2013) The role of uncertainty in climate change adaptation strategies — a Danish water management example. Mitig Adapt Strategies Glob Change 18: 337–359

Rojas R, Feyen L, Bianchi A, Dosio A (2012) Assessment of future flood hazard in Europe using a large ensemble of bias-corrected regional climate simulations. J Geophys Res Atmos 117:D17109, doi:10.1029/2012JD017461

Rosenzweig C, Solecki WD, Blake R, Bowman M and others (2011) Developing coastal adaptation to climate change in the New York City infrastructure-shed: process, approach, tools, and strategies. Clim Change 106:93–127

Sørup HJD, Christensen OB, Arnbjerg-Nielsen K, Mikkelsen PS (2015) Downscaling future precipitation extremes to urban scales using a spatio-temporal Neyman-Scott weather generator. Hydrol Earth Syst Sci Discuss 12: 2561–2605

Sunyer M, Madsen H, Ang PH (2012) A comparison of different regional climate models and statistical downscaling methods for extreme rainfall estimation under climate change. Atmos Res 103:119–128

Sunyer MA, Gregersen IB, Madsen H, Rosbjerg D, Arnbjerg-Nielsen K (2015) Comparison of different statistical downscaling methods to estimate changes in hourly extreme precipitation using RCM projections from ENSEMBLES. Int J Climatol (in press) doi:10.1002/joc.4138

van der Linden P, Mitchell JFB (eds) (2009) ENSEMBLES: Climate change and its impacts. Summary of research and results from the ENSEMBLES project. Met Office Hadley Centre, Exeter. Available at http://ensembles-eu. metoffice.com/docs/Ensembles_final_report_Nov09.pdf (accessed February 2015)

Veijalainen N, Lotsari E, Alho P, Vehviläinen B, Käyhkö J (2010) National scale assessment of climate change impacts on flooding in Finland. J Hydrol (Amst) 391:333–350

Ward PJ, de Moel H, Aerts JCJH (2011) How are flood risk estimates affected by the choice of return-periods? Nat Hazards Earth Syst Sci 11:3181–3195

Wilby RL, Dessai S (2010) Robust adaptation to climate change. Weather 65:180–185

Wilby RL, Harris I (2006) A framework for assessing uncertainties in climate change impacts: low-flow scenarios for the River Thames, UK. Water Resour Res 42:W02419, doi: 10.1029/2005WR004065

Willems P, Vrac M (2011) Statistical precipitation downscaling for small-scale hydrological impact investigations of climate change. J Hydrol (Amst) 402:193–205

Zhou Q, Halsnæs K, Arnbjerg-Nielsen K (2012a) Economic assessment of climate adaptation options for urban drainage design in Odense, Denmark. Water Sci Technol 66:1812–1820

Zhou Q, Mikkelsen PS, Halsnæs K, Arnbjerg-Nielsen K (2012b) Framework for economic pluvial flood risk assessment considering climate change effects and adaptation benefits. J Hydrol (Amst) 414-415:539–549

7

Key drivers and economic consequences of high-end climate scenarios: uncertainties and risks

Kirsten Halsnæs*, Per Skougaard Kaspersen, Martin Drews

Climate Change and Sustainable Development Group, Department of Management Engineering,
Technical University of Denmark, Building 426, Produktionstorvet, 2800 Kgs. Lyngby, Denmark

ABSTRACT: The consequences of high-end climate scenarios and the risks of extreme events involve a number of critical assumptions and methodological challenges related to key uncertainties in climate scenarios and modelling, impact analysis, and economics. A methodological framework for integrated analysis of extreme events and damage costs is developed and applied to a case study of urban flooding for the medium sized Danish city of Odense. Moving from our current climate to higher atmospheric greenhouse gas (GHG) concentrations including a 2°, 4°, and a high-end 6°C scenario implies that the frequency of extreme events increase beyond scaling, and in combination with economic assumptions we find a very wide range of risk estimates for urban precipitation events. A sensitivity analysis addresses 32 combinations of climate scenarios, damage cost curve approaches, and economic assumptions, including risk aversion and equity represented by discount rates. Major impacts of alternative assumptions are investigated. As a result, this study demonstrates that in terms of decision making the actual expectations concerning future climate scenarios and the economic assumptions applied are very important in determining the risks of extreme climate events and, thereby, of the level of cost-effective adaptation seen from the society's point of view.

KEY WORDS: Climate scenarios · Extremes · Risks · Damage and welfare costs · Uncertainties

1. INTRODUCTION

It is well documented that climate change is likely to influence the frequency and severity of some extreme weather and climate events regionally[1] (IPCC 2012). The projected trends in extremes often show a positive correlation with increasing concentrations of atmospheric greenhouse gasses; hence, the most severe changes are projected under high-end scenarios like the RCP8.5 Representative Concentration Pathway (Meinshausen et al. 2011). In this study we address the evaluation of societal risks, recognizing the fact that despite the inherently low probabilities of

extreme events, the economic consequences to society can be very high. Assessing such risks involves specific methodological challenges related to key uncertainties and to economic assumptions. These are again related to the multiple elements involved in climate change impact studies, frequently visualized as a 'cascade' of uncertainties (e.g. Wilby & Dessai 2010). Methodologically, in the cascading picture, uncertainty propagates through the different interlinked steps in a 'top–down' assessment of climate risks, ranging from socio-economic scenarios through emission scenarios, global and regional climate model projections, and impact models to local impacts and possibly adaptation responses. The uncertainties involved are however of a different nature dependent on disciplines, modelling tools, and approaches applied (IPCC 2005, Refsgaard et al. 2013). Thus some of the uncer-

[1]'Extreme events' are here defined as specific outcomes of individual or combinations of climate variables belonging to the tails of a given probability distribution

*Corresponding author: khal@dtu.dk

tainties reflect parameter uncertainties while others are of a more structural character, such as uncertainties related to economic valuation, risk perceptions and preferences (Weitzman 2011). All together this plethora of uncertainties provides a basis for a very wide range of climate change risk estimates.

The present study explores an integrated methodological framework drawn from the cascading picture for assessing the risks of extreme climate events with high consequences, and applies the framework to a real case study of pluvial flood risks in a medium sized Danish city. A systematic assessment is carried out of how risk estimates and uncertainties are related to climate scenario- and impact uncertainties and, in particular, to economic assumptions. Four different climate scenarios are considered: a reference case reflecting current climate conditions, a 2°C and a 4°C scenario corresponding to the RCP4.5 and RCP8.5 scenarios, respectively (IPCC 2013), and finally a special 6°C climate scenario provided by the Danish Meteorological Institute (Christensen et al. 2015, this Special). We investigate the economic consequences of extreme events—these are, as mentioned above, considered as part of an integrated assessment where climate and impact models are linked to economic models. The methodological framework we use for linking physical and economic models are inspired by a paper by Weitzman (2011) describing the role of 'fat-tailed uncertainty in the economics of catastrophic climate change'. Weitzman argues that there are large uncertainties associated with the probability of extreme events as projected by climate models as well as deep structural uncertainties related to economic risk evaluations, including damage cost estimates, discounting, and risk aversion. The latter are key issues in terms of real-life decision making, i.e. how much society should be willing to pay for adaptation in a given future climate scenario, which is often overlooked or severely simplified in many real-life climate change impact assessments. In this study we address the propagation of uncertainties and test critical assumptions in relation to a case study of urban flooding. Through a combination of climate scenarios, urban flood modelling, and economic assumptions, we analyse a total of 32 alternative scenario combinations, highlighting the role of key drivers and economic consequences.

2. METHODOLOGICAL FRAMEWORK

Seen from the perspective of a climate change adaptation decision maker, society should be willing to pay adaptation costs, which are at least equal to the avoided costs of climate change impacts. Adaptation costs should be adjusted for residual damages, up to the point where adaptation costs exceed residual costs. According to this, residual damages are associated with climate change impacts which either have very low damage costs or where adaptation is very expensive.

The avoided costs of climate change in terms of risks depend on damages as well as on the probability of a given event[2]. Adhering to conventional usage of the term, we define climate change risks as the probability × consequence of a climate change event. For high consequence events with low probabilities, the estimated risks will depend on a sort of 'race' between how fast the probabilities of climate events decline, compared with how fast damage costs increase, when we are moving further away from the mean (median) of a climate probability density function.

Climate change impact assessment, e.g. as described by the traditional uncertainty cascade (e.g. Wilby & Dessai 2010), generally involves integrated climate modelling and impact assessment. Specifically, future climate events such as temperature and precipitation extremes, wind storms, droughts, or combinations of these are used as drivers for impact assessments. Subsequently, economic consequence studies address damages to specific sectors, ecosystems, geographical locations, and human assets.

In the present study, damage costs are based on a bottom–up assessment, where cost parameters are assigned to different assets which are expected to be at risk from pluvial flooding; however, the approach could be easily generalized to other types of high impact events. Here assets include buildings, historical values, health, infrastructure, and ecosystems. The costs associated with damages to these assets are transformed to a measure of 'willingness-to-pay' (WTP) reflecting welfare loss, where risk aversion and equity concerns (given by alternative discount rates) are taken into consideration.

Fig. 1 illustrates the different logical steps of the impact assessment. Generally, in terms of describing the risks associated with a specific combination of one or more climate variables like temperature, precipitation, wind or sea level, the probability of a specific (possible compound) event is derived from climate projections. The probability may be expressed

---

[2]A climate event should here be understood as a broad terminology covering particular weather events like hot spells, intensive precipitation, wind storms, etc., which are associated with societal risks

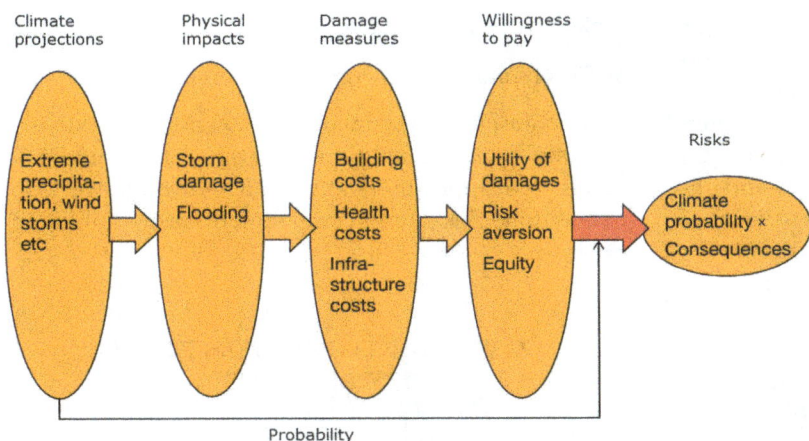

Fig. 1. Example structure of climate change impact assessment and risk analysis. The red arrow shows that the first and last step in the assessment are combined

in the form of a probability density function (pdf), which is typically constructed on the basis of an ensemble of climate models. Principally, the pdf provides a comprehensive description with respect to both frequency and intensity. In practical terms it is far from trivial to construct such a pdf, especially for compound situations where more than one climate variable is involved. Hence, in some cases, more stylized shapes of pdfs are therefore used, e.g. in order to explore the tails of the distribution given certain assumptions about uncertainties (Weitzman 2011).

To exemplify, consider the 2 stylized pdfs illustrated in Fig. 2. The x-axis shows the change in the (anomaly) value of some climate variable, e.g. daily mean temperature, for a future time period with respect to a set control period; the y-axis expresses the probability of this value, e.g. as inferred from single model simulations or an ensemble of climate model projections. In this idealized example both distributions depict a nearly identical central value of an increase in daily mean temperature of 6°C, which could be interpreted as the mean (median) of the cli-

mate model projections, whereas the tails of the pdfs expresses (extreme) values further and further away from the mean. As shown in Fig. 2 the tails of the 2 pdfs differ significantly in their 'fatness'. While the 'thin-tailed' pdf is heavily centred and symmetric around the mean, the red 'fat-tailed' pdf is somewhat skewed and lends higher probability to extremes.

The thin- and fat-tailed distributions could be derived in different ways. One could think of the 2 distributions as being derived from 2 different ensembles of model simulations, e.g. forced by different climate scenarios (a moderate scenario like RCP4.5 versus a high scenario like RCP8.5). For some geographical locations and for some variables like temperature and precipitation extremes many authors (e.g. Christensen & Christensen 2003, Collins et al. 2012) have thus demonstrated that the probabilities of what are considered extremes under present-day conditions are likely to increase significantly under gradually higher levels of global warming. This implies that the tails corresponding to the higher end range of climate scenarios are similarly likely to be relatively 'fat'. Distributions could also be derived from the same ensemble using different methodologies and/or assumptions, or in the case of single or few model simulations, they could be based on entirely different climate models; hence, the difference primarily represents model uncertainty. It could also be a combination of all of the above. Either way, the impact assessment of an extreme event is likely to be heavily compound on the shape of this pdf, which of course introduces a significant uncertainty in terms of determining and quantifying risks.

The perspective of the damage cost assessment in our approach is social welfare[3], where the total damage costs is an aggregate measure of the costs to all individuals of damages to given assets, and total damages are calculated as the sum of damages in all sub categories.

In terms of climate change, the uncertainty surrounding future events and the specific character of extreme events with low probabilities and high consequences suggests that the social welfare function applied to damage cost evaluation is adjusted to reflect society's perspective on uncertain future risks (Heal & Kriström 2002, Weitzmann 2011). One way to

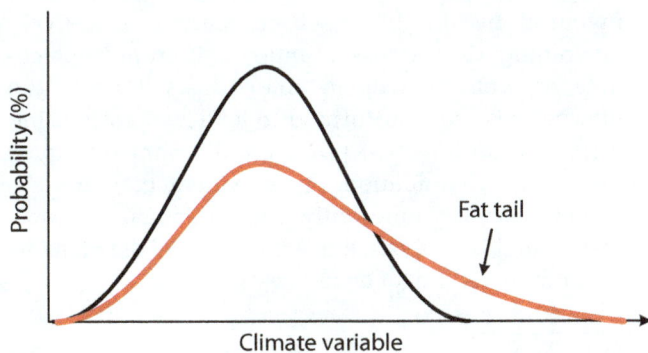

Fig. 2. Stylized representation of 2 alternative climate variable distributions

[3]Social welfare reflects society's perspectives as for example in relation to climate change impacts

include this type of uncertainty in economic analysis is to apply a risk aversion factor. Risk aversion by definition is the reluctance of a person to accept a bargain with an uncertain payoff rather than a bargain with a certain payoff, and as already pointed out, extreme consequences of, for example, high-end climate scenarios are by their very nature uncertain.

As a basis for measuring WTP and following IPCC (Kolstad et al. 2014) we assume a social welfare function (V), where $u(c_t) = V_t$ is the contribution to the social welfare function of generation t consuming c_t. Since c_t is uncertain, we consider the expected value $Eu(c_t)$ of consumption in our social welfare function. The concavity of the function u combines inequality aversion reflected in discount rate and risk aversion to reflect uncertainty:

$$V = \sum_{t=0}^{\infty} Eu(c_t)d(t) \qquad (1)$$

The factor $d(t)$ is a discount factor, which reflects equity in terms of our collective pure time preference for the present versus the future and an ethical parameter reflecting equity among present generations following the prescriptive approach to discounting reflecting equity concerns (IPCC 2014, section 3.6.2 therein).

We assume a risk aversion factor as defined by Arrow (1965):

$$A(v) = -U''(v)/U'(v) \qquad (2)$$

where $A(v)$ is the risk aversion associated with a given social welfare change, and the utility of the social welfare change is:

$$U(v) = \sum_{t=0}^{\infty} u(c_t) \qquad (3)$$

where $U'(v)$ and $U''(v)$ are the first and second order derivatives of $U(v)$, respectively.

In the case of a utility function, which is a polynomial of order n, the form of the risk aversion factor reduces to the expression:

$$A(v) = nc^{n-1} \qquad (4)$$

Hence the risk aversion is a constant. There are to the authors' knowledge no specific climate change risk attitude studies suggesting what the level of risk aversion should be, so instead we consider 2 different risk aversion factors (i.e. high and low risk aversion) based on an approach developed by Heal & Kriström (2002), who suggest to use risk aversion values between 1 and 6 based on risk preferences revealed among investors.

The risks of climate change impacts may now be calculated from:

$$\text{Risks} = \text{WTP to avoid event} \times \text{probability of event} \qquad (5)$$

$$\text{WTP} = \text{damage costs} \times \text{risk aversion factor} \qquad (6)$$

To exemplify how uncertainties and economic assumptions individually and combined influence risk levels, we apply the methodological framework to assess flood risks due to very high intensity rainfall in Odense.

3. PLUVIAL FLOOD RISKS IN THE CITY OF ODENSE

Odense is the largest city on Funen and the third-largest city in Denmark. It has ~172 000 inhabitants and is an eclectic mix of residential housing, enterprises, industry, recreational areas, historical buildings, etc. The city is located next to Odense Stream and close to Odense Fjord, thus making the city centre vulnerable to different kinds of flooding. Recently, the risk of flooding due to heavy rainfall was assessed by the local government as the first step in a large decision-making framework aimed at developing a detailed climate change adaptation strategy and action plan[4] (Odense kommune 2014, pers. comm.).

To identify major risk drivers and illustrate the role of uncertainties as well as the importance of climate scenario assumptions, damage cost functions, risk aversion, and discount rates we carry out a sensitivity analysis, constructing all possible combinations of these factors (illustrated in Fig. 3). Moving radially out from the centre of the circle, our starting point is the choice of climate scenario. The next step is to combine the climate scenario with 2 different damage assessment approaches. We then apply risk aversion factors of 3 and 1, respectively, where a factor of 1 implies risk neutrality, i.e. cost estimates are not adjusted by the risk perception. The factor of 3 represents a 'middle-of-the-road' perspective often favoured by real-life decision makers, effectively 'averaging' risks across a range of different (replaceable as well as irreplaceable) assets. Finally, the alternatives are transformed to levelized costs using a (moderate) 3 % or a (low) 1 % discount rate for a total of 32 combinations. It is evident that the levelized costs are inherently dependent on all parameters in this analysis, e.g. a higher level of risk aversion increases the levelized costs.

---

[4]All Danish local governments are obliged to develop local adaptation plans, which in the first phase until the end of 2014 are focussing on flood risks

3.1. Data

The following physical and socio-economic data are used in the assessments:

- Downscaled climate projections from Arnbjerg-Nielsen et al. (2015, this Special)
- Flood maps for Odense based on urban flood modelling using MIKE Urban/MIKE Flood software (MIKE By DHI 2014), wherein the city's topography and urban drainage system is included; this was supplied by the municipality of Odense (Odense kommune 2014, pers. comm.)
- GIS land cover data for Odense from the Danish Ministry of the Environment (Miljøministeriet 2014)
- Damage cost estimates for roads, railways and irreplaceable assets from Odense kommune (2014, pers. comm.)
- Damage cost estimates for houses, basements and other buildings from Arnbjerg-Nielsen & Fleischer (2009), Zhou et al. (2012) and Forsikring & Pension (2014)
- Since no unit damage cost estimates exists for the service and industry sectors, these were estimated based on insurance claims (Forsikring & Pension 2014). Likewise, unit damage costs for health and waterbodies were estimated from Zhou et al. (2012).

3.2. Climate projections

Heavy rainfall intensities corresponding to 3 different climate scenarios as well as present day conditions have been reported by Christensen et al. (2015) and Arnbjerg-Nielsen et al. (2015). The first 2 scenarios were inferred from regional climate projections of the RCP4.5 and RCP8.5 scenarios for the period 2071–2100 and correspond to a global mean surface warming at the end of the 21st century of ~2°C and 4°C. Conversely, the last scenario represents an arbitrary future 30 yr time slice, where a global mean surface warming of 6°C is realized (Christensen et al. 2015). Based on the 3 time slices we calculate the annual probability of rainfall events of a particular intensity for the different climate scenarios (Fig. 4).

For the scenarios associated with the higher global mean temperature changes, the probability of specific high intensity rainfall events is clearly seen to increase relatively, as do the maximum intensities. This implies that if we consider the frequency of specific events then the distributions derived from the higher end scenarios are effectively 'fat-tailed' as compared to the 'thin-tailed' distributions derived from lower scenarios or present-day conditions (Fig. 2).

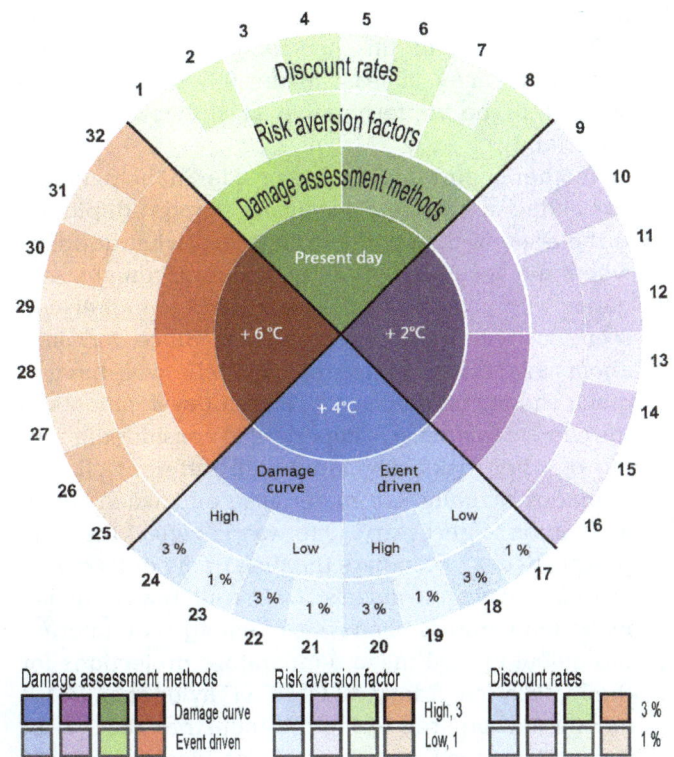

Fig. 3. Structure of a sensitivity analysis applied to the pluvial flooding case; 32 different scenario combinations are shown

Evidently, this is not the whole story, since both global and regional climate projections are influenced by a wide range of further uncertainties, including bias, model uncertainty and internal variability, whose relative importance varies with prediction lead time and with spatial and temporal averaging scale (e.g. Hawkins & Sutton 2009, 2011). Similarly, empirical-statistical downscaling of precipitation is also affected by considerable uncertainties and critical assumptions (Sunyer et al. 2014). In this study, as in many real-life impact assessments, we do not have sufficient information to strictly decompose the variance of the climate projections. For simplicity we instead use the full range of climate scenarios discussed above as sort of a proxy for assigning specific probabilities to specific precipitation intensities across scenarios. In general for precipitation Hawkins & Sutton (2011) identify model uncertainty as the predominant source of uncertainty more or less independently of lead time, which suggests that the climate uncertainty used in this analysis may be deflated. Recent work by Gregersen et al. (2014) on the other hand finds that the spread of the projections for Denmark used herein exceeds the observed spread in a comparable ensemble of regional climate projections from ENSEMBLES (van der Linden & Mitchell 2009), indicating that in this specific case the scenario uncertainty may at least to the 0th order be considered to be representative of the total climate uncertainty.

Another important point to consider in the perspective of decision making on climate change adaptation is the issue of time and learning, e.g. what happens when the focus moves from low-end climate scenarios with possible moderate impacts to higher-end scenarios with more severe impacts. Since decisions about adaptation measures typically will have a much shorter time perspective than developments in the climate system, it is important to consider the timing of when risks associated with different climate scenarios actually can materialise and when given adaptation is necessary. One way to reflect the time perspective is to compare the time frame determined by different climate models of when alternative global mean temperature changes can emerge. Comparing the global annual mean temperature projections for the RCP8.5 scenario of 38 CMIP5 (Taylor et al. 2012) members compared to the pre-industrial 1881–1910 period, Christensen et al. (2015) for example showed that around the year 2100 is the 'earliest' time when

a 6°C global mean temperature change could be achieved. Conversely, the same study finds that most of the projections achieve 6°C before 2130. It is evident that adapting to such high temperature levels will not be needed until far into the future. If we are instead to consider high-end scenarios within the time frame of the 22nd century, it makes sense to focus on the risks associated with moving from a 4°C scenario to a 6°C scenario rather than only focusing on the highest climate scenario.

3.3. Flooded assets

To estimate the damage costs we combine detailed modelling of the topography and the urban drainage system with geographical information in a GIS format. The geographical information includes data on vulnerable assets in terms of buildings, roads, railways, cultural and historical values, ecosystems, and human health, and is based on a static picture of present day city activities and structure. State-of-the-art maps describing the likely location and extent of flooding following alternative extreme precipitation events have recently been produced for Odense using the MIKE Urban/MIKE Flood (MIKE By DHI 2014) modelling tools. For some assets like damage cost estimates of buildings in this assessment step are based on very aggregate general categories of private houses and commercial buildings. Urban ecosystems or recreational areas are not included in the assessment, and neither is discomfort to people due to stress associated with the event or loss of working hours for cleaning up after the flooding. Both of these could potentially be associated with significant costs. The same is the case with losses associated with disruptions in industrial activities and business, which have not been taken into consideration because only very small-scale industrial activities are located in the central city of Odense. Flooding would therefore not have a large economic impact on these activities. Similarly, losses in business activities and shopping are not included, which like in the case of industrial activities could tend to cause an underestimation of damage costs. It could however be expected that many business and shopping activities would be postponed for a few days due to flooding, and that economic losses thereby would be small.

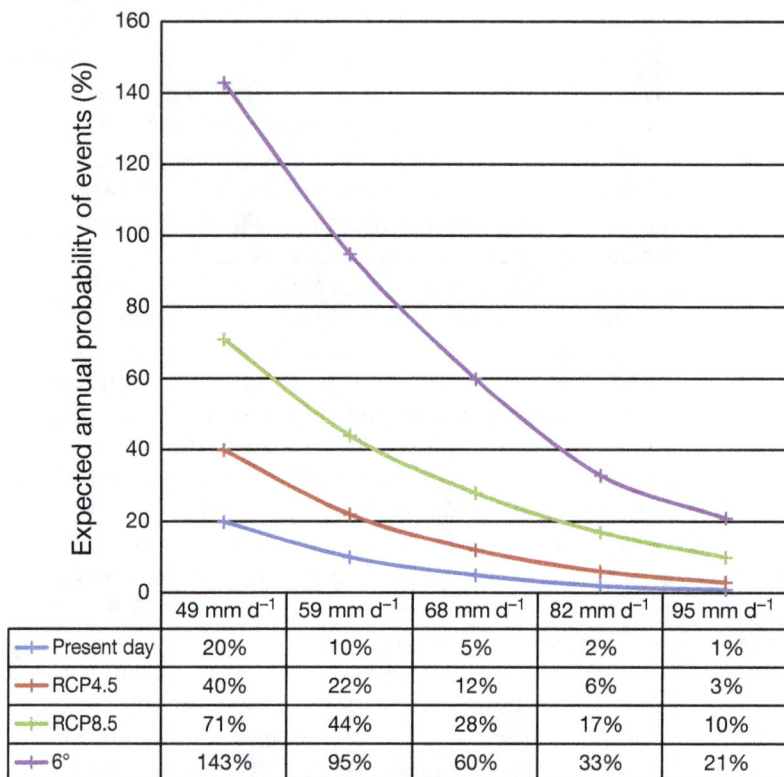

	49 mm d⁻¹	59 mm d⁻¹	68 mm d⁻¹	82 mm d⁻¹	95 mm d⁻¹
Present day	20%	10%	5%	2%	1%
RCP4.5	40%	22%	12%	6%	3%
RCP8.5	71%	44%	28%	17%	10%
6°	143%	95%	60%	33%	21%

Fig. 4. Expected annual probability of intense precipitation events for the different climate scenarios: present day, regional climate projections (RCP) RCP4.5 and RCP8.5, and 6°C increase

It could be argued that basing our damage cost assessment on a static picture of the city would tend to underestimate costs because the value of damaged assets would increase over time. This will certainly be the case, but there is currently no good methodology available that can be used to make a detailed projection of city activities, which can merge the details in our flooding calculations. The present study uses location-specific information about houses, roads and other assets, and we cannot project these. A possibility could be to add a general factor to reflect increases in the value of city assets over time, and this would in general work as a multiplier on the costs and thereby increase the damage estimates. We have chosen not to do this because the major point of our study is not to provide accurate cost estimates, but rather through a sensitivity analysis to demonstrate the importance of key economic assumptions. We are not really expecting high population growth in Danish cities, but the value of assets would increase if the current trend of city development continues.

The total number of buildings and other assets likely to be affected by different flooding events in Odense are compiled in Table 1.

We note from Table 1 that the total number of buildings and basements flooded is clearly increasing with increasing precipitation intensities (maximum event intensities). The same is the case for roads, railways, etc. In the case of health impacts and irreplaceable assets such as historical buildings, a particularly high number of incidences are seen to appear for precipitation events exceeding a threshold of 30 mm h^{-1}.

3.4. Damage cost approach

Damage cost estimates are based on a number of different data sources, implying that unit cost data for different assets may be uncertain. Unit cost data have been adjusted to common measurement standards (Tables A1 & A2 in the Appendix).

The damage cost assessments are based on 2 alternative methods to exemplify differences in real-life damage assessments. In the traditional 'damage curve' (DC) approach, the modelled surface water depth is used directly as a measure of severity. Using the DC approach, damage costs per asset flooded increases linearly as the water depth increases, until it reaches a predefined level, where it is assumed that maximum possible damages occur (Table A1 in the Appendix). This water depth by assumption varies between assets and is set to be 50 to 70 cm in the present study. The selected water levels for maximum damage were based on European and American findings in accordance with Jongman et al. (2012) and Davis & Skaggs (1992).

Table 1. Flooded assets for rainfall events with different intensities. For simplicity, buildings and basements are treated as uniform categories

| Flooded assets | Water depth flooding threshold (cm) | Event intensity (mm h^{-1}) | | | | | Unit |
		20	25	30	35	40	
Buildings							
Service and Industry	20	59	94	174	231	278	No. of buildings
Multistorage residential	20	38	56	95	103	123	
Houses	20	87	201	398	472	576	
Leisure house	20	108	229	419	490	605	
Basements	5	178	311	573	569	726	No. of basements
Health effects from basement flooding	0.3	475	757	1356	1397	1659	No. of people affected
Roads	5	120	212	374	434	533	1000 m^2
Railways	5	2	5	9	12	15	1000 m^2
Waterbodies flooded in the city with mixed surface and sewage water	20	29	31	41	43	46	No. of waterbodies
Irreplaceable assets							
Ancient monuments	20	–	–	1	2	2	Buildings
Churches	20	–	–	1	1	1	Buildings
Conservation worthy buildings	20	10	22	58	60	74	Buildings
Clergy buildings middle age	20	–	–	–	–	–	Buildings
Statues and sculptures	20	2	2	5	5	7	Buildings
Museums	20	–	–	4	4	5	Buildings

In the 'event-driven' (ED) approach, unit damage cost is kept constant for all water levels exceeding a certain water depth threshold. As assets have different susceptibility towards the water level required to cause damages, a water depth threshold is defined for each asset type to represent a given asset damage cost, and this threshold is constant for all precipitation events (see Table A2). Damage unit costs are related to the intensity and total amount of precipitation during a precipitation event. As the intensity of the precipitation events increases so does the total number of assets flooded and the unit cost per damage. The logic behind this approach is that the likelihood of assets being flooded with water levels above the defined threshold increases with the amount of precipitation increase. This relationship has been confirmed by available data for insurance claims from flooding during high-intensity precipitation events in Denmark in the period 2006–2013 (Forsikring & Pension 2014), where it can be seen that the average insurance claim was 3 to 10 times higher for damages from high-intensity precipitation events compared with low-intensity precipitation events.

As shown in Fig. 5 the range of damage costs span from about 17 million EUR for the smallest precipitation event, to over 300 million EUR for the most intensive event. Cost estimates derived using the ED approach start below the DC cost level, but increase more steeply and pass the DC based costs for precipitation events of more than 30 mm h^{-1}. This implies that using the ED approach will generate higher risk

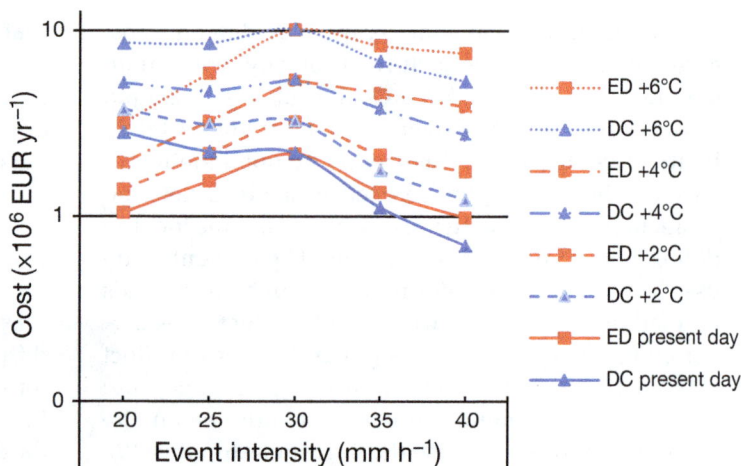

Fig. 6. Levelized costs of flood damage over a 100 yr period for different climate change scenarios (+2°, +4°, and +6°C) using a 3% discount rate. ED: event-driven, DC: damage curve

estimates for very intense precipitation, which may be more likely in higher-end climate scenarios. It is important to recognize here that both approaches depend on the availability of reliable damage cost data, and that such data in most real-life cases is likely to be sparse. Likewise, both approaches ignore the indirect costs of pluvial flooding, which in absolute terms may be considerable, but which for the purpose of a sensitivity analysis makes them equally good (or bad).

The damage costs are transformed to risk estimates by multiplying the estimated costs with the probability of an event happening at a different point in time. From this we calculate net present values and corresponding levelized costs[5]. These risks are illustrated in Fig. 6 for alternative climate scenarios. The levelized costs of the damages are seen to increase for higher precipitation intensities in the ED approach, peaking at 30 mm h^{-1} precipitation. In the case of the 6°C scenario, levelized costs are 3 times higher than for the 2°C scenario. Furthermore, despite the lower inherent probability for very intensive precipitation events of 40 mm h^{-1} as compared to events of 30 mm h^{-1} levelized damages are almost at the same level in both cases under the ED approach. In the DC approach, where damages increase until a maximum threshold level, the levelized costs of the damages reach a maximum already at 20 mm h^{-1}, after which they decrease faster than in the ED approach.

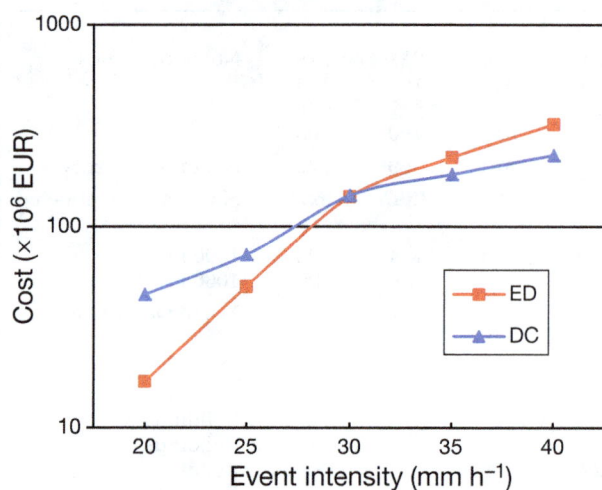

Fig. 5. Estimated total damage costs due to high-intensity precipitation events in Odense, Denmark, using the damage curve (DC) and event-driven (ED) approach

---

[5]The levelized costs are the net present value transformed to constant annual costs by integrating over a time frame of 100 yr

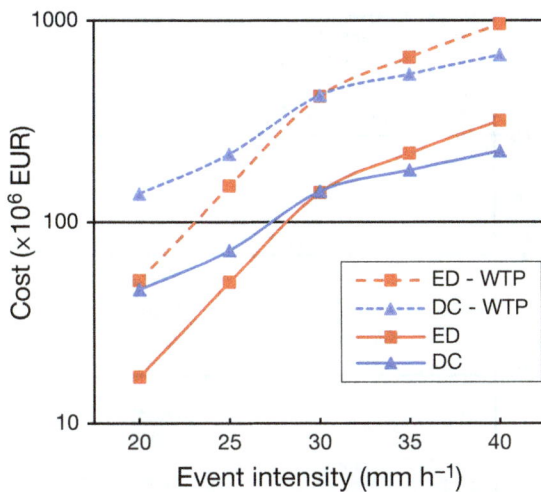

Fig. 7. Total damage costs during high-intensity precipitation events in Odense, Denmark. WTP (willingness-to-pay) = event-driven (ED) or damage curve (DC) costs with a risk aversion factor of 3

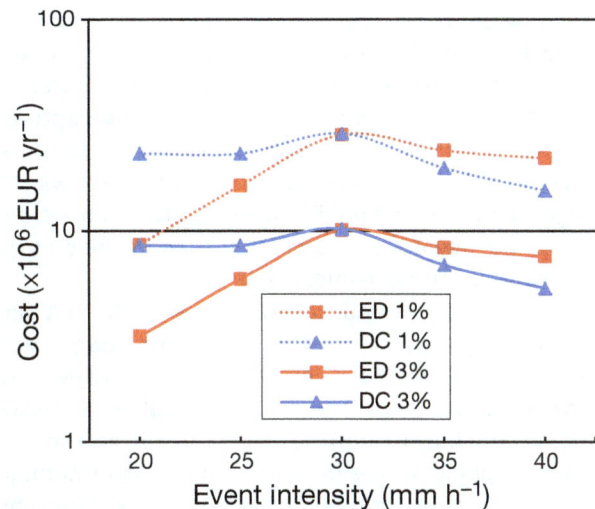

Fig. 8. Levelized costs of flood damage over a 100 yr period under a 6°C scenario and discount rates of 1 and 3% under the damage curve (DC) and event-driven (ED) approach

3.5. Risk aversion and discount rates

Adding a risk aversion factor as suggested by Weitzman (2011) to reflect people's attitudes towards risk will increase the costs (Fig. 7). We apply an absolute risk aversion factor of 3. In terms of WTP this also triples the costs, and the same upwards shift in costs is observed in the levelized costs.

Fig. 8 shows the levelized costs for the 2 damage functions under the 6°C scenario applying both a low (1%) and a medium high (3%) discount rate[6]. Levelized costs are almost 3 times higher again, depending on a 1 versus a 3% discount rate.

In this way the actual level of risks associated with flooding from extreme precipitation in Odense can vary significantly depending in near equal parts on climate scenario assumptions, damage cost approach, and cost assumptions. The importance of these factors is assessed systematically in the next section.

3.6. Sensitivity analysis

The risks measured as levelized costs for all the 32 scenario combinations are shown in Fig. 9. The costs for the lowest and the

highest risks vary from about 85 million EUR down to less than 1 million EUR. In terms of decision making, it is however important to notice that most of the combinations of economic assumptions and climate scenarios assess the risk to be between 7 and 30 million EUR yr^{-1}, while only 4 out of the 32 combinations really stand out and go far beyond a 30 million EUR yr^{-1} risk level. The high risk cases exclusively correspond to the high-end 4 and 6°C climate scenarios, a risk aversion factor of 3 and a low discount rate of 1%.

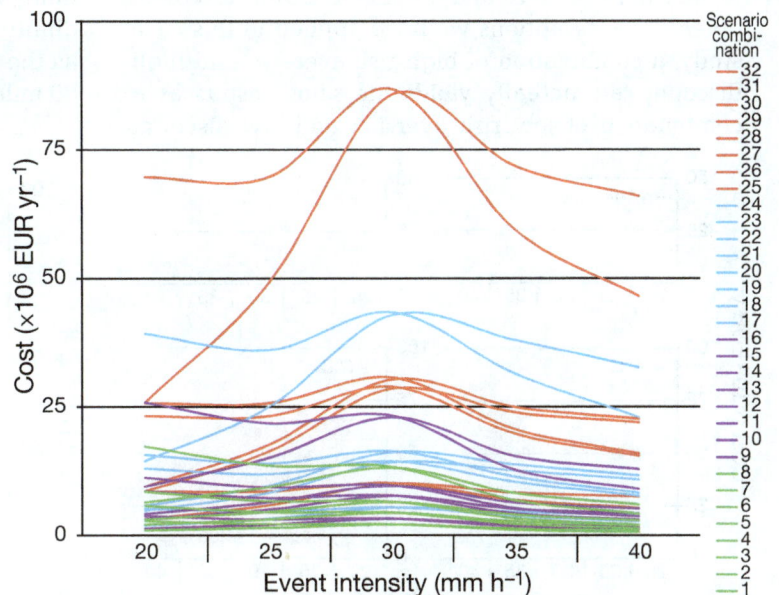

Fig. 9. Risks represented by levelized costs over a 100 yr period calculated for all 32 scenario combinations (see Fig. 3). Red: +6°, blue: +4°, and purple: +2°C climate scenario combinations; green = present day

---

[6]Discount rates between 1 and up to 6% have been suggested for climate change costing studies based on different theoretical arguments; See Arrow et al. (1996) for a detailed discussion

The wide range of risk estimates as presented in Fig. 9 is a result of a combination of climate scenarios and economic assumptions; below we separately examine the importance of these 2 set of assumptions in order to further shed light on key uncertainties. Starting with the climate scenarios, Fig. 10 shows the range of risk estimates by scenario. It is here clear that going beyond a 2°C climate scenario has large implications on risk estimates.

As previously stated, it is important from a decision-making point of view to consider the magnitude and uncertainties of damage estimates when we move from, for example, a 2°C to higher end scenarios, and timing here is important in relation to planning perspectives of adaptation. Recent climate simulations suggest that a 4°C increase could be achieved already around 2050 if current high GHG emission pathways continue. Hence depending on the timeframe of the actual adaptation considered, it may be highly relevant, within a timeframe of up to 2100, to assess options in the context of risks when moving from a 4°C scenario to a 6°C scenario.

Applying alternative economic assumptions to the damage cost assessment expands the range of risk assessment for the climate scenarios. We further examine the role of the economic assumptions keeping the climate scenario constant at the 6°C level. As exemplified in Fig. 11, the choice of discount rate and risk aversion factor can both have a high impact on risk levels. For example, for a precipitation intensity of 30 mm h^{-1} the risks are found to vary between ~10 and 85 million EUR. Moreover, given the assumptions we have applied in this case study, a combination of high risk aversion and high discount rate actually yields the same results as a combination of low risk aversion and low discount

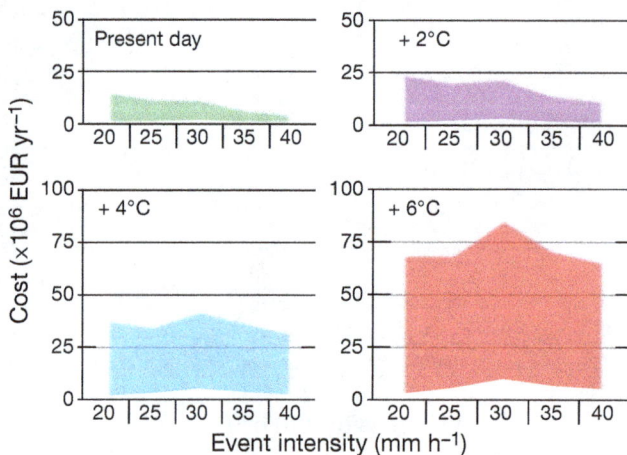

Fig. 11. Levelized costs of risks for the 6°C scenario with risk aversion factors 1 and 3 (risk), and 1 and 3 % discount rates (DR)

rate. This is a coincidence based on the choice of assumptions. From Fig. 11 only the combinations of a high risk aversion factor and a low 1% discount rate result in risks above the 30 million EUR yr^{-1} level, which as previously stated is the maximum level for most of the scenario combinations that are included in the full range of the sensitivity analysis as shown in Fig. 9.

In conclusion it can be said that the alternative climate scenarios, as included in Fig. 10, show a variability of the risk estimates from ~15 million EUR yr^{-1} as the highest estimate for the 2°C scenario to about 80 million EUR yr^{-1} for the 6°C scenario. Keeping the

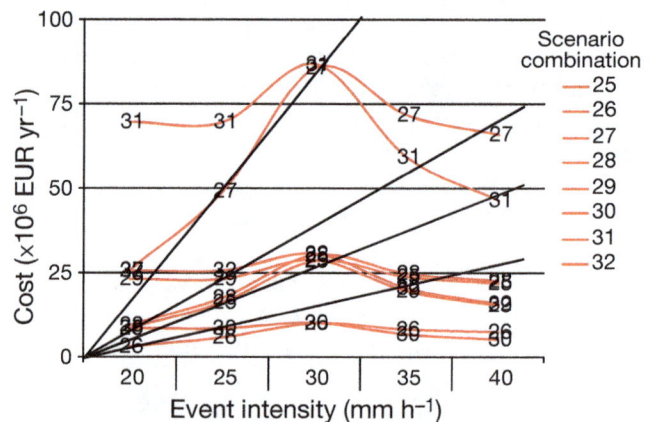

Fig. 10. Range of levelized costs given different climate scenarios and precipitation levels

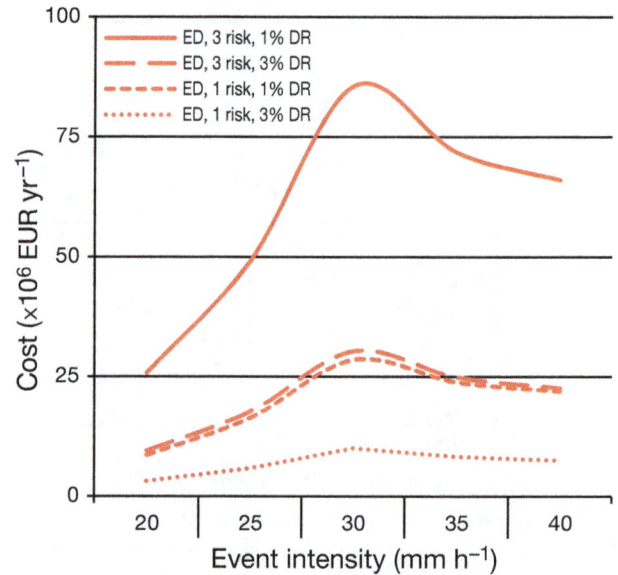

Fig. 12. Levelized climate risks and stylized climate risk reduction curves by adaptation for the 6°C climate scenario. Scenario numbers: from the 32 combinations in Fig. 3

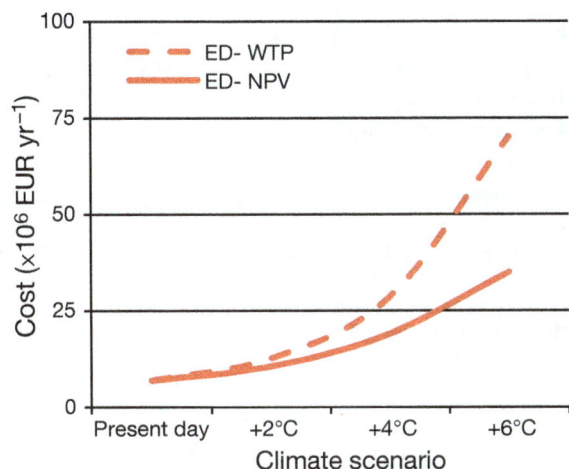

Fig. 13. Levelized costs for different climate scenarios using the event-driven (ED) damage assessment method. The risk aversion factor increases from 1 to 2 as climate change increases in the willingness-to-pay (WTP) measure, with a 3% discount rate. NPV: net present values

6°C scenario constant and then alternatively varying the economic assumptions on risk aversion and discount rate as shown in Fig. 11 provides an almost similar range of risk estimates; so given our assumptions it can be concluded that the set of climate scenarios and economic assumptions influence the risk estimates in a very similar way.

4. DISCUSSION

In the present study we frame climate change risk assessments in terms of how much society should be willing to invest in adaptation measures based on willingness-to-pay measures. The focus here is on a local geographical area, defined by specific climate change risks as exemplified by the case study discussed in the previous sections. Since decision makers in a local context cannot through their own adaptation actions influence atmospheric GHG concentrations significantly, we can assume that they have to consider, at a given point in time, climate change scenarios as a reality. In this construction of the decision-making issues, it is relevant to compare the costs of adaptation with the risk reduction achieved by adaptation assuming that a given climate scenario is emerging. The objective function for decision making can then be formulated as:

$$\text{Climate risks} = \text{adaptation costs} + \text{residual damages after adaptation} \quad (7)$$

where the right-hand side of the equation represents climate risk reductions by adaptation. Residual dam-

ages are included in the calculation in order to reflect that the costs of adaptation at some point can increase to a level where the benefit of risk reduction by adaptation is smaller than the costs. Using the same format as in Fig. 9, picturing the risk of all 32 scenario combinations, the decision-making issue for a given climate scenario objective, e.g. a 6°C scenario, could be as illustrated in Fig. 12. Adaptation costs should be less than or equal to the avoided damages (represented by the risk curves), and the decision maker can then compare adaptation cost curves with the risk curves. When adaptation costs intersect the risk curves, the benefit of implementing adaptation to protect against a high event intensity level is less than the adaptation costs. The straight lines exemplify stylized alternatives of climate risk reduction curves by adaptation and are merely for illustrative purposes.

The recommended risk management levels will of course depend on the exact shape of the adaptation cost and residual damage curves, which were not estimated in this study. As drawn in Fig. 12 in most cases the optimal risk management level will be at a precipitation level of around 30 mm h^{-1}. It is primarily with assumptions of high risk aversion and low discount rate that the recommended safety level exceeds this.

To put the decision-making perspective into a larger context, i.e. in terms of climate change mitigation perspectives, the risk reduction in terms of urban flooding can also be seen as a measure of the benefits of avoiding the consequences of alternative climate change scenarios. For illustrative purposes, Fig. 13 shows risk estimates for moving from no climate change to a 2°, 4°, and 6°C climate change scenario. We assume here that the risk aversion factor applied to the willingness-to-pay assessment increases linearly from 1 to 2 when we are on a trajectory to a 6°C climate scenario. However, one might also argue for the risk aversion factor to increase with global mean temperature change due to ambiguity in relation to future uncertain high consequence events (Weitzmann 2011).

Confronting the climate change risk estimates with the mitigation issues illustrates that by considering different levels of temperature change, the risk function will be convex in shape, while adding a risk aversion factor, which is increasing with temperature, clearly results in a much faster increase in the risk curves. Applying similar assumptions in a global decision-making context would thus point to the conclusion that a more ambitious level of climate change mitigation should be implemented.

That said, the actual shape of damage curves as well as risk aversion factors for different vulnerable assets will of course vary.

5. CONCLUSIONS

A methodological framework for integrated assessment of climate change impacts and welfare consequences has been developed and applied to a case study of urban flooding due to extreme precipitation in Odense. The approach distinguishes climate scenario uncertainties related to climate signals as such and to the probability of tail events with high consequences, while combining this information with alternative economic assumptions for damage functions, risk aversion, and equity as reflected in discount rates.

A systematic sensitivity analysis including 32 scenario combinations demonstrates that alternative climate scenario assumptions as well as economic assumptions together result in risk estimates with a very large variation. We find that a major source of uncertainty relates to the climate scenario uncertainty, in particular related to the probability of tail events associated with high consequences to society. The economic assumptions, particularly on risk aversion factor and discount rate, are both very important and contribute to a very large variation of risk estimates. Furthermore, the actual level of damage costs associated with different levels of precipitation intensity is important in determining the risk levels. The latter is a challenge to impact modellers, and the accuracy of damage cost studies could benefit from the availability of more context-specific studies on impacts on physical assets, human welfare, and risk perception, and on how the full range of economic activities in the city could be affected.

In the context of uncertainty and decision making, the results of the sensitivity analysis seen from a climate modelling perspective and from an economic perspective can be interpreted in different ways. Uncertainties related to the climate scenarios reflect both the state of current climate modelling and statistical downscaling approaches applied to the case study, as well as more general uncertainties related to global decision making on climate change mitigation and future temperature levels. In terms of adequately eliciting these uncertainties in an integrated framework, an ensemble of comprehensive model experiments, specifically designed to decompose the variance, which take into account key factors such as the scenario and model uncertainty is required. The uncertainties related to the economic estimates are more related to different theoretical concepts of risk aversion and discounting, and the sensitivity analysis illustrates what the consequences of these different uncertainties could be.

Acknowledgements. The present study was funded by a grant from the Danish Strategic Research Council for the Centre for Regional Change in the Earth System (CRES) under contract no: DSF-EnMi 09–066868. CRES is a multidisciplinary climate research platform, including key Danish stakeholders and practitioners with a need for improved climate information.

LITERATURE CITED

▶ Arnbjerg-Nielsen K, Fleischer HS (2009) Feasible adaptation strategies for increased risk of flooding in cities due to climate change. Water Sci Technol 60:273–281

▶ Arnbjerg-Nielsen K, Leonardsen L, Madsen H (2015) Evaluating adaptation options for urban flooding based on new high-end emission scenario regional climate model simulations. Clim Res 64:73–84

Arrow KJ (1965) The theory of risk aversion. In: Aspects of the theory of risk bearing, Yrjö Jahnssonin Säätiö, Helsinki. Reprinted in: Arrow KJ (1971) Essays in the theory of risk bearing, Markham, Chicago, IL, p 90–109

Arrow KJ, Cline WR, Maler KG, Munasighe M, Squitieri R, Stiglitz JE (1996) Intertemporal equity, discounting, and economic efficiency. In: Bruce JP, Lee H, Haites EF (eds) Climate change 1995: economic and social dimensions of climate change. Contribution of Working Group III to the Second Assessment Report of the Intergovernmental Panel on Climate Change, Cambridge University Press, Cambridge, p 127–144

▶ Christensen JH, Christensen OB (2003) Severe summertime flooding in Europe. Nature 421:805–806

▶ Christensen OB, Yang S, Boberg F, Fox Maule C and others (2015) Scalability of regional climate change in Europe for high-end scenarios. Clim Res 64:25–38

Collins M, Chandler RE, Cox PM, Huthnance JM, Rougier J, Stephenson DB (2012) Quantifying future climate change. Nat Clim Change 2:403–409

Davis SA, Skaggs LL (1992) Catalog of residential depth-damage functions used by the army corps of engineers in flood damage estimations. US Army Corps of Engineers, Institute for Water Resources Report 92-R-3, Springfield, VA

Forsikring & Pension (Danish Insurance Association) (2014) Erstatning for vandskader. Available at www.forsikringogpension.dk/presse/Statistik_og_Analyse/statistik/forsikring/erstatninger/Sider/Erstatninger_for_vandskader.aspx (accessed 20 January 2014)

Gregersen IB, Madsen H, Linde JJ, Arnbjerg-Nielsen K (2014) Opdaterede klimafaktorer og dimensionsgivende regnintensiteter. Spildevandskomiteen, Skrift 30. Available at https://ida.dk/sites/prod.ida.dk/files/svk_skrift30_0.pdf (accessed 20 March 2015)

▶ Hawkins E, Sutton R (2009) The potential to narrow uncertainty in projections in regional climate predictions. Bull Am Meteorol Soc 90:1095–1107

▶ Hawkins E, Sutton R (2011) The potential to narrow uncertainty in projections of regional precipitation change. Clim Dyn 37:407–418

▶ Heal G, Kriström B (2002) Uncertainty and climate change. Environ Resour Econ 22:3–39

IPCC (Intergovernmental Panel on Climate Change) (2005) Guidance notes for lead authors of the IPCC Fourth Assessment Report on addressing uncertainties. Available at www.ipcc-wg1.unibe.ch/publications/supportingmaterial/uncertainty-guidance-note.pdf

IPCC (2012) Managing the risks of extreme events and disasters to advance climate change adaptation. In: Field CB, Barros V, Stocker TF, Qin D and others (eds) A Special Report of Working Groups I and II of the Intergovernmental Panel on Climate Change. Cambridge University Press, Cambridge, and New York, NY

IPCC (2013) Annex II: climate system scenario tables (Prather M, Flato G, Friedlingstein P, Jones C, Lamarque JF, Liao H, Rasch P [eds]). In: Stocker TF, Qin D, Plattner G-K, Tignor M and others (eds) Climate change 2013: the physical science basis. Contribution of Working Group I to the Fifth Assessment Report of the Intergovernmental Panel on Climate Change. Cambridge University Press, Cambridge, and New York, NY, p 1395–1445

▶ Jongman B, Kreibich H, Apel H, Barredo JI and others (2012) Comparative flood damage model assessment: towards a European approach. Nat Hazards Earth Syst Sci 12:3733–3752

Kolstad C, Urama K, Broome J, Bruvoll A and others (2014) Social, economic and ethical concepts and methods. In: Edenhofer O, Pichs-Madruga R, Sokona Y, Farahani E and others (eds) Climate change 2014: mitigation of climate change. Contribution of Working Group III to the Fifth Assessment Report of the Intergovernmental Panel on Climate Change. Cambridge University Press, Cambridge, and New York, NY, p 207–282

▶ Meinshausen M, Smith SJ, Calvin KV, Daniel JS and others (2011) The RCP greenhouse gas concentrations and their extension from 1765 to 2300. Clim Chang 109:213–241

MIKE By DHI (2014) www.MIKEbydhi.com (accessed 29 September 2014)

Miljøministeriet (Danish Ministry of the Environment) (2014) Kortforsyningen www.kortforsyningen.dk (accessed 15 November 2014)

▶ Refsgaard JC, Arnbjerg-Nielsen K, Drews M, Halsnæs K and others (2013) The role of uncertainty in climate change adaptation strategies—a Danish water management example. Mitig Adapt Strategies Glob Change 18: 337–359

▶ Sunyer MA, Hundecha Y, Lawrence D, Madsen H and others (2014) Inter-comparison of statistical downscaling methods for projection of extreme precipitation in Europe. Hydrol Earth Syst Sci Discuss 11:6167–6214

▶ Taylor KE, Stouffer RJ, Meehl GA (2012) An overview of CMIP5 and the experiment design. Bull Am Meteorol Soc 93:485–498

van der Linden P, Mitchell JFB (eds) (2009) ENSEMBLES: climate change and its impacts: summary of research and results from the ENSEMBLES project. Met Office Hadley Centre, Exeter

Weitzman ML (2011) Fat-tailed uncertainty in the economics of catastrophic climate change. Rev Environ Econ Policy 5:275–292

▶ Wilby RL, Dessai S (2010) Robust adaptation to climate change. Weather 65:180–185

▶ Zhou Q, Mikkelsen PS, Halsnæs K, Arnbjerg-Nielsen K (2012) Framework for economic pluvial flood risk assessment considering climate change effects and adaptation benefits. J Hydrol (Amst) 414–415:539–549

APPENDIX

Table A1. Assumptions for flood damage cost calculations using the damage curve (DC) approach

Unit costs	Surface water depth							Unit
Buildings	*10 cm*	*20 cm*	*30 cm*	*40 cm*	*50 cm*	*60 cm*	*70 cm*	
Service and industry	69 418	138 835	208 253	277 670	347 088	416 506	485 923	EUR/building
Multistorage residential	45 561	91 122	136 684	182 245	227 806	273 367	318 928	EUR/building
Houses	16 667	33 333	50 000	66 667	83 333	100 000	116 667	EUR/building
Leisure house	833	1667	2500	3333	4167	5000	5833	EUR/building
	2.5 cm	*5 cm*	*10 cm*	*15 cm*	*20 cm*	*30 cm*	*50 cm*	
Basements	9	18	35	53	70	106	176	EUR/m^2
	2.5 cm	*5 cm*	*10 cm*	*15 cm*	*20 cm*	*30 cm*	*50 cm*	
Roads	9	18	35	53	70	106	176	EUR/m^2
Railways	44	88	176	264	352	528	881	EUR/m^2
	0.15 cm	*0.3 cm*	*1 cm*	*5 cm*	*10 cm*	*20 cm*	*50 cm*	
Health	11	22	72	361	722	1444	3610	EUR/person
	10 cm	*20 cm*	*30 cm*	*40 cm*	*50 cm*	*60 cm*	*70 cm*	
Waterbodies flooded in the city	16 667	33 333	50 000	66 667	83 333	100 000	116 667	EUR/waterbody
Irreplaceable assets	*10 cm*	*20 cm*	*30 cm*	*40 cm*	*50 cm*	*60 cm*	*70 cm*	
Ancient monuments	33 333	66 667	100 000	133 333	166 667	200 000	233 333	EUR/building
Churches	333 333	666 667	1 000 000	1 333 333	1 666 667	2 000 000	2 333 333	EUR/building
Conservation worthy buildings	33 333	66 667	100 000	133 333	166 667	200 000	233 333	EUR/building
Clergy buildings middle age	33 333	66 667	100 000	133 333	166 667	200 000	233 333	EUR/building
Statues and sculptures	33 333	66 667	100 000	133 333	166 667	200 000	233 333	EUR/building
Museums	333 333	666 667	1 000 000	1 333 333	1 666 667	2 000 000	2 333 333	EUR/building

Table A2. Assumptions for flood damage cost calculations using the event-driven (ED) approach

Unit costs	Water depth threshold (cm)	Maximum event intensity (mm h^{-1})					Unit
		20	25	30	35	40	
Buildings							
Service and industry	20	87 972	182 821	277 670	372 520	467 369	EUR/building
Multi-storage residential	20	45 689	113 967	182 245	250 523	318 801	EUR/building
Houses	20	43 718	55 192	66 667	78 141	89 616	EUR/building
Leisure house	20	1315	2324	3333	4342	5352	EUR/building
Basements	5	33	47	67	87	100	EUR/m^2
Roads	5	167	233	333	433	500	EUR/m^2
Railways	5	33	47	67	87	100	EUR/m^2
Health	0.3	446	624	892	1159	1337	EUR/person
Waterbodies flooded in the city	20	33 333	46 667	66 667	86 667	100 000	EUR/waterbody
Irreplaceable assets							
Ancient monuments	20	66 667	93 333	133 333	173 333	200 000	EUR/building
Churches	20	666 667	933 333	1 333 333	1 733 333	2 000 000	EUR/building
Conservation worthy buildings	20	66 667	93 333	133 333	173 333	200 000	EUR/building
Clergy buildings middle age	20	66 667	93 333	133 333	173 333	200 000	EUR/building
Statues and sculptures	20	66 667	93 333	133 333	173 333	200 000	EUR/building
Museums	20	666 667	933 333	1 333 333	1 733 333	2 000 000	EUR/building

8

Adaptation of rice to climate change through a cultivar-based simulation: a possible cultivar shift in eastern Japan

Ryuhei Yoshida[1,6,*], Shin Fukui[2,3], Teruhisa Shimada[4], Toshihiro Hasegawa[2],
Yasushi Ishigooka[2], Izuru Takayabu[5], Toshiki Iwasaki[1]

[1]Graduate School of Science, Tohoku University, Sendai 980-8578, Japan
[2]Agro-Meteorology Division, National Institute for Agro-Environmental Sciences, Tsukuba 305-8604, Japan
[3]Faculty of Human Sciences, Waseda Univeristy, Tokorozawa 359-1192, Japan
[4]Graduate School of Science and Technology, Hirosaki University, Hirosaki 036-8561, Japan
[5]Meteorological Research Institute, Tsukuba 305-0052, Japan

[6]*Present address:* Faculty of Symbiotic Systems Science, Fukushima University, Fukushima 960-1296, Japan

ABSTRACT: As surface warming threatens rice production in temperate climates, the importance of cool regions is increasing. Cultivar choice is an important adaptation option for coping with climate change but is generally evaluated with a single metric for a few hypothetical cultivars. Here, we evaluate adaptation to climate change based on multiple metrics and cultivars in presently cool climates in Japan. We applied the outputs of a global climate model (MIROC5) with a Representative Concentration Pathways 4.5 scenario, dynamically downscaled to a 10 km mesh for the present (1981–2000) and future (2081–2099) climate conditions. The data were input into a rice-growth model, and the performances of 10 major cultivars were compared in each mesh. With the present-day leading cultivars, the model predicted reduced low-temperature stress, a regional average yield increase of 17%, and several occurrences of high-temperature stress. The most suitable cultivars in each grid cell changed dramatically because of climate change when a single metric was used as a criterion, and the yield advantage increased to 26%. When yield, cold, and heat stress were taken into account, however, the currently leading cultivars maintained superiority in 64% of the grid cells, with an average regional yield gain of 22%, suggesting a requirement for developing new cultivars by pyramiding useful traits. A trait such as low sensitivity to temperature for phenology helps in ensuring stable growth under variable temperatures. Increasing photoperiod sensitivity can be an option under future climates in relatively warmer regions.

KEY WORDS: Rice cultivar · Yield · High-temperature stress · Low-temperature stress · Climate change

1. INTRODUCTION

Recent and projected surface warming has the potential to negatively affect crop growth (Porter et al. 2014); therefore, climate change related impact assessments are required for ensuing food security. Climate change scenarios from global climate models are widely used as input in crop-growth models, and global impact assessments have been conducted for major crops, such as wheat, maize, and rice (Parry et al. 2004, Lobell & Field 2007, Deryng et al. 2011, 2014, Rosenzweig et al. 2014). Regional impact assessments for major crop species have also been reported in various regions (reviewed by Porter et al. 2014) considering their respective agro-ecosystems and projected climatic changes. In Asia, rice is the most important food crop and impact assessments for rice have been carried out in various regions (e.g.

*Corresponding author: yoshida@sss.fukushima-u.ac.jp

Tao et al. 2008, Iizumi et al. 2011, Kim et al. 2013, Soora et al. 2013, Yu et al. 2014).

Rice production is sensitive to considerable variation in various climatic factors, including atmospheric CO_2 concentrations, precipitation, solar radiation, and temperatures, all of which are projected to change in the future. Among these factors, high- and low-temperature extremes in the reproductive growth phases are serious concerns (Wassmann et al. 2009) because they cause floret sterility, which reduces grain yield (Satake 1976, Satake & Yoshida 1978). In temperate rice-growing countries such as Japan, cool-summer damages have been a major yield constraint for many years (Satake 1976), but heat-induced sterility is also emerging as a result of recent hot-summer events (Hasegawa et al. 2011) and will become a serious threat to rice production in Japan (Nakagawa et al. 2003).

The effects arising from predicted changes in climate will differ, even within Japan. According to Iizumi et al. (2011), the probability of a decrease in rice yield in the 2090s relative to the 1990s is >20% for western Japan, whereas it is <10% in eastern Japan, where current temperatures are lower. Rice quality is currently deteriorating in western Japan because of high temperatures, and this is projected to continue under future climate conditions (Okada et al. 2011). While high-temperature stress is not presently evident in eastern Japan, climate change would bring this stress to presently cool regions as well (Nemoto et al. 2012). In eastern Japan, cold damage caused by a local northeasterly wind known as Yamase is a more serious threat to crop growth than heat damage (Shimono 2011). The frequency of Yamase airflows is projected to decrease in the future, but the event will still occur (Kanno et al. 2013) and cool-summer damages to rice are projected to persist under future climate conditions (Kanda et al. 2014). However, because surface warming caused by climate change has the potential to negatively affect rice growth in currently warm regions, such as in low-latitude or western Japan (e.g. Iizumi et al. 2007), rice production in presently cool regions (e.g. eastern Japan) is becoming more significant for a stable food supply (Nakagawa et al. 2003, Easterling et al. 2007, Shimono 2011).

Climatic conditions at each site affect farmers' choice of cultivars, which is one of the most important options for agricultural management. The choice of cultivars will become even more important in the future because they are considered to be the most effective adaptation measures against climate change (Porter et al. 2014). For this reason, a number of simulations were conducted to examine the impact of climate change on currently planted cultivars and cultivars adapted to projected climate change conditions based on a single metric, namely grain yield (reviewed by Porter et al. 2014). These simulations most commonly use one or a few hypothetical cultivars that match growth duration in new climates based on the altered thermal time requirements, assuming a linear response of developmental rate to temperature (e.g. Challinor et al. 2009, Yu et al. 2014, Kassie et al. 2015). Because most crop models predict shorter growth duration due to warmer climate, which leads to less biomass accumulation, and thus reduction in yield, varieties with longer thermal time requirements (i.e. late maturation) than the current varieties are expected to be more productive in warmer climates. However, rice is a short-day crop, and phenological traits of cultivars are determined by a combination of different degrees of photoperiod and temperature sensitivities. Replacement of the currently planted cultivars with those planted in warmer climates is tempting, but whether this simple solution is an alternative should be examined based on the realistic representation of the phenological traits of the cultivars. Previously, we evaluated phenological traits of various major cultivars, which cover more than 80% of the current rice harvest area in Japan (Fukui et al. 2015). By incorporating these parameters into a rice growth model, we can evaluate the adaptability of major rice cultivars to new environments.

In this study, we aimed to determine the suitability of the response of rice cultivars to climate change in the presently cool regions in Japan. To objectively determine the spatial distribution of the most suitable cultivar, we simulated grain yields of 10 major rice cultivars for 10 km gridded cells under current and projected future climatic conditions. Because both low and high temperatures are concerns in the current and future climates, we introduce these metrics as criteria for cultivar rankings in addition to grain yield. A consideration of multiple metrics together, such as yield and stress, for a multiple-cultivar choice would offer useful information on the efficacy of cultivar replacement and traits of cultivars conferring adaptability to climate change.

2. METHODS

2.1. Downscaling of the climate change scenario

We used the outputs of the global climate model (GCM) Model for Interdisciplinary Research on Climate (MIROC5; Watanabe et al. 2010) using the Representative Concentration Pathways (RCP) 4.5, because the climate data for Asia have focused on the Yamase airflow (Kanno et al. 2013). The original 150 km grid mesh resolution is insufficient for resolving regional differences in the meteorological elements of eastern Japan; therefore, we used the 20 km mesh MIROC5 output dataset for Japan based on the method of Ishizaki et al. (2012; Fig. 1). We downscaled the data to a 10 km mesh for eastern Japan in the JMA-NHM regional climate model (RCM; Saito et al. 2007). The JMA-NHM model utilizes the Kain-Fritsch scheme for convective parameterization (Kain 2004) and the improved Mellor-Yamada Level 3 scheme for turbulent parameterization (Nakanishi & Niino 2004). The downscaling was conducted for the key growing period, from May 28 to August 31, for 1981–2000 (present climate) and 2081–2099 (future climate), and the first 4 d were used as a spin-up period and excluded from the analysis.

Outputs from GCMs and RCMs are generally biased by the observed values (e.g. Yoshida et al. 2012a). To avoid biases, we used changes in the climatological mean (20 yr mean for present climate and 19 yr mean for future climate), rather than the original, downscaled climate data. Based on previous studies (Kimura & Kitoh 2007, Iizumi et al. 2010, Yoshida et al. 2012b), the following procedures were conducted to compose the climate datasets. First, the gridded, observed dataset from the Automated Meteorological Data Acquisition System (Mesh-AMeDAS; Seino 1993) for 1981–2000 was defined as the present climate. Then, the future climate was calculated from climate differences derived from the downscaled data and the Mesh-AMeDAS data. Here, differences (i.e. difference = future − present) were calculated for the daily maximum, mean, and minimum temperatures, and the multiplying ratios (ratio = future/present) were calculated for downward shortwave radiation, relative humidity, and wind speed. This method was used to account for the impacts of summer climate change on rice production.

2.2. Rice yield simulation

To account for cultivar-based rice production, we modified the Hasegawa/Horie rice-growth model (H/H model; Hasegawa & Horie 1997, Nakagawa et al. 2005, Fukui et al. 2015) to incorporate responses to extreme temperatures and CO_2 based on climate change studies, which are outlined below.

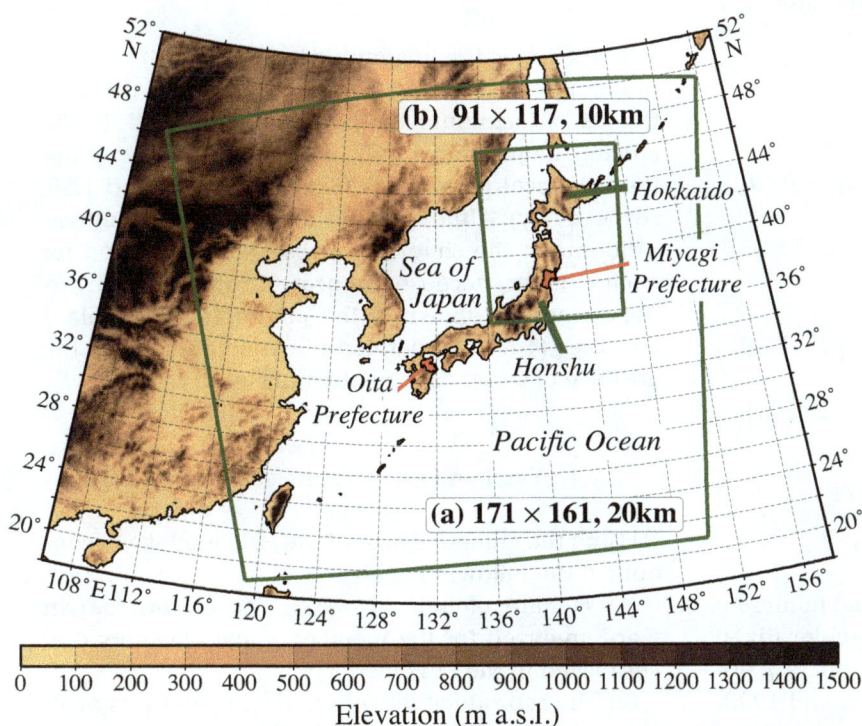

Fig. 1. Calculation domain used for the Japan Meteorological Agency nonhydrostatic model (JMA-NHM) simulations. (a) Outer domain: 171 × 161 grids with 20 km spacing for the dataset based on the method in Ishizaki et al. (2012). (b) Inner domain: 91 × 117 grids with 10 km spacing for the JMA-NHM model simulation

2.2.1. Stomatal conductance

Rising atmospheric CO_2 concentrations will enhance photosynthesis and reduce stomatal conductance. In this model, we used a combination of the Farquhar-von Caemmerer-Berry (FvCB) photosynthesis model (Farquhar et al. 1980) and the Ball-Berry (BB) stomatal conductance model (Collatz et al. 1991) to account for the CO_2 response. Two key parameters in the FvCB model, the maximum rate of Rubisco activity (V_{cmax}, µmol m^{-2} s^{-1}) and potential rate of electron transport (J_{max}, µmol m^{-2} s^{-1}), were expressed as linear functions of specific leaf nitrogen (SLN, g m^{-2}) derived from leaf-level gas exchange measurements in controlled chamber experiments (data presented in Hasegawa et al. 2015):

$$V_{cmax} = \max[86 \times (SLN - 0.5), 0] \quad (1)$$

$$J_{max} = \min \{\max[138 \times (SLN - 0.4), 0], 210\} \quad (2)$$

The other parameters for the FvCB model were from Medlyn et al. (1999). Stomatal conductance (g_c, mol m^{-2} s^{-1}) was then given by the BB model:

$$g_c = \frac{b_1 \times A_f \times \mathrm{RH}}{C_a} + b_0 \qquad (3)$$

where A_f is the assimilation rate obtained from the FvCB model, RH is the relative humidity (dimensionless), C_a is the atmospheric CO_2 concentration, and b_0 and b_1 are empirical parameters (0.00001 and 5.9; Katul et al. 2000).

2.2.2. Phenology

Rice phenology is simulated by a development index (DVI; 0 = seeding emergence, 1 = panicle initiation, 2 = heading, and 3 = maturity). The DVI value increases daily, and its development rate (DVR) is controlled by the air temperature and photoperiod (Nakagawa et al. 2005) as follows:

$$\mathrm{DVR} = \begin{cases} \dfrac{f_1(T)g(L)}{G_1}, & (0 \le \mathrm{DVI} < 2) \\[2mm] f_2(T), & (2 \le \mathrm{DVI} \le 3) \end{cases} \qquad (4)$$

where T indicates the average temperature between the daily maximum and minimum temperatures, L denotes the photoperiod, G_1 is the parameter. Each respective development function was expressed as follows:

$$f_1(T) = \begin{cases} \left\{ \left(\dfrac{T-T_{\min}}{T_o-T_{\min}}\right)\left(\dfrac{T_{\max}-T}{T_{\max}-T_o}\right)^{\frac{T_{\max}-T_o}{T_o-T_{\min}}} \right\}^{\alpha}, & (T_{\min} < T < T_{\max}) \\[3mm] 0, & (T < T_{\min} \text{ or } T_{\max} < T) \end{cases} \qquad (5)$$

$$g(L) = \begin{cases} \left\{ \left(\dfrac{L}{L_{\min}}\right)\left(\dfrac{L_{\max}-L}{L_{\max}-L_{\min}}\right)^{\frac{L_{\max}-L_{\min}}{L_{\min}}} \right\}^{\beta}, & (L_{\min} < L < L_{\max}) \\[3mm] 1, & (L < L_{\min}) \end{cases} \qquad (6)$$

$$f_2(T) = \begin{cases} \dfrac{1}{G_2}[1.0 - e^{-A(T-T_c)}], & (T_c < T) \\[2mm] 0, & (T \le T_c) \end{cases} \qquad (7)$$

where $T_{\max}$ and $T_{\min}$ are the maximum and minimum temperatures in terms of the growth thresholds (fixed at 42 and 8°C) (Yin et al. 1997). $L_{\max}$ and $L_{\min}$ are maximum and minimum photoperiods (24 and 10 h), and α, β, A, G_2, T_o and T_c are cultivar-dependent parameters. We analyzed 10 Japanese cultivars developed in the cool-temperate and temperate climates (Table 1).

2.2.3. Yield and temperature stress

Rice growth is sensitive to the daily maximum and minimum temperatures after initiation of the panicle (Horie et al. 1999):

$$Y = \frac{H_v}{0.85}\{M_a \times [1 - f(\mathrm{HDD})] \times [1 - f(\mathrm{CDD})]\} \qquad (8)$$

where Y is the yield (unhulled grain weight expressed at the 15% moisture content, g m^{-2}), H_v is the potential harvest index (0.5), M_a is the mass above the ground (g m^{-2}), and $f(\mathrm{HDD})$ and $f(\mathrm{CDD})$ are the high- and low-temperature stress functions (i.e. fertility under the high-temperature stress and sterility under the low-temperature stress; value range: 0 = free to 1 = stress). Each stress function was calculated according to Horie et al. (1999):

$$f(\mathrm{HDD}) = (1 + e^{k1 \times \mathrm{HDD}})^{-1} \qquad (9)$$

$$\mathrm{HDD} = 36.6 - \frac{1}{n}\sum_{\mathrm{DVI}=1.6}^{2.2} T_x \qquad (10)$$

$$f(\mathrm{CDD}) = \frac{\max[100 - (\gamma_0 + k_2 \times \mathrm{CDD}^y), 0]}{100 \times [1 + e^{-6.2 \times (\mathrm{DVI}-2.29)}]} \qquad (11)$$

$$\mathrm{CDD} = \sum_{\mathrm{DVI}=1.5}^{2.2} \max(22 - T, 0) \qquad (12)$$

where HDD is the heating degree-days (°C), n is the number of days during the flowering period (d), T_x is the daily maximum temperature (°C), CDD is the cooling degree-days (°C), and k_1, γ_0, k_2, and y are the empirical parameters (8.53, 4.6, 0.054, and 1.56, respectively). Although this scheme excludes cultivar differences, the simulated stress values differed for each cultivar, because the growth rate was cultivar-dependent, as shown in the parameters in Table 1 and Eqs. (4–7). We defined the yield attained in the absence of stress as the potential yield.

2.2.4. Yield and temperature stress simulation

Using the climate data and the H/H model, we conducted the following analyses:
(1) climate change scenarios for eastern Japan were analyzed for the meteorological elements that were used in the H/H model.
(2) The reproducibility of the H/H model was evaluated by simulating rice yields using the Mesh-AMeDAS data for the present climate (1981–2000). In this simulation, we used the current leading cultivar of each prefecture (Japanese administrative dis-

Table 1. Cultivars and parameters used in the H/H phenology scheme (Fukui et al. 2015). G_1 = minimum number of days required from emergence to heading under optimum conditions; G_2 = insensitiveness to temperature after heading; α and β = sensitiveness to temperature and photoperiod before heading; A = multiplying factor of temperature impact on growth after heading; T_o = optimum temperature before heading; T_c = minimum temperature required for growth. These parameters were estimated based on the days to heading observed in the nationwide variety trials (Fukui et al. 2015). Because dates for panicle initiation were not recorded in the database, development index (DVI) was assigned as 1 at heading and 2 at maturity. To conform to the Hasegawa/Horie (H/H) rice-growth model, which assumes DVI = 1 at panicle initiation, 2 at heading and 3 at maturity, the estimated DVI values using the above parameters were corrected to the 0–3 system scale assuming that panicle initiation occurs at DVI = 0.64 in the 0–2 system

ID	Cultivar	G_1 (d)	G_2 (d)	α	β	A ($\times 10^{-2}$)	T_o (°C)	T_c (°C)
1	Hitomebore	56.7	23.3	0.93	0.95	3.5	32.9	0.4
2	Kirara397	45.3	22.0	1.25	0.07	3.3	32.1	1.9
3	Hinohikari	30.9	38.0	1.25	7.34	15.9	30.6	9.9
4	Asahinoyume	30.9	35.8	1.07	6.70	9.9	31.3	4.3
5	Akitakomachi	54.6	24.7	1.24	0.13	5.8	34.3	7.4
6	Aichinokaori	30.7	29.9	1.48	7.90	6.2	28.3	4.9
7	Haenuki	55.1	24.0	1.68	1.12	3.7	30.0	1.5
8	Koshiibuki	59.0	23.6	1.20	0.45	6.1	32.3	10.0
9	Koshihikari	36.6	29.1	1.11	3.42	5.3	34.6	0.2
10	Kinuhikari	30.2	26.5	1.50	4.32	6.7	33.3	9.2

trict) obtained from Crop Statistics of the Ministry of Agriculture, Forestry, and Fisheries (MAFF, data as of 2009, www.maff.go.jp/j/tokei/kouhyou/kensaku/bunya2.html) for simulating yields in the grid cells for the 1981–2000 period. For all grid cells, the current leading cultivar corresponds to 1 of the 10 cultivars listed in Table 1. The amount of nitrogen fertilizer each year was given as follows. We obtained consumption of nitrogenous fertilizers for rice production from the Rice and Wheat Production Cost Statistics (MAFF 1981–2000, www.e-stat.go.jp/SG1/estat/List.do?bid=000001014632&cycode=0) and calculated the elemental weight of N applied, which was then divided by the rice-planted area. Because this statistic provides only aggregated data for the regions covering several prefectures, N supply in the simulation was almost homogenous over the entire study area, as was the soil fertility. The time of transplanting by prefecture was fixed for each grid cell based on the Crop Statistics (MAFF, data as of 2000) for each prefecture. Nitrogen fertilizer was spilt-applied thrice during the growing season; 58% of the total N was applied immediately before planting and the remaining 42% in 2 equal splits during the panicle development (21% each at about the spikelet differentiation stage and reduction division stage of the pollen mother cell), as commonly practiced by farmers in

the study regions. The H/H model simulates rice growth under no water limitation, and the paddy was assumed to be maintained in flooded condition, which is supported by the fact that >99% of the rice-growing areas are fully irrigated. We compared the simulated yields with those obtained from the crop data on a sub-prefectural scale, as summarized by MAFF. Because the crop statistics by MAFF provided the hulled grain yield, we converted the yield to unhulled data multiplying it by 1.25 as reported by Yoshida (1981).

(3) Climate change impacts on rice yields and temperature stresses for the current cultivars were estimated using the projected future climate data. Planting times and N fertilizer amount were kept constant at the year 2000 values. The timing of the split-application of N fertilizer was determined according to the predicted phenology.

(4) Potential yield and temperature stresses under climate change conditions in all the grid cells were also examined for all 10 cultivars listed in Table 1. On the basis of simulated yields, we selected a 'top cultivar' for each grid cell that provided the maximum yield among the 10 cultivars. We also counted the total number of top grid cells by cultivars, which was divided by the total number of analyzed grid cells (i.e. 1307). This ratio was defined as the share of each cultivar. For all crop simulations, grid cells with paddy-field ratios of <1% in the National Land Numerical Information database (MLITT 2012) were defined as non-paddy areas and excluded from the analysis. The geographical distribution of paddy fields in 2006 (the latest available year) was applied and fixed throughout the analysis period.

3. RESULTS

3.1. Effects of climate change on meteorological elements

We estimated the present and future meteorological elements required for the H/H model (Fig. 2). Temperature variables (daily mean, maximum, and minimum temperatures) showed similar geographical distributions, and surface warming was more evident in the northern area on the Pacific Ocean side (Fig. 2a–i). A regional average of ~3°C of surface

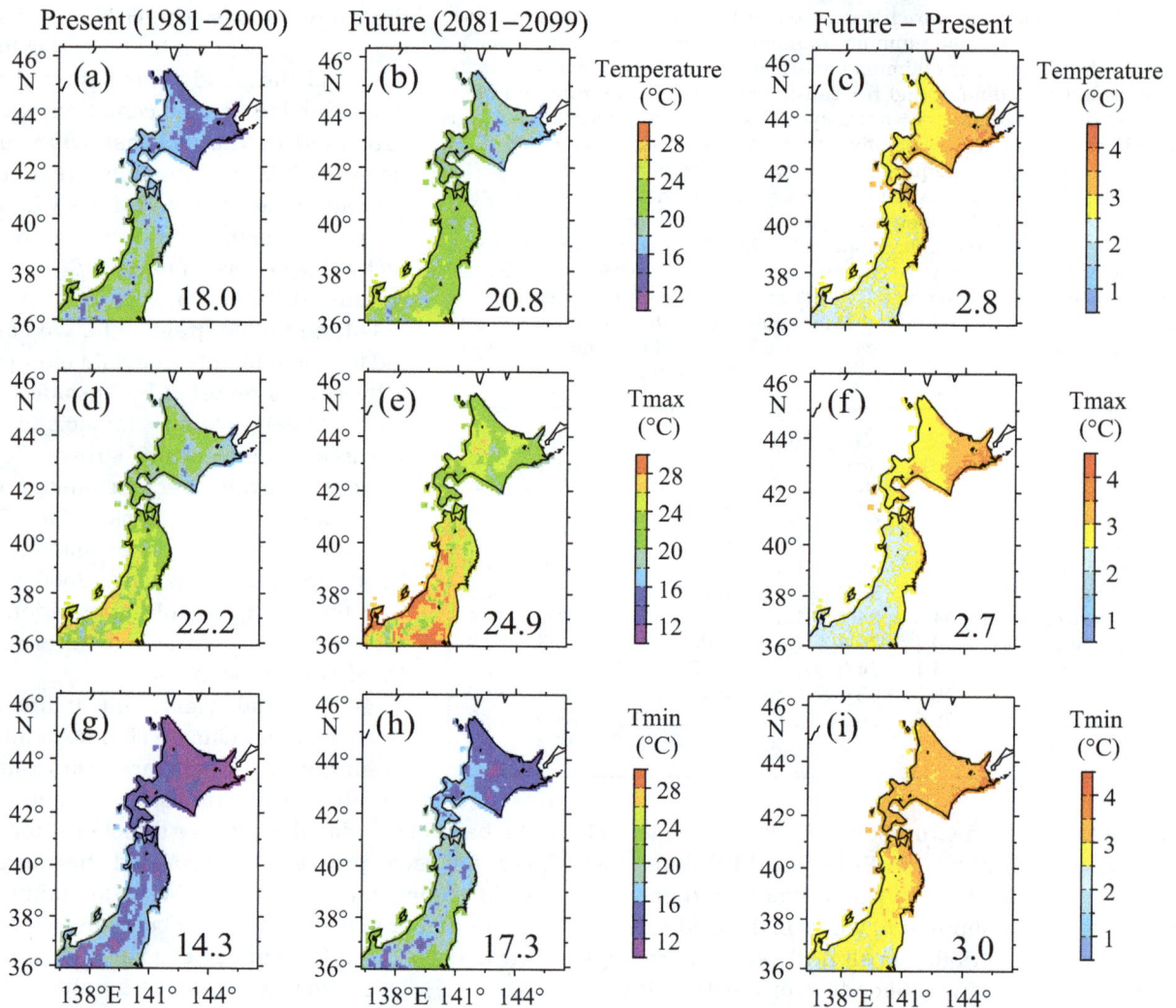

Fig. 2. (cont. on next page). Geographical distributions of the downscaled summer meteorological factors in the present and future climates, and their differences. Daily (a–c) mean, (d–f) maximum and (g–i) minimum temperatures

warming was obtained across the land surface, which fell within the 5 to 95 % ranges for the global land-surface warming projected by the global climate models in the Coordinated Modeling Intercomparison Project Phase 5 (1.3 to 3.4°C) (Collins et al. 2013). Similar east–west geographical patterns were indicated for downward shortwave radiation in both present and future climates (Fig. 2j,k). A small increase (<1 % of the regional average) occurs over most of the area, and a relatively large increase (~10 %) occurs in eastern Hokkaido (the northern island in the 10 km mesh domain, see Fig. 1) and the southern part of the analyzed area (Fig. 2l). Relative humidity also showed similar geographical patterns but decreased slightly with climate change (Fig. 2m,n). The decrease appeared in most grid cells (80.0 % of the total land grid cells), but a small increase (<1 %)

was also estimated for a part of the northern area (Fig. 2o). Future changes in wind speed were estimated to be the smallest among the elements analyzed. Although intensified wind speed was found in the western part of Hokkaido, its contribution was negligible compared with changes at the regional scale (Fig. 2p–r). These meteorological elements were utilized in the H/H model to simulate the yield and temperature stresses.

3.2. Reproducibility of the rice growth model

We first examined the performances of the H/H model by comparing the simulated yields with historical yield records between 1981 and 2000. Fig. 3 depicts the inter-annual variation of the observed

Present (1981–2000)   Future (2081–2099)   Future – Present

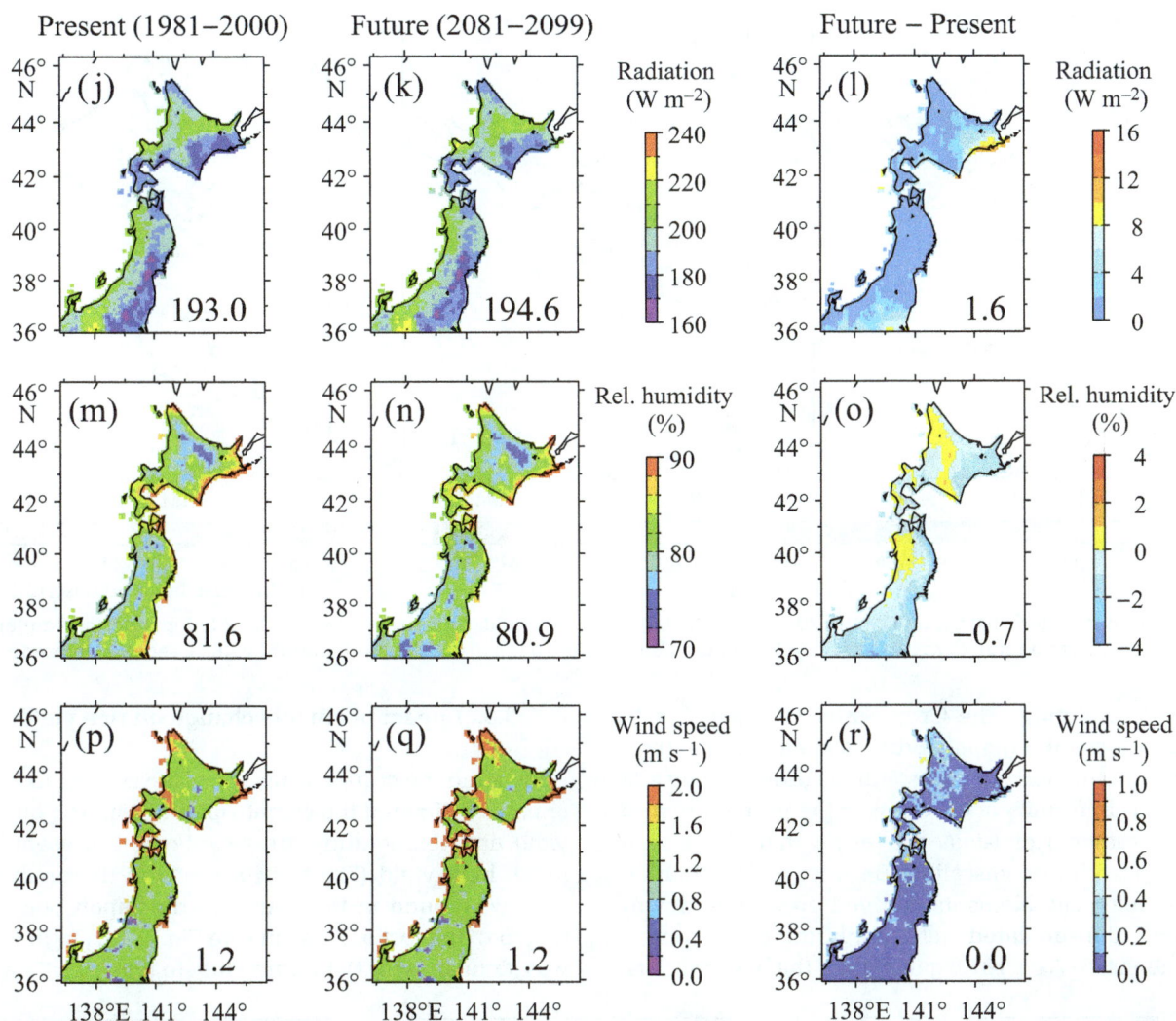

Fig. 2. (cont.). (j–l) Downward shortwave radiation at the surface; (m–o) relative humidity; and (p–r) wind speed. Bottom right value in each figure: regional average

and simulated yields for the period as an average over the grid cells with paddy fields ≥1%. The H/H model reasonably reproduced the inter-annual variation in the observed yield (Fig. 3a). The correlation coefficient (r) was 0.80 and the average bias –1.8%. The yield dropped sharply in 1993 because of a strong Yamase airflow, which resulted in severe cold damage to rice. The model simulated the large 1993 reduction in yield, reflecting a high f(CDD) value (Fig. 3b). The simulation also predicted relatively low yields in other years, such as 1981, 1983, 1995, 1996, and 1997, when f(CDD) exceeded 0.3, although the predicted yield reduction tended to be larger than the observed yield reduction.

The geographical distribution (Fig. 4) of the observed yield showed a small regional difference (mean ± SD: 628.5 ± 67.9 g m^{-2}) because of the coarse

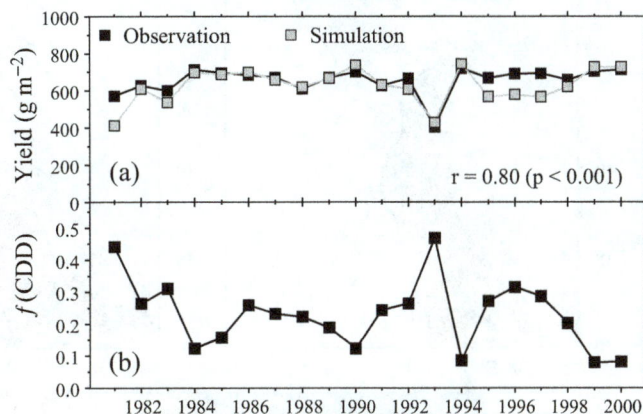

Fig. 3. (a) Interannual variation of observed and simulated yields from 1981 to 2000 averaged over the analyzed area. r: Pearson's correlation coefficient. (b) Same as (a) but for f(CDD), the simulated low-temperature stress value

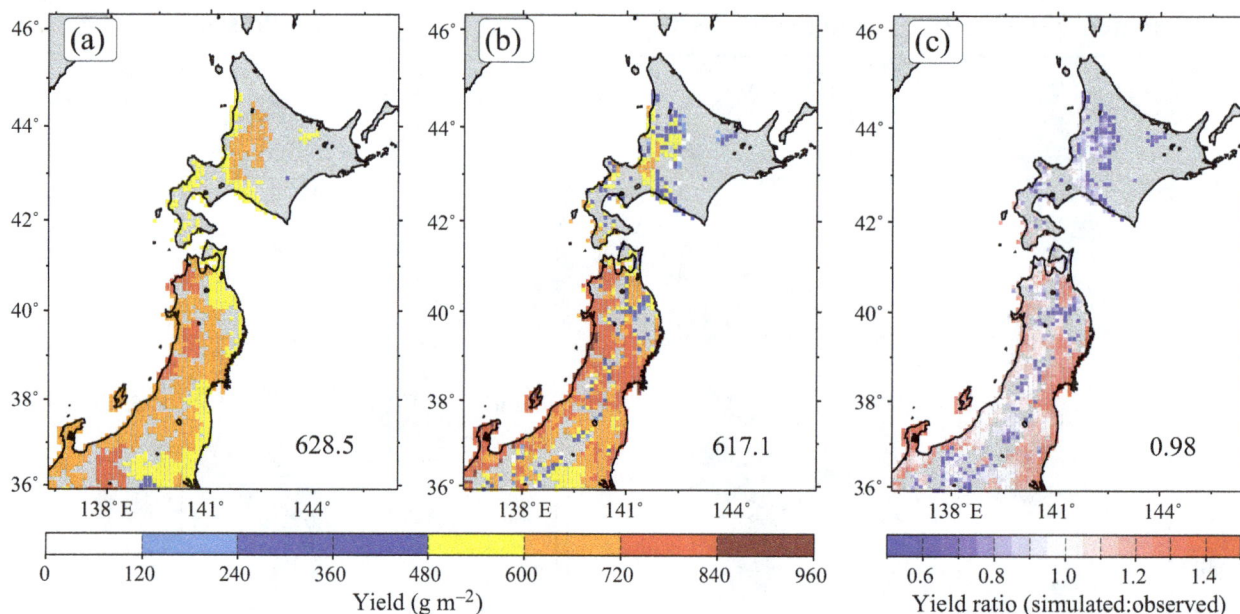

Fig. 4. Geographical distribution of the 20 yr (1981–2000) averaged yield. (a) Observation, (b) simulation, and (c) simulated: observed yield ratio. Gray: areas with <1 % paddy fields in the 10 km grid cells. Bottom right values: regional average

spatial resolution. The Crop Statistics reports yield data by several regional divisions in each prefecture. Although the horizontal resolutions differ because of the non-uniformity of the areas of the prefectures, all of the regional divisions are larger than the 10 km mesh used for downscaling. Because of the detailed meteorological elements derived from the downscaling, the simulated yield had larger geographical variations (617.1 ± 187.8 g m^{-2}) than the observations.

3.3. Impact of climate change on rice yield

By fixing the cropped cultivar as the current leading one, we estimated the climate change impacts on rice yield and temperature stresses. For the present climate, high yield (720 to 840 g m^{-2}) in the analyzed area was found in the plains on the Japan Sea side (Fig. 5a). In most of Honshu (see Fig.1), this high yield was simulated for the future climate (Fig. 5b). The in-

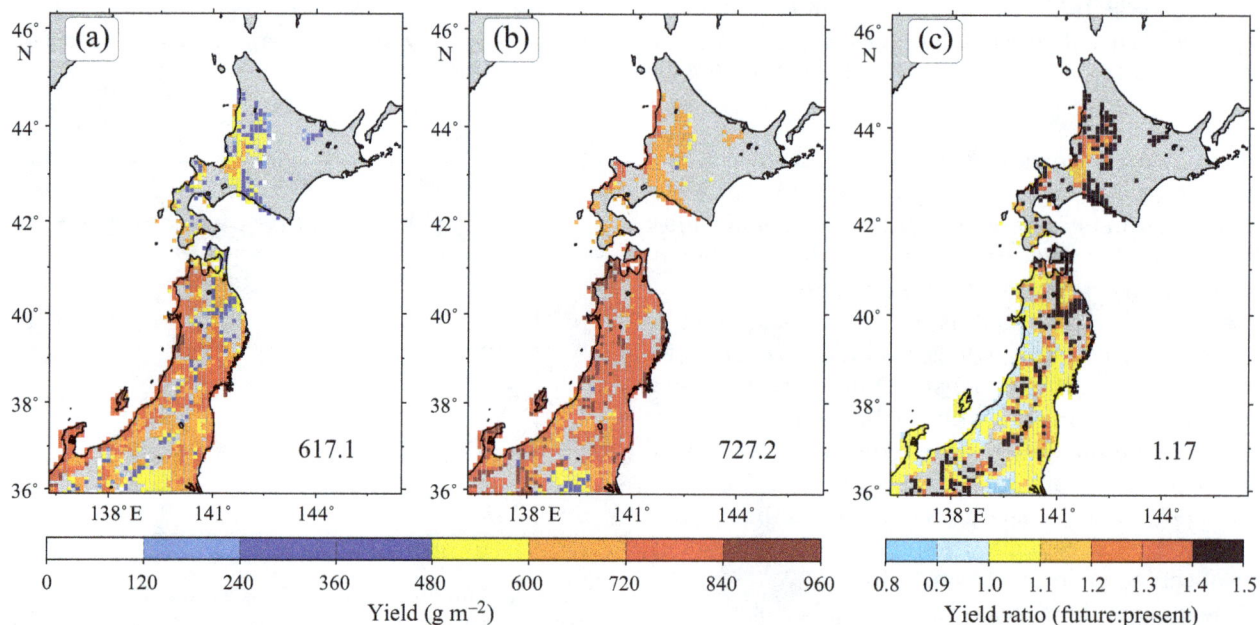

Fig. 5. Simulated yield for (a) present (1981–2000) and (b) future (2081–2099) climates, and (c) future:present yield ratio with the current leading cultivar. Bottom right values: regional average. Gray shading as in Fig. 4

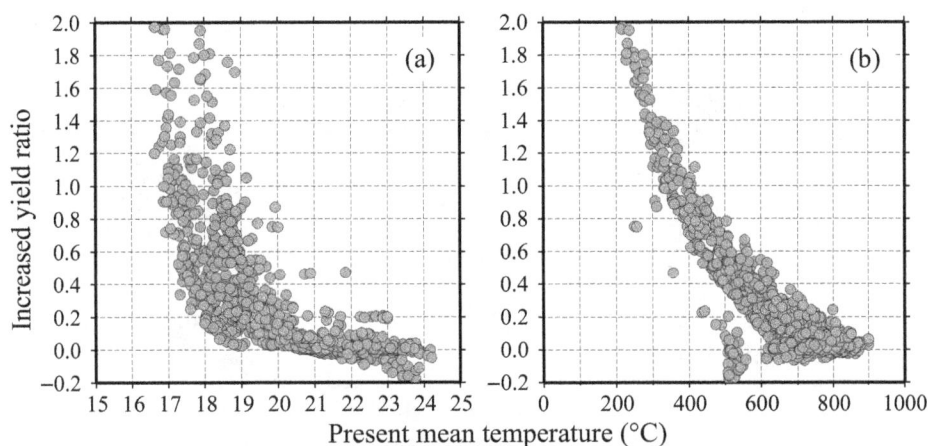

Fig. 6. Relationship between daily (a) mean temperature and (b) rice yield in the present climate and the increased yield ratio from the present to the future climate

creased-yield ratio (i.e. future yield/present yield) averaged 17%, but differed between regions. It was ~10% in the plains and >40% in the mountainous areas and on Hokkaido (Fig. 5c), whereas a negative impact was found in several areas of the Japan Sea side and in a southern area with a higher estimated surface temperature (Fig. 2e). A high increased-yield ratio was found in the northern and high-elevation areas, so we focused on the relation between the increased-yield ratio and the present mean temperature. The negative correlation (r = −0.71) illustrated the high increased-yield ratio for low temperature (Fig. 6a). Present yield was more strongly related to the ratio (r = −0.85), whereas a high ratio was estimated for the lower-yield area (Fig. 6b).

In the present climate, the rice crop was not exposed to high-temperature stress in eastern Japan (Fig. 7a), but the high-temperature stress became obvious in the future climate (Fig. 7b). In particular, the increase was more evident in the southern plains (Fig. 7c). The increased ratio of the future high-temperature stress normalized by the present stress (i.e. [future − present]/present) was 551% on average regionally. The ratio is large because of the small value in the present climate. On the other hand, the geographical distribution of low-temperature stress in the present climate followed the north–south gradient (Fig. 7d). Climate change had positive effects from the perspective of reducing cold damage, as the regional average of low-temperature stress in the future climate decreased to 36% of the present climate (Fig. 7e). In contrast to the situation for high-temperature stress, low-temperature stress was reduced in the northern or mountainous area (Fig. 7f). When we focused on the balance between high and low-temperature stresses, high-temperature stress was negligibly small in the present climate [f(HDD):f(CDD) = 1:71]. However, the intensity of the high-temperature stress increased a quarter of the value of the low-temperature stress under the future climate [f(HDD):f(CDD) = 1:3.9]. High-temperature stress was projected to increase in southern Japan and low-temperature stress was projected to decrease in the northern area.

3.4. Potential cultivar shift due to climate change

Fig. 8 shows a meridional distribution of the cultivar with the higher yield. More than one top cultivar is occasionally presented within the same latitudinal zones, because the best cultivars vary, depending on the variation in climatic conditions between east and west. The location of the top cultivar generally shifted northward in the future climate (Fig. 8a). In particular, Kirara397 (cultivar ID: 2), which was widely distributed as the top cultivar in the present climate, shifted and became limited to high latitudes or high altitudes in the future climate (corresponding to a sparse distribution in the south). A similar shift was found for Akitakomachi (ID: 5).

Hitomebore (ID: 1) was predominant in both climates (Fig. 8b). One of the northward-shifting cultivars, Kirara397 (ID: 2), occupied the highest share in the present climate (37.6%), but decreased to 11.3% in the future climate. The decreased-share cultivars, such as Kirara397 or Akitakomachi, are sensitive to changes in the daily surface air temperature. This characteristic is expressed as the smaller value of G_2 and the larger value of α (Table 1). In contrast, the photoperiod-sensitive cultivars (larger β value), such as Asahinoyume (ID: 4) or Aichinokaori (ID: 6), increased their share by ~20%.

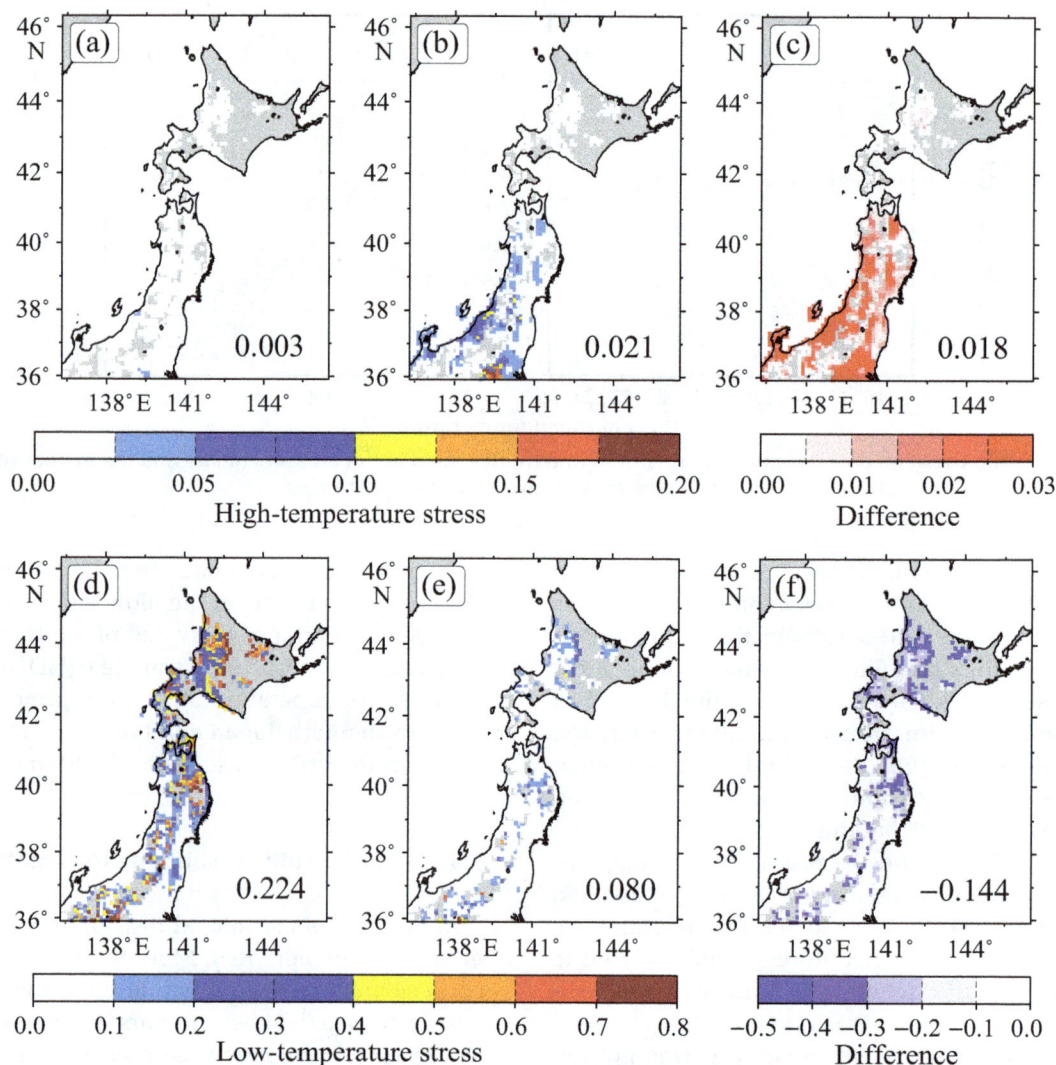

Fig. 7. Distribution of simulated temperature stress for the (a,d) present climate (1981–2000), (b,e) future climate (2081–2099), and (c,f) their difference (future climate – present climate): Function values of (a–c) high-temperature stress and (d–f) low-temperature stress. Gray shading: areas with <1% paddy fields in 10 km grid cells. Bottom right values: regional average

For the 3 major cultivars in the future climate (i.e. Hitomebore, Asahinoyume, and Aichinokaori), their distribution is shown for the present and future in Fig. 9. The Hitomebore grid cells are distributed throughout the study area except Hokkaido in the present climate, and they shift toward the north or the mountainous areas in the future climate. Asahinoyume and Aichinokaori, the photoperiod-sensitive cultivars, barely exist in the present climate, but become dominant in the southern plains in the future climate.

Yields averaged for the top cultivars in future climates were 26% higher than in the current climate. High yields were recorded for Asahinoyume and Aichinokaori, which are more photoperiod-sensitive than Hitomebore (Table 2). Growth duration, in days from transplanting to maturity, of these top cultivars

were close to the current growth duration (MAFF, data as of 2009, www.maff.go.jp/j/study/suito_sakugara/h2204/pdf/ref_data3-3.pdf), suggesting that these phenological traits match well to local future climates.

Based on the projected yield and temperature stress of the current leading cultivars (Figs. 5 & 7), we analyzed the cultivars that provided more favorable outcomes (i.e. higher potential yield or lower temperature stress) than the current leading cultivars (Fig. 10). Additionally, we extracted the grid cells for which the current leading cultivar was ranked 6 or lower (i.e. 6 of the current non-leading cultivars surpassed the leading one) and designated them as reversed grid cells. For example, 53.9% of the paddy grid cells were selected as reversed grid cells when

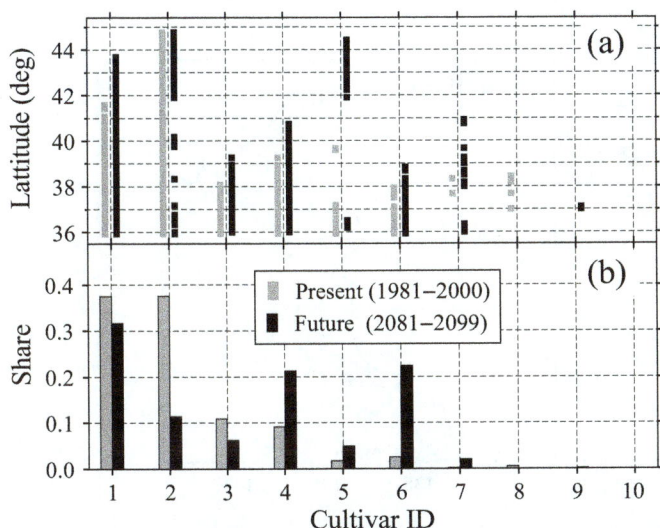

Fig. 8. (a) Meridional distribution of cultivars that provide the maximum yield among the 10 cultivars and (b) their share of the entire analyzed area. Gray/black = present/future climate, respectively. Cultivar names for each ID are presented in Table 1

Table 2. Future yield and growing period averaged over the grid cells for the 3 major simulated cultivars (i.e. maximum yield value among 10 cultivars)

	Yield (g m^{-2})	Growing duration (d)
Hitomebore (ID: 1)	828	127
Asahinoyume (ID:4)	886	144
Aichinokaori (ID: 6)	937	154

we focused on a potential yield (Fig. 10a). This tendency was more evident in the high-temperature stress case, with 86.8% of the total paddy grid cells (Fig. 10b). However, the ratio dropped to 10.4% for the low-temperature stress case (Fig. 10c).

Next, we performed the multi-condition assessment. Multi-condition assessment means that two or three conditions are considered concurrently for judgment of reversed grid cells. For the case of higher potential yield and less high-temperature stress, the southern area was expected to have a large proportion of reversed grid cells (Fig. 10d; 29.2%), whereas the northern area had fewer because of difficulties in maintaining higher yield. As the decrease in low-temperature stress was more difficult for the current non-leading cultivars compared to high-temperature stress (Fig. 10b,c), the reversed grid cells under the condition of higher potential yield and less low-temperature stress were rarely found in eastern Japan (Fig. 10e; 5.2%). Therefore, good performance under all 3 conditions (higher potential yield, less high-temperature, and less low-temperature stresses, Fig. 10f) was limited to only a few grids and cultivars. In approximately 64% of the grid cells, the current leading cultivar was the top cultivar.

4. DISCUSSION

When the current leading cultivars were maintained in the future (2081–2099) climate, rice yield was projected to increase by an average of 17% compared with the baseline climate (1981–2000). The positive effect was mainly for 2 reasons: (1) increases in atmospheric CO_2 concentrations (C_a) and (2) rise in temperatures from sub-optimal levels. In the simulation, mean C_a for the baseline period was 354 µmol mol^{-1}, but in the future period it reached 534 µmol mol^{-1} under RCP 4.5 (Meinshausen et al. 2011). A meta-analysis of the previous free-air CO_2 enrichment (FACE) studies showed a 12% increase in grain yield under the C_a range of 500 to 599 µmol mol^{-1} (Ainsworth 2008). A recent rice model intercomparison study demonstrated that current rice models also predict a yield increase of ~14 to 15%, with an increase in C_a by 180 µmol mol^{-1} (Li et al. 2015). These findings support that rising C_a is an important factor for the increase in rice yields.

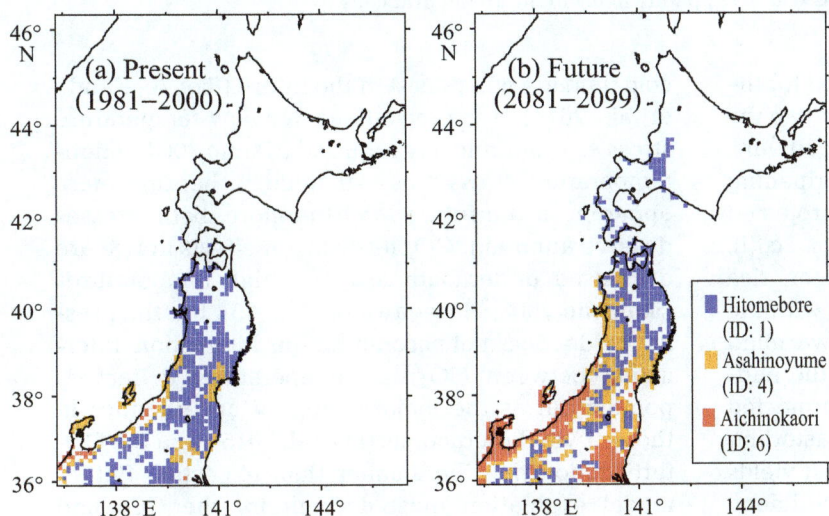

Fig. 9. Geographical distribution of the top cultivar for Hitomebore (blue), Asahinoyume (orange), and Aichinokaori (red) in (a) present and (b) future climates. Non-paddy area is unmasked to clarify the figure

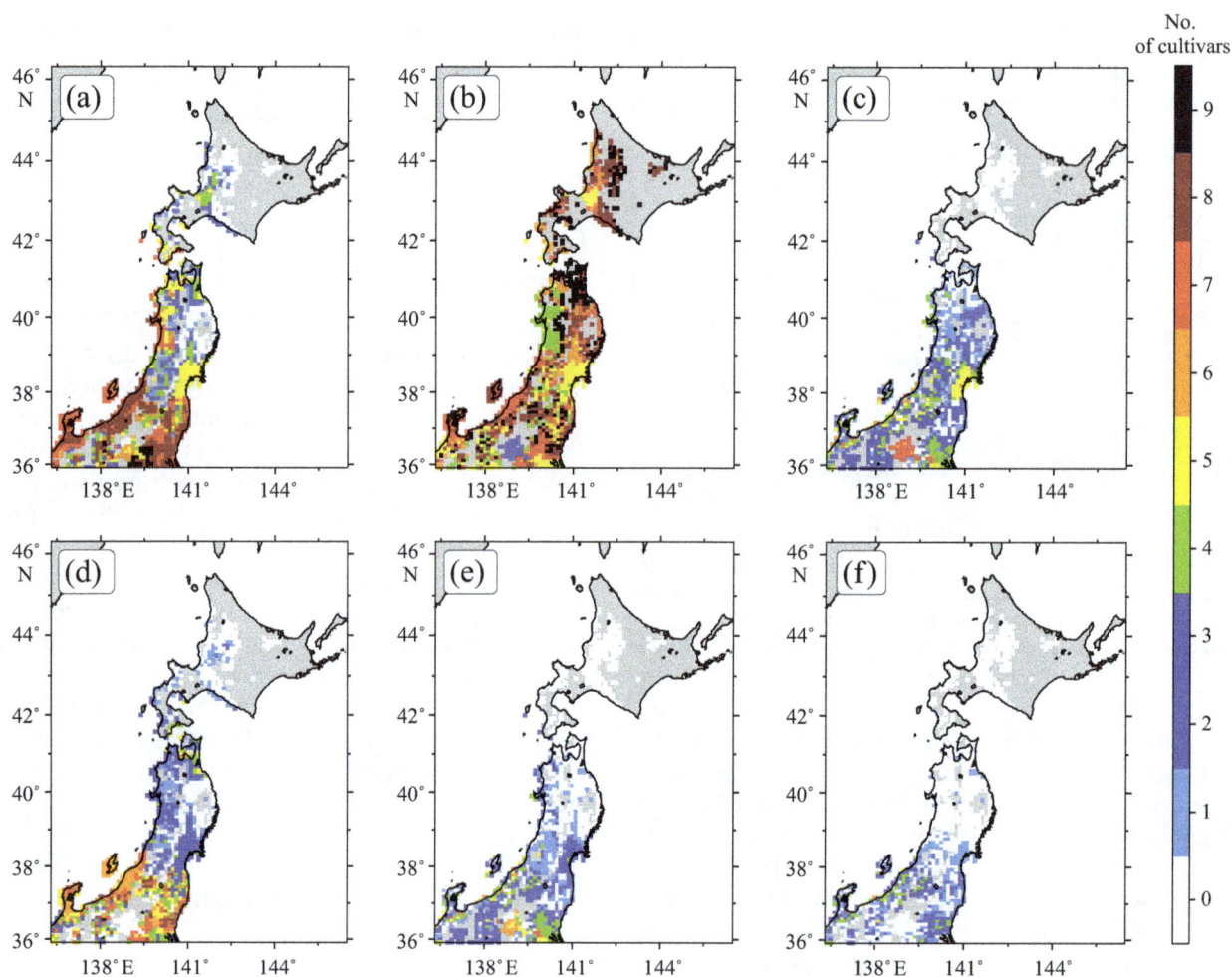

Fig. 10. Number of cultivars that provide more favorable rice production in the future climate than the current leading cultivar (i.e. higher potential yield, less high-temperature stress, and less low-temperature stress): (a) higher yield, (b) less high-temperature stress, (c) less low-temperature stress, (d) higher yield + less high-temperature stress, (e) higher yield + less low-temperature stress, and (f) higher yield + less high-temperature stress + less low-temperature stress. Gray shading: non-rice-cropped areas with <1 % paddy fields in the 10 km grid cells

Summer air temperature of 18.0°C averaged for the baseline period (Fig. 2a) is apparently cooler than the optimum, even for temperate japonica cultivars, whose optimal temperature range for the ripening process is 20 to 22°C (Yoshida 1981). The projected surface warming in this study was ~3°C (Figs. 2c,f,i), and thus provided favorable conditions for rice growth. Our results for yields are consistent with the previous studies that demonstrated how a warming of 1 to 3°C brings favorable rice growth in the mid-latitudes in contrast to the negative impact projected in the low latitudes (Nakagawa et al. 2003, Easterling et al. 2007). An observational study found that yields are currently increasing in northern Japan (Shimono 2008).

Several concerns remain despite these optimistic projections for the impacts of climate change. Our projections and other studies suggest that the risk of

cold damage will persist in the future (Fig. 7e; Kanda et al. 2014), so precautions for low-temperature stress are continuously required (Kanno 2004). High-temperature stress was estimated to become more apparent in the future, and therefore, both stresses deserve attention. CO_2 fertilization effects in rice are decreased under both low and high temperatures (Shimono 2008, Hasegawa et al. 2015), but the present model does not account for this interaction. Interaction between CO_2 and temperature, in fact, is pointed out as the major source of uncertainties in the rice-yield predictions (Li et al. 2015), and the CO_2 fertilization may be smaller than expected. Future model evaluation must account for the CO_2 and temperature interaction.

The present study is unique in that it examines whether cultivar shifts occur as a result of climate change among the widely planted cultivars in Japan at present based on 3 metrics: (1) potential yield, (2) cold stress, and (3) heat stress. A major decline was predicted for Kirara397 (ID: 2), an early maturing cultivar widely planted in Hokkaido at present. This cultivar was replaced with longer, but weakly photoperiod-sensitive cultivars, such as Hitomebore (ID: 1) and Akitakomachi (ID: 5) (Fig. 8a). Hitomebore (ID: 1) maintained its high share values in terms of yield under climate change conditions (Fig. 8b) by becoming the leading cultivar in the future climate (i.e. it was selected in most grid cells; Fig. 10a), while giving away several current 'winning' grid cells. This cultivar was developed in Miyagi Prefecture (38° N, 140° E on the Pacific Ocean side, Fig. 1) in 1991 as a highly cold tolerant cultivar with a high eating quality. It is currently the second most popular cultivar in Japan and is widely planted across different regions, including southern prefectures such as Okinawa (24°–27° N) and Oita prefecture (33° N, Fig. 1), with 13.7 % of the total paddy area in 2009 (MAFF 2014). This wide adaptability could in part be attributed to a low temperature sensitivity of phenology, as evidenced by the smallest α value (Table 1), i.e. growth duration is relatively unaffected by the temperature change. This could be one of the key traits for stable rice production in the future climate. Two longer-duration cultivars, Asahinoyume (ID: 4) and Aichinokaori (ID: 6) (Fig. 8b), gained share in the future climates. These cultivars are currently not major winners but are projected to become dominant in the southern plains (Fig. 9), suggesting that photoperiod sensitivity helps to ensure growth duration in relatively warmer regions.

When we consider all of the higher potential yields and reduced high- and low-temperature stresses, cultivar replacement was projected to occur in 35.6 % of the paddy grid cells, whereas the current leading cultivars kept their superiority in the remaining 64.4 % grid cells (Fig. 10f). The regional yield average for the top cultivars, based on the multiple metrics, was 22 % higher than the baseline. Of the new cultivars, those selected in each grid cell were developed in the more southern areas. When they are grown in the current climate, they take too long to mature, but under warmer conditions, they grow sufficiently long to ensure higher yield but without temperature stress. Thus, introducing the 'southern cultivars' would be an effective way to adapt to the future climate in approximately 36 % of the study area. An observational study also suggested the introduction of a southern cultivar to mitigate the high-temperature damage projected for the current leading cultivars (Nemoto et al. 2011).

In approximately 64 % of the grid cells, however, presently leading cultivars remained the winners if the 3 metrics are taken into account (Fig. 9f). This finding suggests that cultivar replacement is not an 'easy' option. Southern cultivars often failed to win despite many reverse grid cells based on a single metric (Figs. 9a–c). A trait such as low sensitivity to temperature for phenology observed for Hitomebore also helps current winners under climate change conditions because of its smaller variation in growth duration. This trait could also be a valuable trait under variable weather conditions. The results also suggest that breeding new cultivars for the future by utilizing or pyramiding higher yield potentials and stress tolerances is essential. Phenology is the first trait that needs to be considered for climate change adaptation, but keeping growth duration long is not the only way to improve productivity. In the context of climate change, enhancing the CO_2 fertilization effect is an option for the future (Ziska et al. 2012). A previous FACE study demonstrated a large genotypic variability in the grain-yield response to elevated C_a, ranging from 3 to 36 % (Hasegawa et al. 2013).

The current study is an initial attempt to account for multiple metrics to visualize a possible cultivar shift under climate change conditions. In reality, farmers' cultivar choices are more complex, with many more metrics taken into account. Marketability, eating, and appearance quality, in addition to pest, diseases, and lodging resistance, all count when choosing cultivars, all of which can be potentially influenced by climate change. Future work is required to link these metrics, as they directly affect the profitability of farmers.

5. CONCLUSION

This study evaluated adaptation to climate change for crop production based on multiple metrics and multiple choices of existing cultivars, taking the presently cool climate as an example for rice production in Japan. Projection based on the MIROC5 global climate model under RCP4.5 and a rice model (H/H) demonstrated that climate change in the 2081–2099 period would increase grain yield by a regional average of 17 % compared with the baseline climate (1981–2000), even without any cultivar shift. The simulation also predicted that low-temperature stresses would be reduced but that high-temperature stresses would become apparent, suggesting that countermeasures against both heat and cold damages will be required in the future climate.

Comparisons of simulations among 10 major Japanese cultivars demonstrated that top cultivars change dramatically with climate change if only a single metric, such as yield, is accounted for. The predicted yield averaged for the yield 'winners' was 26% higher than the baseline yield. When all 3 metrics (yield, cold and heat stress) are taken into account, however, cultivar replacement is projected to occur in ~36% of the grid cells, and the current leading cultivars maintained their superiority in ~64% of the total grid cells. The resultant yield advantage compared with the baseline was 22%. This finding suggests a requirement for developing new cultivars by pyramiding useful traits for the future climates. A trait such as low sensitivity to temperature for phenology helps in ensuring growth duration under variable temperatures. Increasing photoperiod sensitivity can be an option in the future climates in relatively warmer regions.

Future studies must include other important metrics, such as quality and profitability, in combination with management practices, including crop calendar and fertilizer managements. Uncertainties from climate projections and crop models should also be combined with the benchmark testing of these adaptation measures.

Acknowledgements. This study was supported by JSPS KAKENHI (Grant No.: 25892004), Asahi Group Foundation, the Research Program on Climate Change Adaptation (RECCA) of the Ministry of Education, Culture, Sports, Science, and Technology of Japan, the Environment Research and Technology Development Fund (S-8-1) of the Ministry of the Environment, Japan, and the Cross-ministerial Strategic Innovation Promotion Program of the Cabinet Office, Government of Japan. The H/H model improvement was conducted through a research project entitled 'Development of technologies for mitigation and adaptation to climate change in agriculture, forestry and fisheries,' funded by the Ministry of Agriculture, Forestry and Fisheries, Japan. We thank Dr. Shinji Sawano for providing N fertilizer input data converted from the Production Cost Statistics.

LITERATURE CITED

ä Ainsworth EA (2008) Rice production in a changing climate: a meta-analysis of responses to elevated carbon dioxide and elevated ozone concentration. Glob Change Biol 14: 1642–1650

▶ Challinor A, Wheeler T, Hemming D, Upadhyaya H (2009) Ensemble yield simulations: crop and climate uncertainties, sensitivity to temperature and genotypic adaptation to climate change. Clim Res 38:117–127

▶ Collatz GT, Ball JT, Grivet C, Berry JA (1991) Physiological and environmental regulation of stomatal conductance, photosynthesis and transpiration: a model that includes a laminar boundary layer. Agric Meteorol 54:107–136

Collins M, Knutti R, Arblaster J, Dufresne JL and others (2013) Long-term climate change: projections, commiments and irreversibility. In: Stocker TF, Qin D, Plattner G-K, Tignor M and others (eds) Climate change 2013: the physical science basis. Contribution of Working Group I to the Fifth Assessment Report of the Intergovernmental Panel on Climate Change. Cambridge University Press, Cambridge

▶ Deryng D, Sacks WJ, Barford CC, Ramankutty N (2011) Simulating the effects of climate and agricultural management practices on global crop yield. Glob Biogeochem Cycles 25:GB2006, doi:10.1029/2009GB003765

▶ Deryng D, Conway D, Ramankutty N, Price J, Warren R (2014) Global crop yield response to extreme heat stress under multiple climate change futures. Environ Res Lett 9:034011,

Easterling WE, Aggarwal PK, Batima P, Brander KM and others (2007) Food, fiber and forest products. In: Parry ML, Canziani OF, Palutikof JP, van der Linden PJ, Hanson CE (eds) Climate change 2007: impacts, adaptation and vulnerability. Contribution of Working Group II to the Fourth Assessment Report of the Intergovernmental Panel on Climate Change. Cambridge University Press, Cambridge, p 273–313

▶ Farquhar GD, von Caemmerer S, Berry JA (1980) A biochemical model of photosynthesis CO_2 assimilation in leaves of C_3 species. Planta 149:78–90

Fukui S, Ishigooka Y, Kuwagawa T, Hasegawa T (2015) A methodology for estimating phenological parameters of rice cultivars utilizing data from common variety trials. J Agr Meteorol 71:77–89, doi:10.2480/agrmet.D-14-00042

Hasegawa T, Horie T (1997) Modelling the effect of nitrogen on rice growth and development. In: Kropff MS, Teng PS, Aggarwal PK, Bouma J, Bouman BAM, Jones JW, van Laar HH (eds) Applications of systems approaches at the field level. Kluwer, Dordrecht, p 243–257

▶ Hasegawa T, Ishimaru T, Kondo M, Kuwagawa T, Yoshimoto M, Fukuoka M (2011) Spikelet sterility of rice observed in the record hot summer of 2007 and the factors associated with its variation. J Agric Meteorol 67:225–232

▶ Hasegawa T, Sakai H, Tokida T, Nakamura H and others (2013) Rice cultivar responses to elevated CO_2 at two free-air CO_2 enrichment (FACE) sites in Japan. Funct Plant Biol 40:148–159,

Hasegawa T, Sakai H, Tokida T, Usui Y and others (in press) (2015) Rice FACE studies to improve assessment of climate change effects. In: Hatfield JL, Fleisher D (eds) Improving modeling tools to assess climate change effects on crop response. Adv Agric Syst Model Ser, American Society of Agronomy, Madison, WI, doi: 10.2134/advagricsystmodel7.2014.00158

Horie T, Nakagawa H, Centeno HGS, Fropff MJ (1999) The rice crop simulation model SIMRIW and its testing. In: Matthews RB, Kropff MJ, Bachelet D, van Laar HH (eds) Modeling the impact of climate change on rice production in Asia. International Rice Research Institute and CAB International, Wallingford, p 51–66

▶ Iizumi T, Hayashi Y, Kimura F (2007) Influence on rice production in Japan from cool and hot summers after global warming. J Agric Meteorol 63:11–23

▶ Iizumi T, Nishimori M, Ishigooka Y, Yokozawa M (2010) Introduction to climate change scenario derived by statistical downscaling. J Agric Meteorol 66:131–143 (in Japanese with English abstract)

▶ Iizumi T, Yokozawa M, Nishimori M (2011) Probabilistic evaluation of climate change impacts on paddy rice productivity in Japan. Clim Change 107:391–415,

▶ Ishizaki NN, Takayabu I, Oh'izumi M, Sasaki H and others (2012) Improved performance of simulated Japanese climate with a multi-model ensemble. J Meteorol Soc Jpn 90:235–254

▶ Kain JS (2004) The Kain-Fritch convective parameterization: an update. J Appl Meteorol 43:170–181

Kanda E, Kanno H, Okubo S, Shimada T, Yoshida R, Kobayashi T, Iwasaki T (2014) Estimation of cool summer damage in the Tohoku region based on the MRI AGCM. J Agric Meteorol 70:187–198, doi:10.2480/agrmet.D-14-00004

▶ Kanno H (2004) Five-year cycle of north–south pressure difference as an index of summer weather in northern Japan from 1982 onwards. J Meteorol Soc Jpn 82:711–724

▶ Kanno H, Watahabe M, Kanda E (2013) MIROC5 predictions of Yamase (cold northeasterly winds causing cool summers in northern Japan). J Agric Meteorol 69:117–125

Kassie B, Asseng S, Rotter R, Hengsdijk H, Ruane AC, Van Ittersum MK (2015) Exploring climate change impacts and adaptation options for maize production in the Central Rift Valley of Ethiopia using different climate change scenarios and crop models. Clim Change 129:145–158, doi:10.1007/s10584-014-1322-x

▶ Katul GG, Ellsworth DS, Lai CT (2000) Modeling assimilation and intercellular CO_2 from measured conductance: a synthesis of approaches. Plant Cell Environ 23:1313–1328

▶ Kim HY, Ko J, Kang S, Tenhunen J (2013) Impacts of climate change on paddy rice yield in a temperate climate. Glob Chang Biol 19:548–562

Kimura F, Kitoh A (2007) Downscaling by pseudo global warming method. In: The Final Report of the ICCAP. Research Institute for Humanity and Nature (RIHN), Kyoto

▶ Li T, Hasegawa T, Yin X, Zhu Y and others (2015) Uncertainties in predicting rice yield by current crop models under a wide range of climatic conditions. Glob Chang Biol 21: 1328–1341

Lobell DB, Field CB (2007) Global scale climate–crop yield relationships and the impacts of recent warming. Environ Res Lett 2:014002, doi:1088/1748-9326/2/1/014002

MAFF (Ministry of Agriculture, Forestry and Fisheries) (2000) Crop statistics. Association of Agriculture and Forestry Statistics, Tokyo

MAFF (2014) Crop statistics. MAFF, Tokyo

▶ Medlyn BE, Badeck FW, De Pury DGG, Barton CVM and others (1999) Effects of elevated $[CO_2]$ on photosynthesis in European forest species: a meta-analysis of model parameters. Plant Cell Environ 22:1475–1495

Meinshausen M, Smith SJ, Calvin K, Daniel JS and others (2011) The RCP greenhouse gas concentrations and their extensions from 1765 to 2300. Clim Change 109:213–241

MLITT (Ministry of Land, Infrastructure, Transport and Tourism) (2012) National Land Numerical Information. MLITT, Tokyo

Nakagawa H, Horie T, Matsui T (2003) Effects of climate change on rice production and adaptive technologies. In: Mew TW, Brar DS, Peng S, Dawe D, Hardy B (eds) Rice science: innovations and impact for livelihood. IRRI, Chinese Academy of Engineering, and Chinese Academy of Agricultural Sciences, p 635–658

▶ Nakagawa H, Yamagishi J, Miyamoto N, Motoyama M, Yano M, Nemoto K (2005) Flowering response of rice to photoperiod and temperature: a QTL analysis using a phenological model. Theor Appl Genet 110:778–786

▶ Nakanishi M, Niino H (2004) An improved Mellor-Yamada level-3 model with condensation physics: its design and verification. Boundary-Layer Meteorol 112:1–31

▶ Nemoto M, Hamasaki T, Sameshima R, Kugawai E, Ohno H,

Wakiyama Y, Maruyama A, Ozawa K (2011) Assessment of paddy rice heading date under projected climate change conditions for Hokkaido region based on the field experiment. J Agric Meteorol 67:275–284

▶ Nemoto M, Hamasaki T, Sameshima R, Kumagai E and others (2012) Probabilistic risk assessment of the rice cropping schedule for central Hokkaido, Japan. J Appl Meteorol Climatol 51:1253–1264

▶ Okada M, Iizumi T, Hayashi Y, Yokozawa M (2011) Projecting climate change impacts both on rice quality and yield in Japan. J Agric Meteorol 67:285–295

▶ Parry ML, Rosenzweig C, Iglesias A, Livermore M, Fischer G (2004) Effects of climate change on global food production under SRES emissions and socio-economic scenarios. Glob Environ Change 14:53–67

Porter JR, Xie L, Challinor AJ, Cocharane K and others (2014) Food security and food production systems. In: Field CB, Barros VR, Dokken DJ, Mach KJ, and others (eds) Climate change 2014: impacts, adaptation, and vulnerability. Part A: global and sectoral aspects. Contribution of Working Group II to Fifth Assessment Report of the Intergovernmental Panel on Climate Change. Cambridge University Press, Cambridge, p 485–553

▶ Rosenzweig C, Elliott J, Deryng D, Ruane AC and others (2014) Assessing agricultural risks of climate change in the 21st century in a global gridded crop model intercomparison. Proc Natl Acad Sci (USA) 111:3268–3273,

▶ Saito K, Ishida J, Aranami K, Hara T, Segawa T, Narita M, Honda Y (2007) Nonhydrostatic atmospheric models and operational development at JMA. J Meteorol Soc Jpn 85B:271–304

Satake T (1976) Sterile-type cool injury in paddy rice plants. In: Climate and rice. International Rice Research Institute, Los Baños, p 281–300

▶ Satake T, Yoshida S (1978) High temperature-induced sterility in indica rices at flowering. Jpn J Crop Sci 47:6–17,

▶ Seino H (1993) An estimation of distribution of meteorological elements using GIS and AMeDAS data. J Agric Meteorol 48:379–383

▶ Shimono H (2008) Impact of global warming on yield fluctuation in rice in the northern part of Japan. Jpn J Crop Sci 77:489–497 (in Japanese with English abstract),

ä Shimono H (2011) Earlier rice phenology as a result of climate change can increase the risk of cold damage during reproductive growth in northern Japan. Agric Ecosyst Environ 144:201–207

Shimono H, Okada M, Yanakawa Y, Nakamura H, Kobayashi K, Hasegawa T (2008) Rice yield enhancement by elevated CO_2 is reduced in cool weather. Glob Change Biol 14:276–284

ä Soora N, Aggarwal PK, Saxena R, Rani S, Jain S, Chauhan N (2013) An assessment of regional vulnerability of rice to climate change in India. Clim Change 118:683–699

ä Tao F, Hayashi Y, Zhang Z, Sakamoto T, Yokozawa M (2008) Global warming, rice production, and water use in China: developing a probabilistic assessment. Agric Meteorol 148:94–110

Wassmann R, Jagadish SVK, Heuer S, Ismail A and others (2009) Climate change affecting rice production: the physiological and agronomic basis for possible adaptation strategies. Adv Agron 101:59–122

ä Watanabe M, Suzuki T, O'ishi R, Komuro Y and others (2010) Improved climate simulation by MIROC5: mean states, variability, and climate sensitivity. J Clim 23: 6312–6335

ä Yin X, Kropff MJ, Horie T, Nakagawa H, Centeno HGS, Zhu D, Goudriaan J (1997) A model for photothermal responses of flowering in rice I. Model description and parameterization. Field Crops Res 51:189–200

Yoshida S (1981) Fundamentals of rice crop science. International Rice Research Institute, Los Baños

➤ Yoshida R, Iizumi T, Nishimori M (2012a) Inter-model differences in the relationships between downward shortwave radiation and air temperatures derived from dynamical and statistical downscaling models. J Meteorol Soc Jpn 90B:75–82

Yoshida R, Iizumi T, Nishimori M, Yokozawa M (2012b) Impacts of land-use changes on surface warming rates and rice yield in Shikoku, western Japan. Geophys Res Lett 39:L22401, doi:10.1029/2012GL053711

➤ Yu Y, Zhang W, Huang Y (2014) Impact assessment of climate change, carbon dioxide fertilization and constant growing season on rice yields in China. Clim Change 124:763–775

➤ Ziska LH, Bunce JA, Shimono H, Gealy DR and others (2012) Food security and climate change: on the potential to adapt global crop production by active selection to rising atmospheric carbon dioxide. Proc R Soc B 279: 4097–4105

9

Effects of climate change on heating and cooling degree days and potential energy demand in the household sector of China

Ying Shi[1], Xuejie Gao[2,*], Ying Xu[1], Filippo Giorgi[3], Deliang Chen[4]

[1]National Climate Center, China Meteorological Administration, Zhongguancun Nandajie 46, Haidian District, Beijing 100081, PR China

[2]Climate Change Research Center, Institute of Atmospheric Sciences, Chinese Academy of Sciences, Huayanli 40, Chaoyang District, Beijing 100029, PR China

[3]The Abdus Salam International Centre for Theoretical Physics, PO Box 586, Trieste 34100, Italy

[4]Department of Earth Sciences, University of Gothenburg, PO Box 460, 40530 Gothenburg, Sweden

ABSTRACT: Future changes of heating and cooling degree days (HDD and CDD) in the 21st century over mainland China are projected with a regional climate model to investigate the potential effects of climate change on energy demand in the household sector. Validation of the model shows a good performance in reproducing the spatial distribution, magnitude and interannual variability of the present day HDD and CDD. Significant decreases in HDD and increases in CDD are projected under the warming. These are further weighted by population projections for a first-order assessment of future changes in energy demand. A larger decrease in population-weighted regional mean HDD compared to the increase in CDD is projected, indicating a decrease of about 15% in potential energy demand for different periods and scenarios in the future. In addition, the simulations show a marked spatial heterogeneity in the change in energy demand. Specifically, we find increases in both heating and cooling demand in parts of northern China due to the increased population there, an increase in cooling demand in the south and decreases in heating demand in the northernmost and western regions. Furthermore, a seasonal shift occurs, with increasing demand in summer and a decrease in winter. Finally, when the future reference temperatures for household heating and cooling change from standards currently used in China to values closer to those in Europe and the USA, potentially large increases in energy demand (~80%) are expected, illustrating the importance of policy decisions concerning household heating and cooling.

KEY WORDS: Climate change · Regional climate model · Heating and cooling degree days · China

1. INTRODUCTION

Warming of the climate system is unequivocal, globally ~0.85°C (0.65 to 1.06°C) over the period of 1880 to 2012 (IPCC 2013). Continued emissions of greenhouse gases (GHGs) will cause further warming, with a projected temperature increase of 0.3 to 4.8°C by the end of the century (2081–2100) relative to the present day (1986–2005) (IPCC 2013). Significant warming in the last decades has also occurred in China, as in other parts of the world, at a larger rate than the global mean (e.g. Zhai & Pan 2003, CCSNARCC 2011).

Changes in temperature strongly affect energy consumption for heating and cooling in the household sector, which accounts for nearly 20% of the total energy consumption in China (SSB 2013). Warmer climate conditions are expected to lead to a decrease in energy demand in winter and an increase in summer. The simplest way to estimate the

*Corresponding author: gaoxuejie@mail.iap.ac.cn

relationship between household energy consumption and temperature is through the concept of degree day, and specifically heating and cooling degree days (HDD and CDD, respectively, e.g. Quayle & Diaz 1980). This concept has been widely used to estimate energy demand in climate change studies at both global and regional scales (e.g. Kadioğlu & Şen 1999, Christenson et al. 2006, Isaac & van Vuuren 2009, Wang et al. 2010, Rehman et al. 2011). Analysis of HDD and CDD has also been conducted for China, mostly for individual cities or small areas (Chen et al. 2007, Li et al. 2010, Tan et al. 2012). Limited studies have been carried out covering the whole country, in particular for investigating future changes in energy consumption under global warming conditions (Ren et al. 2009, You et al. 2014).

Regional climate models (RCMs) can be especially useful tools to study the response of HDD and CDD to global warming over China because they can simulate the regional climate detail associated with local complex topographical features and the unique weather and climate systems of the region (e.g. the East Asian monsoon) (e.g. Gao et al. 2001, 2012, Zhou & Yu 2006, Ju et al. 2007, Li & Zhou 2010, Yu et al. 2010). Therefore, in this study, we present an analysis of the future changes in HDD and CDD over China based on climate projections conducted with a high-resolution RCM under 2 representative concentration pathways (RCPs) (the high-level RCP8.5 and mid-level RCP4.5; Moss et al. 2010). We analyze the 21st century future periods 2046 to 2065 (mid-century) and 2080 to 2099 (end-of-century) with respect to the 'present day' period 1986 to 2005, and in particular, we focus on the mid-21st century under RCP4.5 and end of the century under RCP8.5 to present low and high ends of the range of future changes in our simulations.

As a further step in the study, the climate data are weighted by population density data at 0.5° by 0.5° degree resolution. This enables us to assess not only the total future energy demand in the household sector according to the climatological indicators (HDD and CDD) but also its modulation by the present and projected population distribution. Such analysis can indeed provide important indications of significant spatial and seasonal shifts in energy demand within China, which is important information for the development of future energy management strategies and policies in the country. We stress that our study represents the first attempt in the literature to investigate changes in heating/cooling energy demand over China based on high-resolution climate and population data.

2. DATA AND METHODS

2.1. Data

The first step in our analysis is an assessment of the model performance when reproducing observed CDD and HDD over China under present-day conditions. The observational dataset employed for this purpose is CN05.1 developed by Wu & Gao (2013), which is an augmentation of CN05 (Xu et al. 2009) including a greater number of station observations. The interpolation from station to grid data essentially follows the same approach used in generating the Climatic Research Unit dataset (CRU; New et al. 2002), whereby a gridded climatology is first calculated by thin-plate smoothing splines, and then a gridded anomaly derived via an angular weighting method is added to obtain the final data. CRU, CN05 and CN05.1 employ about 200, 750 and 2400 stations over China, respectively. Comparison of CN05 and CN05.1 with CRU shows basic similarities, with differences mostly in the areas where new and denser station data were introduced (e.g. western China and the Tibetan Plateau). Different spatial resolutions of 0.25°, 0.5° and 1° are available in CN05.1, and here, we use the 0.5° resolution to match the resolution of the RCM employed in the study. CN05 and the updated CN05.1 are becoming increasingly popular in model validation analyses over China (e.g. Gao et al. 2011, Wu et al. 2012, Guo & Wang 2013, Sui et al. 2014).

The model simulations are conducted with the Abdus Salam International Centre for Theoretical Physics (ICTP) Regional Climate Model v.4, RegCM4 (Giorgi et al. 2012), driven by the global model BCC_CSM1.1 (Beijing Climate Center Climate System Model v.1.1; Wu et al. 2010, Xin et al. 2013). The RegCM4 domain covers continental China and surrounding areas with a grid spacing of 50 km, and the simulation covers the period 1951–2005 for the present day (with observed GHG concentrations) and 2006–2099 for the future under the mid-level RCP4.5 and high-level RCP8.5 scenarios, respectively (Moss et al. 2010). A comparison of the RegCM4 and driving GCM simulations can be found in Gao et al. (2013), which shows remarkable improvements by RegCM4 in reproducing the present day climate over the region compared to the driving GCM. Specifically, the GCM simulation exhibits a general cold bias in the range of 1.0 to 2.5°C in eastern China and up to 10.0°C cold or warm biases in the complex topography areas of western China. Conversely, in the RegCM4 simulations, the temperature bias in eastern China decreases to values mostly within ±1.0°C, and only a cold bias of ~2.5°C is found over the Tibetan Plateau. The model data are bilinearly interpolated into a regular 0.5° by 0.5° latitude-longitude grid to match the observation grid.

2.2. Degree days

A degree day is defined as the difference between mean daily temperature and a given reference temperature. For metrics designed to reflect the energy demand to heat/cool a building, the reference temperature is considered to be a human comfort temperature. While a reference temperature values of 18°C for heating and 22°C for cooling are widely used (e.g. Roltsch et al. 1999, Isaac & van Vuuren 2009, Spinoni et al. 2015), these references vary across countries depending on the level of economic development, general characteristics of the buildings, climatic conditions, etc. In China, the current policy for the starting/ending dates of room heating (DSH and DEH), which can be dated back to the 1950s, states that only rooms in northern China are heated to a reference temperature of 18°C, and the DSH and DEH are decided by the local governments in the cites and counties at the provincial level. Thus, tens of DSH and DEH may exist in the country.

The heating policy can be summarized as follows: (1) only areas with >90 d of temperature <5°C can be heated (to the reference value of 18°C); (2) the heating starts/ends when the temperature is lower/greater than 5°C for a continuous 5 d period in an annual cycle from 1 September to 31 August. Here, we use the above standards to define the areas and periods of heating for all provinces and municipalities and assume that these will remain unchanged in the future. (Note that the calculation of HDD follows the annual cycle from 1 September to 31 August instead of the calendar year.) Conversely, cooling days occur when the out-door air temperature exceeds 26°C, following the Code for Design of Heating Ventilation and Air Conditioning (GB50019-2003) (MCPRC 2003). In this case, we assume that cooling lowers the room temperature to the reference value of 26°C.

Thus, HDD and CDD in a given year are defined as follows:

$$\begin{cases} HDD = \sum_{i=1}^{n} rd(T_{b1} - T_i) & \text{(if DEH} - \text{DSH} > 90) \\ CDD = \sum_{i=1}^{n} rd(T_i - T_{b2}) \end{cases} \qquad (1)$$

where n is the number days in the year (365 or 366), T_i is daily mean temperature for day i, T_{b1} and T_{b2} are the reference temperatures for heating (18°C) and cooling (26°C), respectively, and rd is equal to 1 if T_i is lower than T_{b1} (reference temperature for heating) or higher than T_{b2} (reference temperature for cooling) and is equal to 0 otherwise.

Note that the HDD (CDD) calculated following these policies are substantially lower (higher) than for policies adopted in Europe (Spinoni et al. 2015) and the USA (Roltsch et al. 1999) (typically with a reference temperature of 18°C for heating and 22°C for cooling and right after the temperature passes this threshold) and will likely change in the future following further economic development in China. For example, currently people in southern China often do not heat their rooms, although temperature may drop to 0°C in winter, mainly because the government, following the policy described above, does not provide subsidies for heating (MCPRC 2003, Chen et al. 2007). Thus, to test the importance of the reference temperature, we also present a sensitivity analysis of the HDD and CDD calculation using a reference temperature of 18°C for heating and 22°C for cooling (hereafter referred to as HDD18 and CDD22, respectively).

2.3. Weighting by population

Energy demand in the household sector is related to both climate and population density. Thus, in order to assess the impact of warming on energy demand, the geographically varying values of HDD and CDD are weighted by the corresponding population in the same half-degree grid box. The population weighted HDD and CDD, referred to as HDDP and CDDP, should thus describe more realistically the regional distribution of energy demand (e.g. Quayle & Diaz 1980, Guttman 1983, Labriet et al. 2015). For present day and future population distribution, we use the dataset developed by the International Institute for Applied System Analysis (IIASA) (Riahi & Nakicenovic 2007; GGI Scenario Database v.2.0, available at www.iiasa.ac.at/Research/GGI/DB/). This dataset provides population density on a 0.5° × 0.5° (longitude-latitude) grid rather than at the country or regional level (e.g. Lutz et al. 2008, Chen & Liu 2009, UNFPA 2010).

The distribution of present-day mean population over China from the IIASA dataset is presented in Fig. 1a. The population is unevenly distributed due to the complex natural features and economic condi-

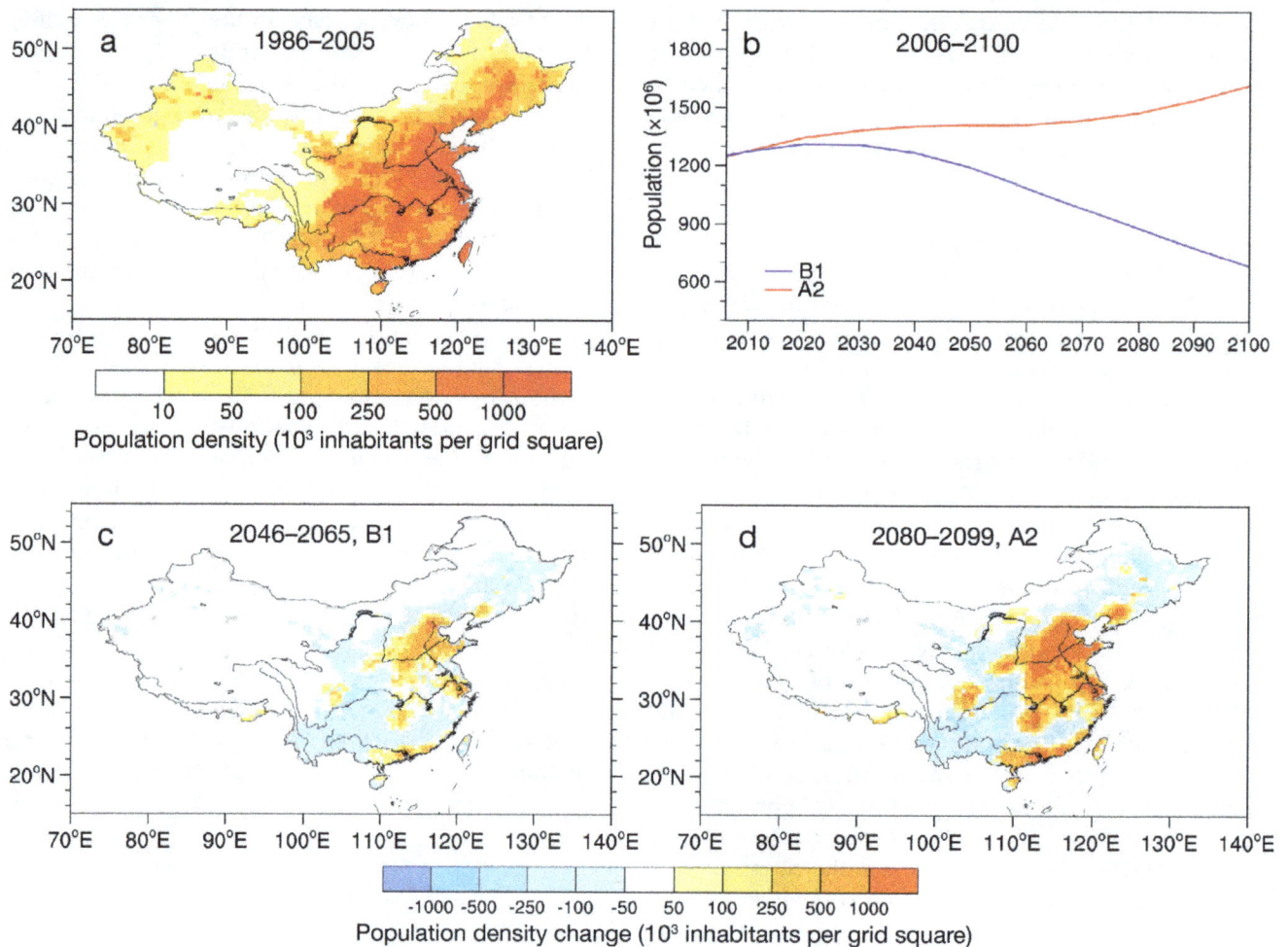

Fig. 1. Population density in China. (a) 'Present day' (1986–2005); (b) mean population in China for the period 2006–2100; (c) population change for the B1 mid 21st century (2046–2065); (d) population change for the A2 end of 21st century (2080–2099). Units are thousands per 0.5° × 0.5° grid in (a), (c) and (d). Data are from the International Institute for Applied Systems Analysis (Riahi & Nakicenovic 2007)

tions of the country, with higher density in the plains of southern and eastern China and lower density in the mountainous west, except the Sichuan Basin in the southwest.

Three scenarios of population growth are provided in the dataset: A2 for high, B2 for medium and B1 for low growth. We consider A2 for our high-end RCP8.5 scenario and B1 for the low-end RCP4.5. Future changes of total population in China are reported in Fig. 1b, showing continuous growth for A2 and growth followed by a decline in the B1. Population totals for China at the end of the 21st century in the A2 and B1 scenarios are 1.689 and 0.716 billion, respectively, slightly higher than the high-variant (1.554 billion) and low-variant (0.612 billion) of the latest United Nation population estimation (United

Nations 2015). The spatial distribution of the population changes in the mid 21st century under B1 and the end of the century under A2 are presented in Fig. 1c,d, respectively. The changes show similar patterns characterized by the shift of population towards the more developed regions of North China, the Yangtze River valley, Sichuan Basin and southern coastal areas.

The HDDP and CDDP in each of the 0.5° grid box are normalized by the total population of China and then added up to obtain a total for its 31 provinces. We assume that, on average, each person needs the same amount of energy based on the HDD or CDD regardless of the individual economic conditions and life styles, an assumption that, admittedly, in some cases could be unrealistic.

3. RESULTS

3.1. Validation of the climate model

Fig. 2 shows the HDD and CDD over China for the period 1986 to 2005 in the observations and simulations, along with the difference between the model simulation and observations (or model bias). RegCM4 generally reproduces the observed spatial distribution and magnitudes of both HDD and CDD over the region, with spatial correlation coefficients between simulated and observed data of 0.96 and 0.92, respectively (both statistically significant at the 95% confidence level). However, some significant biases are also evident. The model underestimates HDD by >500°D over parts of the Northeast and Northwest China (Fig. 2e) because of a warm model bias over high-latitude areas in the cold season (Zhang et al. 2008). HDD is somewhat overestimated over the Tibetan Plateau due to a cold bias there (Gao et al. 2013), but we note that the regions of maximum model error include only very sparse population (Fig. 1a).

Concerning CDD, the simulation captures the observed pattern of 0 values in the Tibetan Plateau and surrounding areas as well as the large values in eastern China and the basins of the Northwest (Fig. 2b,d). Over the most populated eastern China regions, the model shows a positive CDD bias (overestimation of up to 100°D) in the northern portions, extending from North China to the Huanghuai area. Conversely, the CDD is underestimated over southern China (Fig. 2f). In general, the model shows a poorer performance in simulating the high-end warm events compared to the cold events, as reported by the previous studies of Zhang & Shi (2012) and Ji & Kang (2015).

Annual cycles of the regional mean HDD and CDD from observations and simulation over China are presented in Fig. 2g. Because China is mainly located in the mid and high latitudes, much larger HDD values are found than CDD. The HDD values are >500°D during November to February, with a maximum in January, and drop to a few tens of °D in the summer months. The mean CDD over China is much smaller, with values around 10°D in June to August both in the observations and model simulation. The model captures the observed seasonal patterns and monthly values reasonably well, with overestimations of HDD in most months mainly due to the bias over the Tibetan Plateau and its surrounding areas (Fig. 2e).

Fig. 3 compares observed and simulated present-day interannual variability of HDD and CDD, as measured by the interannual standard deviation. Both for HDD and CDD, the model reproduces well the observed patterns of interannual variability, with pattern correlations of 0.92 and 0.90, respectively. Also, the simulated magnitudes are in line with observations, with HDD maxima over northeast China and the Tibetan Plateau and CDD maxima over eastern and northwestern China. The most significant differences between the model simulation and observations include negative (positive) biases over the northern regions (Tibetan Plateau) for HDD and positive (negative) biases in northern China (southern China) for CDD, respectively.

We also calculated the trend in HDD and CDD during the present-day 1986–2005 period. When averaged over the whole China territory, the observations show a trend of –106°D for HDD and 7°D for CDD, while the model simulates a mean trend of –52°D and 1°D, respectively. Therefore, the model captures the sign of the trends but underestimates the magnitude. The pattern correlations between the observed and simulated trends are 0.51 (statistically significant at the 95% confidence level) for HDD and 0.27 (not statistically significant) for CDD.

In summary, the validation analysis presented in this section shows that the model reproduces reasonably well the observed spatial pattern, annual cycles, and interannual variability of both HDD and CDD, although biases over some regions are present. The model also reproduces the spatial pattern of the trends in HDD but shows a poorer performance for the CDD trend.

3.2. Future changes of temperature

The spatial distribution of projected temperature changes in winter (December-January-February, DJF) and summer (June-July-August, JJA) for the RCP4.5 mid-21st century and RCP8.5 end of century are presented in Fig. 4. Annual changes of HDD and CDD are mostly affected by the temperature changes in DJF and JJA, respectively, and, as mentioned, the 2 time slices in Fig. 4 provide the lower and upper ends of the responses analyzed here. The temporal evolution of the regional mean temperature changes in the 21st century is also provided in the figure.

As shown in Fig. 4, substantial warming is found in both seasons and future periods. The warming is mostly in the range of 1.0 to 2.0°C in the mid-21st century under RCP4.5, except for a larger value of 2.0 to 2.5°C over the southern and eastern regions of the Tibetan Plateau in DJF (Fig. 4a,b). Regional mean changes of temperature in DJF and JJA averaged over China for the RCP4.5 mid-21st century are 1.5

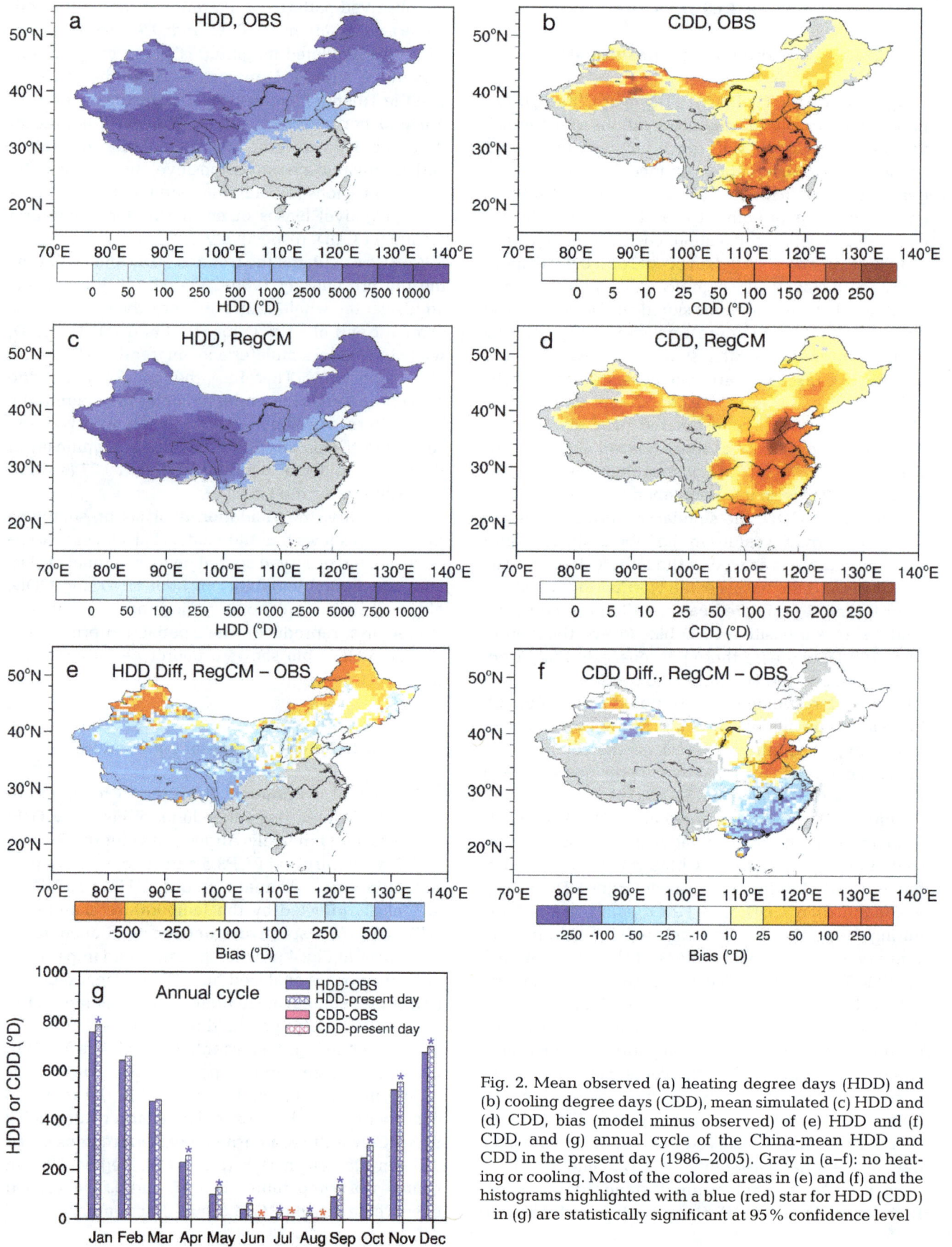

Fig. 2. Mean observed (a) heating degree days (HDD) and (b) cooling degree days (CDD), mean simulated (c) HDD and (d) CDD, bias (model minus observed) of (e) HDD and (f) CDD, and (g) annual cycle of the China-mean HDD and CDD in the present day (1986–2005). Gray in (a–f): no heating or cooling. Most of the colored areas in (e) and (f) and the histograms highlighted with a blue (red) star for HDD (CDD) in (g) are statistically significant at 95 % confidence level

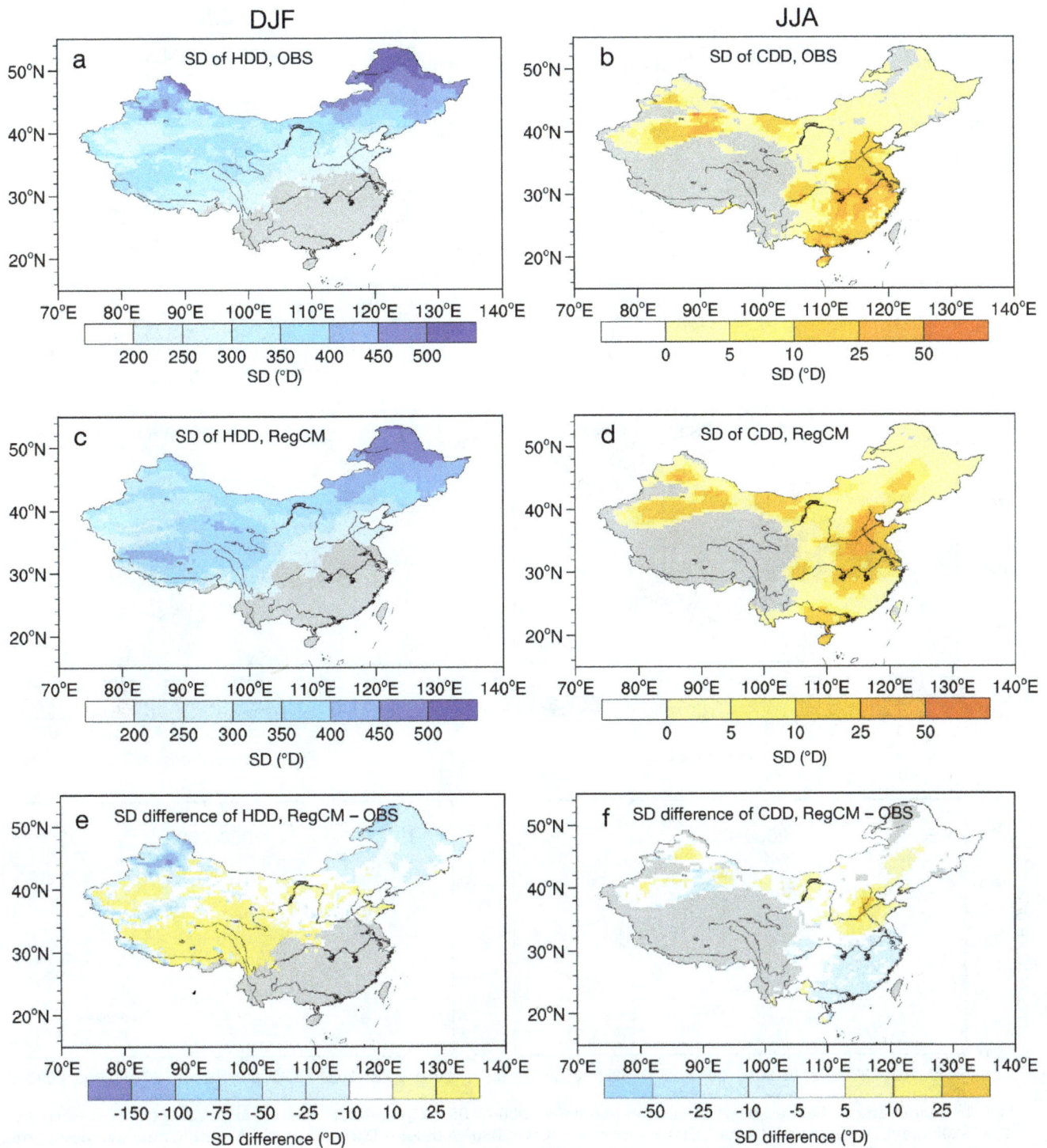

Fig. 3. Interannual variability as measured by the interannual standard deviation for the present day period (1986–2005). Observed (a) HDD and (b) CDD; simulated (c) HDD and (d) CDD; and simulated minus observed (e) HDD and (f) CDD. Gray in (a–f): no heating or cooling

and 1.4°C, respectively. The warming is much more pronounced in the RCP8.5 end of century scenario, ranging from 2.5 to 6°C (Fig. 4c,d). The largest warming (>5°C) is found in northeast China and the southern part of the Tibetan Plateau in DJF, along with the Hetao area in northern China in JJA. Greater warming is found in DJF than JJA, with China mean values of 4.2 and 3.8°C, respectively. In general, larger temperature increases in the high-latitude and high-elevation areas (Tibetan Plateau and the mountains) are

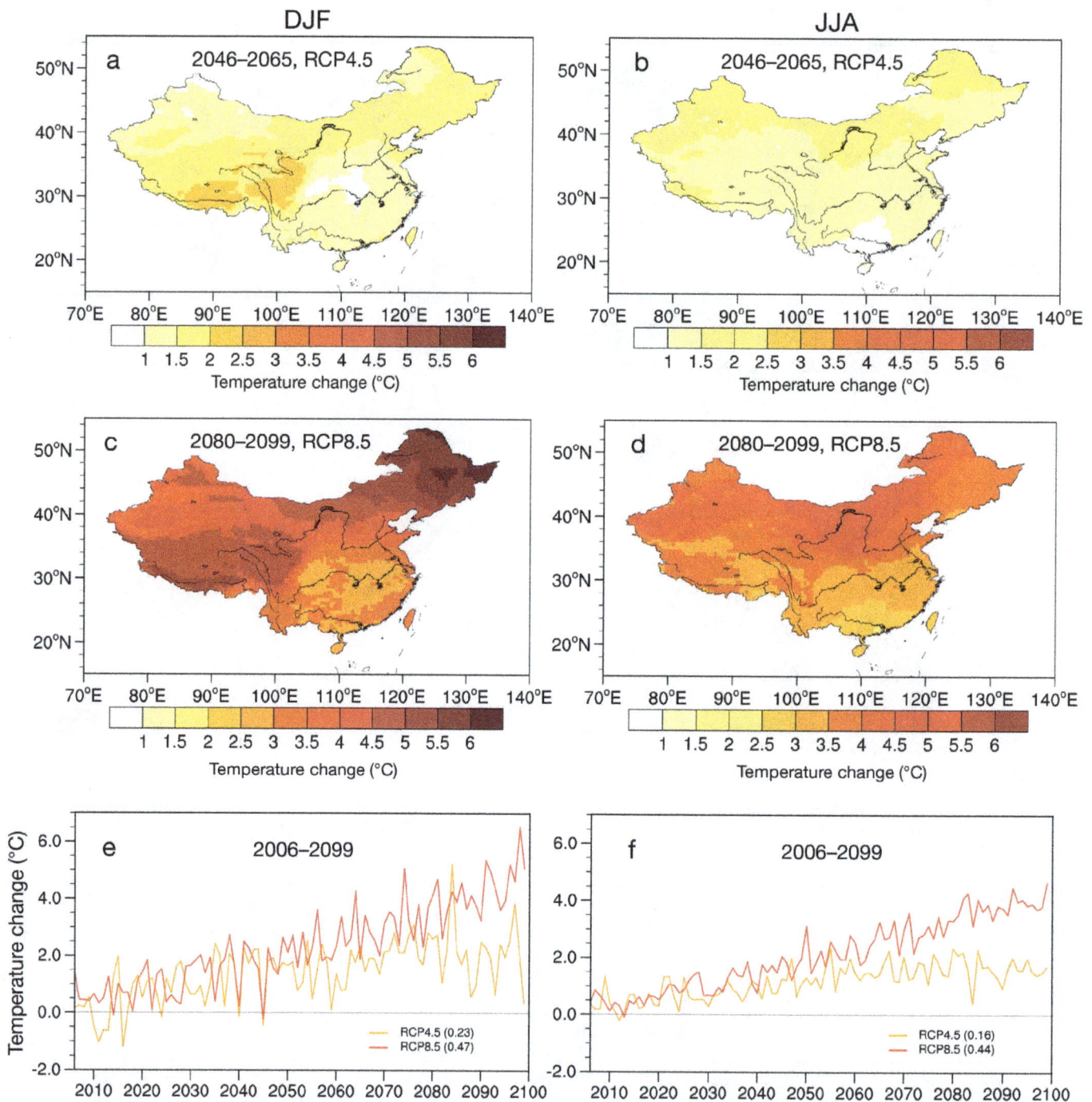

Fig. 4. Temperature changes over China compared to 1986–2005 (°C). (a) DJF and (b) JJA for RCP4.5 mid-century (2046–2065); (c) DJF and (d) JJA for RCP8.5, end of century (2080–2099); (e) DJF and (f) JJA for the China average during the 21st century. Parentheses: trends of changes (unit: °C 10 yr^{-1}). The changes are all significant at 95% confidence level

found in the annual mean changes (figures not shown; see Gao et al. 2013 for more detail).

The greater warming in DJF can also be observed when comparing Fig. 4e,f. The temperature increase under RCP8.5 is almost linear throughout the 21st century, with linear trends for DJF and JJA of 0.47

and 0.44°C decade^{-1}, respectively. A temperature stabilization is found under the RCP4.5 scenario in the second half of the century, with overall century trends of 0.23 (DJF) and 0.16°C decade^{-1} (JJA), respectively. All the changes in Fig. 4 are statistically significant at the 95% confidence level.

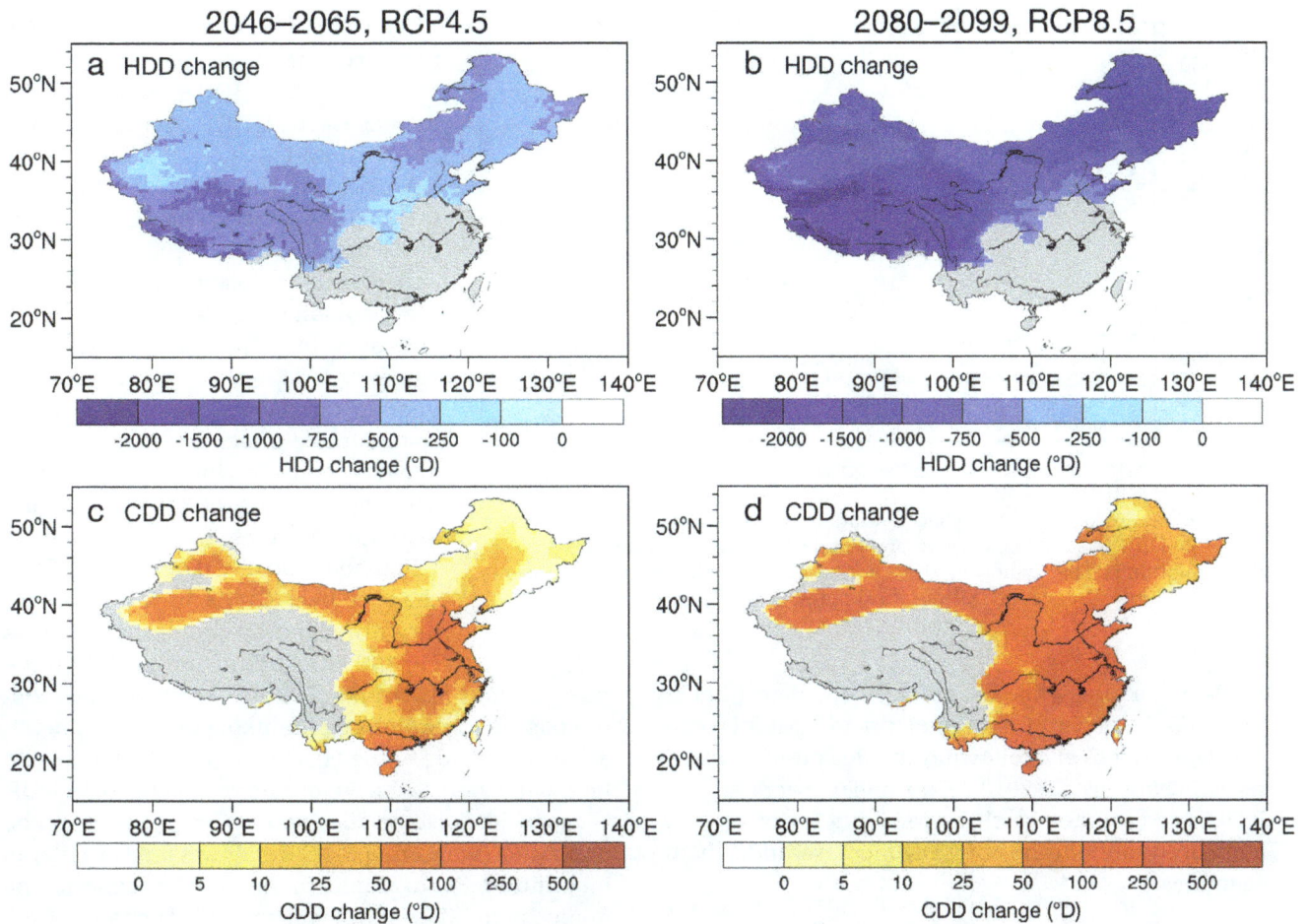

Fig. 5. Change of (a,b) HDD and (c,d) CDD compared to 1986–2005. (a) HDD and (c) CDD for RCP4.5 mid-century (2046–2065); (b) HDD and (d) CDD for RCP8.5 end of century. Gray: no heating or cooling. In all colored areas, the changes are significant at 95% confidence level, except when they are <5°D in (c)

3.3. Future changes of HDD and CDD

The spatial distributions of the projected changes in HDD and CDD for the RCP4.5 mid-21st century and RCP8.5 end of century are presented in Fig. 5. A predominant decrease of HDD is projected in the future, larger in the high-latitude and high-elevation areas with cold climate conditions, consistent with the maximum warming there. The decrease of HDD for the RCP8.5 late 21st century is clearly much larger than the mid-century RCP4.5, e.g. in the range of 500 to 750 (mid-century RCP4.5) and 1000 to 1500°D (end of century RCP8.5) over most of the Northeast. The largest decreases in HDD, with maxima of up to −1000 and −2000°D, are found over the Tibetan Plateau.

An increase of CDD throughout the 21st century is found over most of China except over the cold or cool regions of the Tibetan Plateau and some other mountainous regions (Tianshan Mountain in the northwest and Changbai Mountain in the northeast) (Fig. 5c,d).

In eastern China, CDD increases mostly in the range of 100–250 and 250–500°D for the 2 time slices are found in the area extending from northern China to the southern extent of the Yangtze River Basin and in Hua'nan. Larger increases are mainly found in the basins of western China, most notably in the heavily populated Sichuan Basin.

The temporal evolution of mean HDD and CDD changes over China under the RCP4.5 and RCP8.5 scenarios are presented by the solid lines in Fig. 6. In general, the changes in HDD and CDD show small differences between RCP4.5 and RCP8.5 before the 2040s, in agreement with the low scenario dependence of climate change in the early 21st century (IPCC 2013). The changes are more pronounced under the RCP8.5 compared to RCP4.5 in latter half of the century, and the decrease in HDD is much larger than that of CDD. In the latter part of the century, the decrease and increase in HDD and CDD, respectively, stabilize at ~450°D and 30°D in the

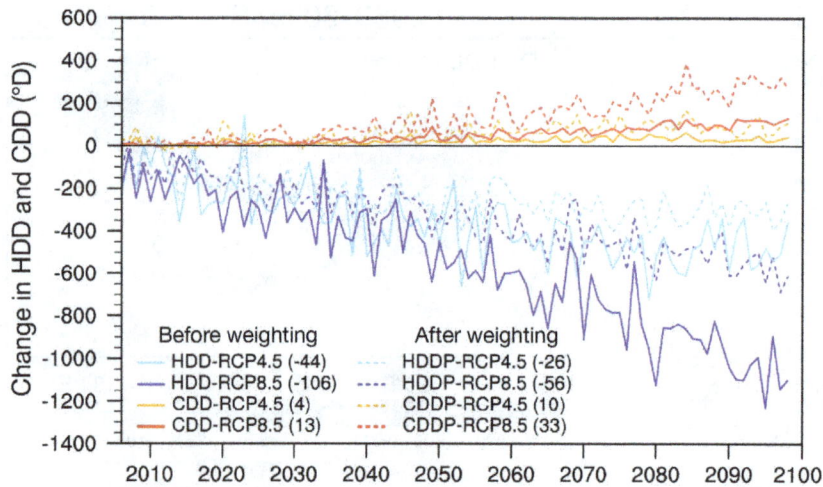

Fig. 6. Changes of HDD and CDD averaged over China in the 21st century (2006–2099). The solid lines are before and dashed lines are after the population weighting. The trends of the changes are presented in parentheses (unit:°D decade^{-1}; all significant at 95 % confidence level)

RCP4.5 scenario, while the changes are almost linear in RCP8.5, reaching values of up to −1100°D and 100°D, respectively, following the design of the scenarios (Moss et al. 2010). The linear trends of the HDD decrease are much larger (by a factor of ~10) compared to the trends of CDD increase under both scenarios.

3.4. Changes of population weighted HDD and CDD

We now turn our attention to the effect of population weighting on the CDD and HDD projections. Spatial distribution of the HDDP and CDDP changes for the mid-century RCP4.5 and late century RCP8.5 are presented in Fig. 7. Large differences can be found when comparing Fig. 7 to Fig. 5. In general, the changes of HDDP and CDDP are more pronounced in the regions with higher population density (Fig. 1). For HDDP, the most noticeable difference compared to HDD is the increase in northern China despite the warming there. This increase is due to the large future population increase projected over the region (Fig. 1c,d), which overwhelms the effect of climate change. Northern China is also the region with largest CDDP increase in both mid-century RCP4.5 and late century RCP8.5. While the changes for HDD over the Tibetan Plateau and CDD over the northwest are large, much lower changes in HDDP and CDDP can be found over these regions characterized by very low population density.

When averaged over the different provinces, in both scenarios, large decreases in HDDP and smaller increases in CDDP dominate over northern and western China (Fig. 8); however, increases in HDDP are found in the northern provinces (cities) of Beijing, Tianjin, Hebei, Shandong and Henan in eastern China, due to the population increases there. The increases in CDDP over these provinces are even more pronounced than in southern China due to the compounded effect of summer warming and population increase, indicating a strong increase of energy demand over these regions.

Regional mean HDDP and CDDP changes over China in the 21st century are given by the dashed lines in Fig. 6. The trends in general follow the un-weighted HDD and CDD (solid lines) ones, but the population weighting dramatically affects the relative magnitude of the HDDP and CDDP changes. In fact, while before the weighting the decrease in HDD is much larger than the increase in CDD, after the population weighting the 2 trends are more similar in magnitude. For example, in the RCP8.5 scenario, the decrease in HDD and increase in CDD change from 1000 and 100°D, respectively, in the un-weighted case to around 500 and 300°D after the weighting. This is because the areas undergoing the increase in CDD, mostly warm areas in the plains, are expected to undergo much more pronounced population growth than those exhibiting a decrease in HDD (cold and mountainous areas).

Changes in HDDP and CDDP for the mid- and late 21st century over the whole of China, expressed in degree days and percent of present day values, are provided in Table 1. The changes in HDDP are 3-fold higher than the changes in CDDP in the mid-century under the 2 scenarios and in the late century under RCP4.5, but only 2-fold higher in the late century RCP8.5. By the end of the century, HDDP decreases by about 23 and 39%, while CDDP increases by factors greater than 1.3 and 3.7 under RCP4.5 and RCP8.5, respectively. Therefore, mitigation from the higher emission scenario RCP8.5 to the lower RCP4.5 shows more effect on CDDP (thus the cooling energy demand) than HDDP, mostly as a result of the changes in population densities. It is interesting to note that the sum of HDDP and CDDP shows only relatively small differences between the 2 scenarios

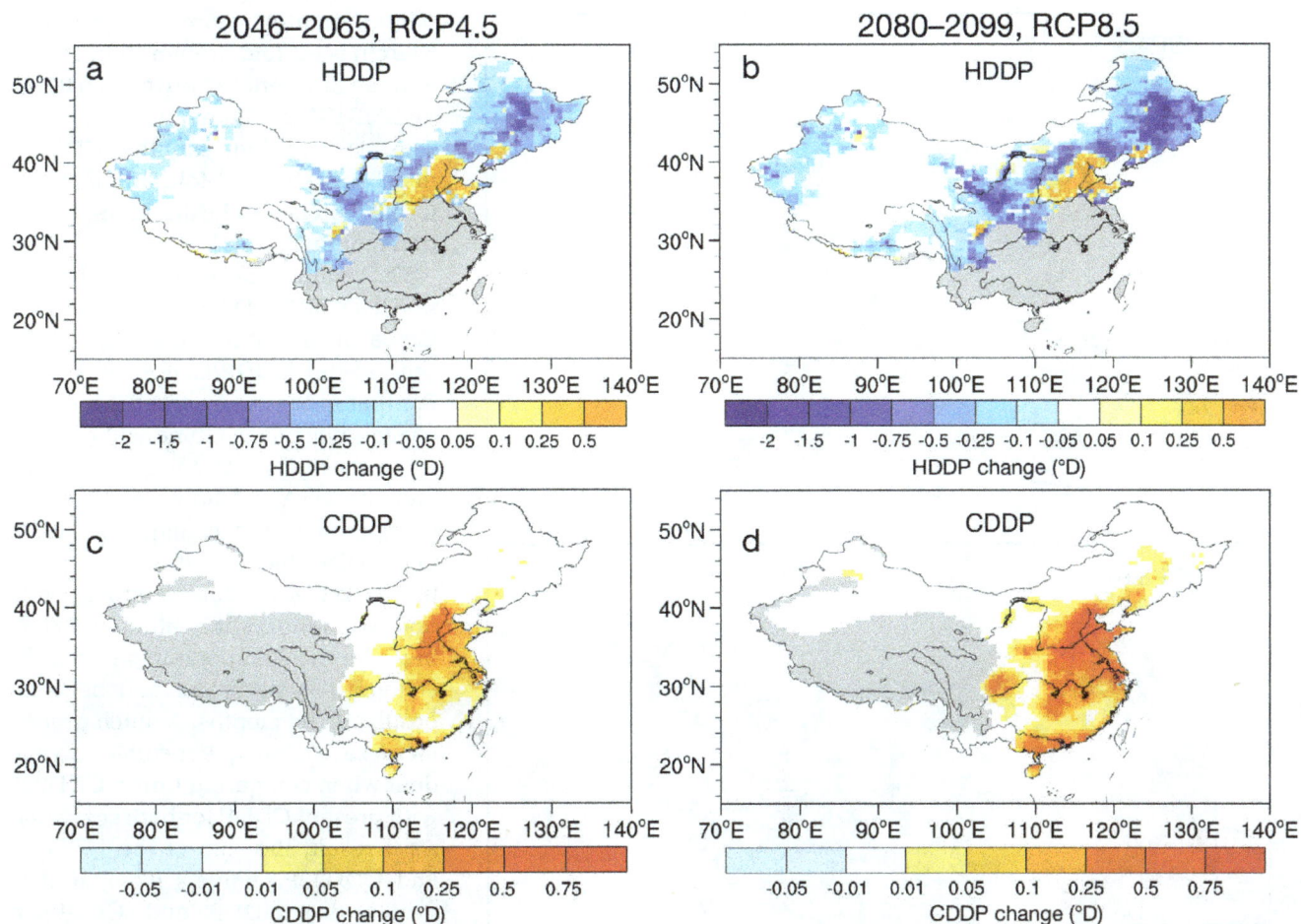

Fig. 7. Same as Fig. 5, but for population weighted HDD and CDD (HDDP and CDDP, respectively). In all colored areas, the changes are significant at the 95% confidence level

and across time slices, with values ranging from −189 to −265°D (or about −15%). The negative value of this sum implies a general reduction in future energy demand in the household sector.

The annual cycle of the changes in HDDP and CDDP averaged over China for mid-century RCP4.5 and late century RCP8.5 as well as their corresponding values in the present day and the 2 future time slices are shown in Fig. 9. A general decrease in HDDP in the cold half of the year and an increase in CDDP in the warm half is found (Fig. 9a). In addition, the decrease in HDDP in winter and the increase in CDDP in summer for the late century RCP8.5 are 2- to 3-fold and 3- to 4-fold larger than for the mid-century RCP4.5, respectively. Compared to the present day (Fig. 9b), the peak of HDDP in winter becomes less pronounced in the future, while the peak of CDDP in summer is amplified, with this shift being more pronounced in the RCP8.5 than the RCP4.5.

3.5. Sensitivity of HDD and CDD to the reference temperature

In this section, we briefly discuss the sensitivity of the calculations for HDD and CDD to the values of reference temperatures. In particular, we test values in line with those presently used in economically advanced countries (USA and Europe), i.e. 18°C for heating and 22°C for cooling (referred to as HDD18 and CDD22, respectively).

The spatial distributions of the present-day HDD18 and CDD22 are similar to those for HDD and CDD but with much larger values, particularly for CDD22 (figures not shown for brevity). Population weighted China-wide annual mean HDD18P and CDD22P are 2562 and 321°D, respectively, which are about 2- and 4.5-fold higher than HDDP and CDDP using the current reference temperatures (1366 and 71°D, respectively). Differences between HDD18P and HDDP can be found in all months (Fig. 10a). HDD18P in the

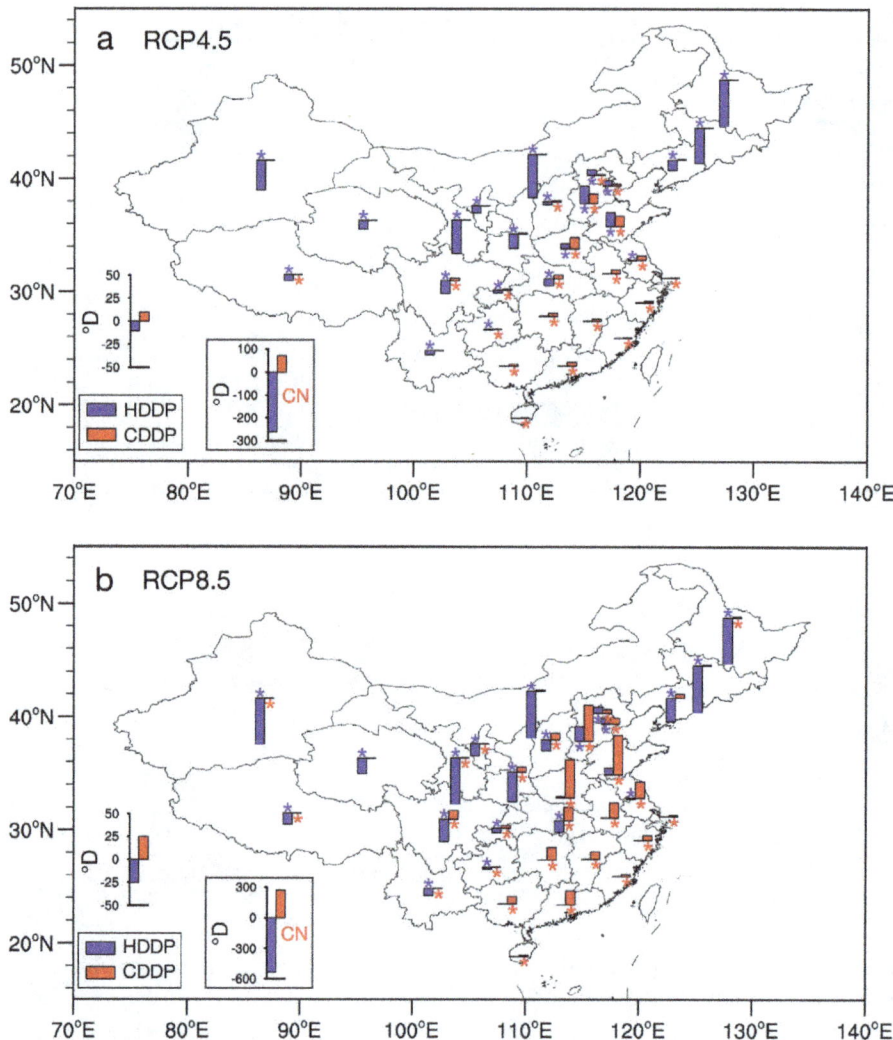

Fig. 8. Changes in HDDP and CDDP in different provinces (scale at left of panel) and the whole country (marked with CN) for (a) RCP4.5 mid-century (2046–2065) and (b) RCP8.5 end of century (2080–2099). Asterisks indicate significance at the 95 % confidence level

Table 1. Changes in the annual mean HDDP, HDD18P and CDDP, CDD22P in the mid (2046–2065) and end (2080– 2099) of the 21st century under RCP4.5 and RCP8.5 relative to the present day (1986–2005) HDDP and CDDP over China

	Mid-21st century (°D/%)		End of 21st century (°D/%)	
	RCP4.5	RCP8.5	RCP4.5	RCP8.5
HDDP	−261/−19	−370/−27	−311/−23	−535/−39
CDDP	72/101	128/179	95/133	270/379
HDDP+CDDP	−189/−13	−242/−17	−216/−15	−265/−19
HDD18P	749/55	590/43	672/49	312/23
CDD22P	422/594	524/738	449/632	745/1049
HDD18P+CDD22P	1171/81	1114/76	1121/78	1077/75

winter months (from November to March) is ~2-fold greater than HDDP, and the differences between HDD18P and HDDP can be even larger in the seasonal transition months. CDD22P in JJA is 4- to 8-fold higher than CDDP and is also larger in the transition months.

The annual cycle of the changes in HDD18P/CDD22P averaged over China in the future compared to the present-day HDDP/CDDP, i.e. assuming that in the future the heating and cooling policies in China will be similar to those currently used in Europe and the USA, is shown in Fig. 10b. Compared to the changes of HDDP and CDDP (Fig. 9a), the most important effect of the future implementation of the different heating and cooling reference temperatures is that the change in heating demand becomes positive in all months. A much greater increase in cooling demand is also evident when comparing future CDD22P with present CDDP for both scenarios. As a result, the sum of the HDD18P and CDD22P changes (compared to present-day HDDP and CDDP) is positive and in fact rather large and with relatively small variations across scenarios and time slices: 1077 to 1121°D (or ~80 %) (Table 1). This illustrative example thus shows how future changes in heating and cooling policies can have dominant effects on the overall energy demand in China.

4. CONCLUSIONS AND DISCUSSION

In this paper, future changes in HDD/ HDDP and CDD/CDDP in the 21st century over mainland China under different GHG forcing scenarios are investigated using the regional climate model RegCM4 driven by the BCC_CSM1.1 global climate model. The climate indicators are further weighted by present day and projected population densities in order to provide a first-order assessment of potential energy demand. Our main conclusions and considerations can be summarized as follows:

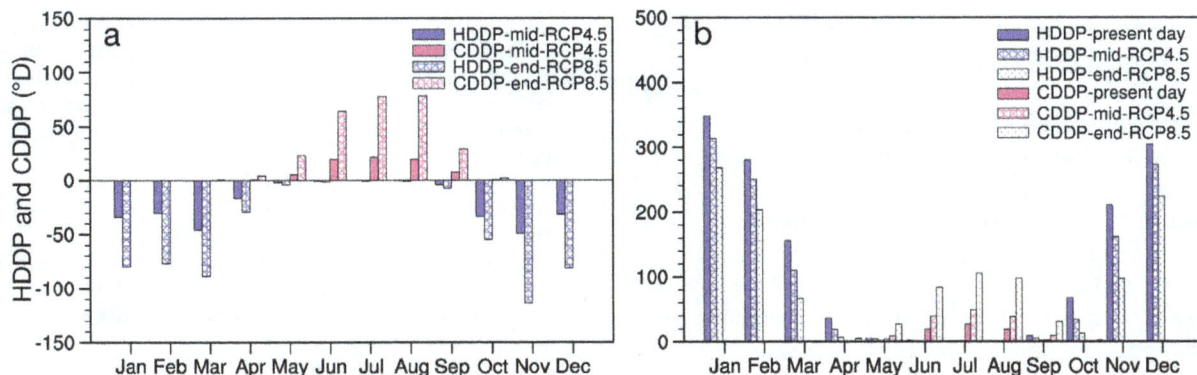

Fig. 9. Annual cycle of (a) China-mean changes of HDDP and CDDP in the mid (2046–2065) and end (2080–2099) of the 21st century and (b) HDDP and CDDP for the present day (1986–2005), RCP4.5 mid-21st century and RCP8.5 end of 21st century. The changes are all significant at the 95 % confidence level

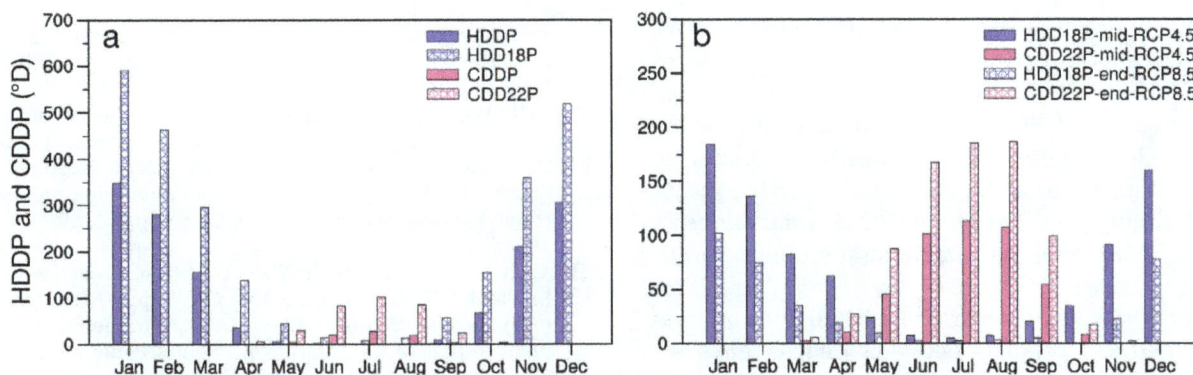

Fig. 10. Annual cycle of (a) China-mean HDDP, HDD18P and CDDP, CDD22P (see text) for the present day period (1986–2005) and (b) future changes of HDD18P and CDD22P compared to present day HDDP and CDDP, respectively, for the RCP4.5 mid-century (2046–2065) and RCP8.5 end of century (2080–2099) periods. The changes are all significant at the 95 % confidence level

(1) The model can reproduce well the present-day mean climatology, seasonal cycle, and interannual variability of HDD and CDD over China, along with the trend of HDD during the present day period. A lower performance is found for the present-day trend of CDD over the region. More generally, a better model performance is found for HDD compared to CDD.

(2) A substantial decrease in HDD and increase in CDD is found in the future in response to the GHG-induced warming, with more pronounced changes under the RCP8.5 than the RCP4.5 scenario, in particular during the second half of the 21st century. When averaged over the whole of China, the decrease in HDD is larger than the increase in CDD; however, there is a distinct regional variability in these trends. For example, over eastern China, the change in HDD shows a latitudinal distribution with a larger decrease in the north. The increase in CDD is more pronounced in the eastern plain areas and the southern coasts.

(3) Population weighting substantially affects the changes in HDD and CDD, even reversing its sign

over some regions of northern China, where both HDDP and CDDP increase. This illustrates the importance of adding a metric of exposure (in this case population density) to estimates of climate change impacts on energy demand.

(4) Assuming the same energy demand for changes in HDDP and CDDP, the future total energy consumption in China in the household sector shows a decrease of about 15 %. However, while most of the cold part of the country benefits from a lower energy use for heating, the southern part experiences an increase in energy demand for cooling. This will lead to the need for additional transport of coal, oil, and electricity, etc., from the north to the south. Energy consumption will also show changes in seasonal distribution, with greater demand in the summer and lower in the winter.

(5) For our calculations of HDDP and CDDP, we used the standards that are currently used in China, with heating restricted to northern China and starting only when the temperature drops to 5°C, and

cooling starting when temperatures exceed 26°C. With economic growth, these reference temperatures will most likely move closer to the values currently used in economically advanced countries, 18°C and 22°C for heating and cooling, respectively. Sensitivity of the HDD18P and CDD22P calculations to the use of these values shows that potentially large increases in energy demand (~80%) are expected in the future in China under this scenario of change in reference temperature, illustrating the dominant role that changes in policy can have regarding energy demand.

Our results have important implications for energy management policies in the country. Clearly, the issue of future energy demand due to changes in HDD and CDD is complex and multifaceted, with a strong dependence on the spatial variability of climate change and population dynamics. Due to the limited scope of this paper, we only focused on a small number of RCM simulations available with 1 model. More studies are thus needed to better address the uncertainties in the climate scenario projections through ensembles of RCM simulations. In addition, we adopted simple assumptions for the relation between degree days, population and energy demand. However, there is the need to measure the direct relation between degree days and energy consumption (Isaac & van Vuuren 2009) and to better account for future changes in economic conditions and population amount and distribution (Isaac & van Vuuren 2009). Future work will thus go in the direction of providing more comprehensive assessments of energy demand within a multi-model framework (e.g. CORDEX; Giorgi et al. 2009) incorporating more detailed socio-economic components.

Acknowledgements. This research was jointly supported by R&D Special Fund for Public Welfare Industry (meteorology) (GYHY201306019) and National Natural Science Foundation of China (41375104).

LITERATURE CITED

CCSNARCC (Committee for China's Second National Assessment Report on Climate Change) (2011) The Second National Assessment Report on Climate Change. Science Press, Beijing

Chen W, Liu JJ (2009) Future population trends in China: 2005-2050. Monash University, Centre of Policy Studies and the Impact Project, Melbourne

ä Chen L, Fang XQ, Li S (2007) Impacts of climate warming on heating energy consumption and southern boundaries of severe cold and cold regions in China. Chin Sci Bull 52: 2854-2858

► Christenson M, Manz H, Gyalistras D (2006) Climate warm-

ing impact on degree-days and building energy demand in Switzerland. Energy Convers Manag 47:671-686

► Gao XJ, Zhao ZC, Ding YH, Huang RH, Giorgi F (2001) Climate change due to greenhouse effects in China as simulated by a regional climate model. Adv Atmos Sci 18: 1224-1230

► Gao XJ, Shi Y, Giorgi F (2011) A high resolution simulation of climate change over China. Sci China Earth Sci 54: 462-472

► Gao XJ, Shi Y, Zhang DF, Wu J, Giorgi F, Ji ZM, Wang YG (2012) Uncertainties of monsoon precipitation projection over China: results from two high resolution RCM simulations. Clim Res 52:213-226

Gao XJ, Wang ML, Giorgi F (2013) Climate change over China in the 21st century as simulated by BCC_CSM1.1-RegCM4.0. Atmos Ocean Sci Lett 6:381-386

Giorgi F, Jones C, Asrar G (2009) Addressing climate information needs at the regional level: the CORDEX framework. WMO Bull 58:175-183

► Giorgi F, Coppola E, Solmon F, Mariotti L, Sylla BM, Bi XQ (2012) RegCM4: model description and illustrative basic performance over selected CORDEX domains. Clim Res 52:7-29

► Guo DL, Wang HJ (2013) Simulation of permafrost and seasonally frozen ground conditions on the Tibetan Plateau, 1981-2010. J Geophys Res Atmos 118:5216-5230

► Guttman NB (1983) Variability of population-weighted seasonal heating degree days. J Clim Appl Meteorol 22: 495-501

IPCC (2013) Climate Change 2013: the physical science basis. In: Stocker TF, Qin DH, Plattner GK, Tignor M and others (eds) Contribution of Working Group I to the 5th Assessment Report of the Intergovernmental Panel on Climate Change. Cambridge University Press, Cambridge

► Isaac M, van Vuuren DP (2009) Modelling global residential sector energy demand for heating and air conditioning in the context of climate change. Energy Policy 37:507-521

► Ji ZM, Kang SC (2015) Evaluation of extreme climate events using a regional climate model for China. Int J Climatol 35:888-902

► Ju LX, Wang HJ, Jiang DB (2007) Simulation of the Last Glacial Maximum climate over East Asia with a regional climate model nested in a general circulation model. Palaeogeogr Palaeoclimatol Palaeoecol 248:376-390

Kadioğlu M, Şen Z (1999) Degree-day formulations and application in Turkey. J Appl Meteorol 38:837-846

► Labriet M, Joshi SR, Vielle M, Holden PB and others (2015) Worldwide impacts of climate change on energy for heating and cooling. Mitig Adapt Strategies Glob Change 20:1111-1136

► Li T, Zhou GQ (2010) Preliminary results of a regional air-sea coupled model over East Asia. Chin Sci Bull 55: 2295-2305

Li XC, Bai ML, Yang J, Yu FM, Di RQ, Ma YF (2010) Impacts of climate warming on energy consumed in heating period in Hohhot. Adv Clim Chang Res 6:29-34 (in Chinese)

► Lutz W, Sanderson W, Scherbov S (2008) The coming acceleration of global population ageing. Nature 451:716-719

MCPRC (Ministry of Construction of the People's Republic of China) (2003) Code for design of heating ventilation and air conditioning (GB50019-2003), 1893-1951. MCPRC, Beijing

► Moss RH, Edmonds JA, Hibbard KA, Manning MR and others (2010) The next generation of scenarios for climate change research and assessment. Nature 463:747–756

▶ New M, Lister D, Hulme M, Makin I (2002) A high-resolution data set of surface climate over global land areas. Clim Res 21:1–25

▶ Quayle RG, Diaz HF (1980) Heating degree day data applied to residential heating energy consumption. J Appl Meteorol 19:241–246

▶ Rehman S, Al-Hadhrami LM, Khan S (2011) Annual and seasonal trends of cooling, heating, and industrial degree-days in coastal regions of Saudi Arabia. Theor Appl Climatol 104:479–488

Ren YY, Ren GY, Qian HS (2009) Change scenarios of China's provincial climate-sensitive components of energy consumption. Geogr Res 28:36–44

Riahi K, Nakicenovic N (eds) (2007) Greenhouse gases—integrated assessment. Technol Forecasting Soc Chang Spec Issue 74:873–1108

▶ Roltsch WJ, Zalom FG, Strawn AJ, Strand JF, Pitcairn MJ (1999) Evaluation of several degree-day estimation methods in California climates. Int J Biometeorol 42:169–176

▶ Spinoni J, Vogt J, Barbosa P (2015) European degree-day climatologies and trends for the period 1951–2011. Int J Climatol 35:25–36

SSB (State Statistical Bureau) (2013) China Energy Statistical Yearbook 2013. China Statistics Press, Beijing

▶ Sui Y, Lang XM, Jiang DB (2014) Time of emergence of climate signals over China under the RCP4.5 scenario. Clim Change 125:265–276

Tan BG, Tian Z, Liu KX, Li MC, Guo J, Xiong MM (2012) Annual variation characteristics of degree-days and the trend analysis in Tianjin. Renew Energy Resour 30:102–105

UNFPA (United Nations Fund for Population Activities) (2010) State of world population. UNFPA, New York, NY

United Nations (2015) World population prospects: the 2015 revision, key findings and advance tables. UN Department of Economic and Social Affairs, Population Division, Working Paper No. ESA/P/WP.241, New York, NY

▶ Wang XM, Chen D, Ren ZG (2010) Assessment of climate change impact on residential building heating and cooling energy requirement in Australia. Build Environ 45:1663–1682

Wu J, Gao XJ (2013) A gridded daily observation dataset over China region and comparison with the other datasets. Chin J Geophys 56:1102–1111

▶ Wu TW, Yu RC, Zhang F, Wang ZZ and others (2010) The Beijing Climate Center atmospheric general circulation model: description and its performance for the present-day climate. Clim Dyn 34:123–147

▶ Wu J, Gao XJ, Giorgi F, Chen ZH, Yu DF (2012) Climate effects of the Three Gorges Reservoir as simulated by a high resolution double nested regional climate model. Quat Int 282:27–36

Xin XG, Wu TW, Li JL, Wang ZZ, Li WP, Wu FH (2013) How well does BCC_CSM1.1 reproduce the 20th century climate change over China? Atmos Oceanic Sci Lett 6:21–26

▶ Xu Y, Gao XJ, Shen Y, Xu CH, Shi Y, Giorgi F (2009) A daily temperature dataset over China and its application in validating a RCM simulation. Adv Atmos Sci 26:763–772

▶ You QL, Fraedrich K, Sielmann F, Min JZ and others (2014) Present and projected degree days in China from observation, reanalysis and simulations. Clim Dyn 43:1449–1462

▶ Yu ET, Wang HJ, Sun JQ (2010) A quick report on a dynamical downscaling simulation over china using the nested model. Atmos Ocean Sci Lett 3:325–329

▶ Zhai PM, Pan X (2003) Trends in temperature extremes during 1951–1999 in China. Geophys Res Lett 30:1913–1916

▶ Zhang DF, Shi Y (2012) Numerical simulation of climate changes over North China by the model RegCM3. Chin J Geophys 55:474–487

Zhang DF, Gao XJ, Ouyang LC (2008) Simulation of present climate over China by a regional climate model. J Trop Meteorol 14:19–23

▶ Zhou TJ, Yu RC (2006) Twentieth century surface air temperature over China and the globe simulated by coupled climate models. J Clim 19:5843–5858

10

Frozen ground temperature trends associated with climate change in the Tibetan Plateau Three River Source Region from 1980 to 2014

Siqiong Luo[1,*], Xuewei Fang[1,2], Shihua Lyu[1,3], Di Ma[1], Yan Chang[1], Minghong Song[1], Hao Chen[1]

[1]Key Laboratory of Land Surface Process and Climate Change in Cold and Arid Regions,
Cold and Arid Regions Environmental and Engineering Research Institute, Chinese Academy of Sciences, Lanzhou 730000,
PR China [2]University of the Chinese Academy of Sciences, Beijing 100049, PR China
[3]Chengdu University of Information Technology, Chengdu 610225, PR China

ABSTRACT: Long-term (1981–2014) trends in soil temperatures at depths ranging from 0–320 cm were used to examine relationships between regional climate change and soil temperatures in the Three River Source Region (TRSR) of the Tibetan Plateau (TP). Air temperature, precipitation, maximum depth of frozen ground and maximum snow depth were also analysed for trends and for correlations with soil temperatures. A significant warming trend was evident in the TRSR during the 35 yr analysed. Significant soil warming was detected, at rates of 0.706°C decade^{-1} for surface soils (0 cm), 0.477°C decade^{-1} for shallow layer soils (5–20 cm) and 0.417°C decade^{-1} for deep layer soils (40–320 cm). There was a clear effect of air temperature on soil temperature, as seen from the concurrent significant increases in air and soil temperature trends and the strong correlation between them. The relationship between precipitation and soil temperatures was complicated. Higher levels of precipitation on the ground also resulted in higher soil temperatures during the summer in the frozen soil, while the process of freezing and thawing had an inhibitory effect on the soil temperature increase. The warming trends in soil temperature are expected to continue with the degradation of frozen soil in the TRSR.

KEY WORDS: Soil temperature trends · Frozen ground · Air temperature · Three River Source Region · Tibetan Plateau

1. INTRODUCTION

The important modulating role played by frozen ground on the land surface is strongly affected by climate. Observations and simulations have indicated that the frozen ground conditions of the earth are currently experiencing rapid changes in response to global warming associated with climate change — particularly in the Tibetan Plateau (TP), a region that encompasses extensive areas of permafrost and seasonal frozen ground. Due to its unique geographical location and high altitude, the frozen ground on the TP differs from that of other high latitude regions.

The frozen ground of TP is relatively warm and thin compared with high latitude frozen ground in both North America and Russia, and thus is more sensitive to climate changes and surface conditions (Cheng 1998). Therefore, more reliable evidence and rapid changes can be anticipated in these regions in response to climate change (Cheng & Jin 2013). Meteorological observatories have demonstrated that the TP has experienced statistically significant warming since the mid-1950s, especially during the winter (Liu & Chen 2000). Areas of permafrost and seasonally frozen ground have decreased (Lawrence et al. 2012, Ran et al. 2012, Cheng & Jin 2013, Guo &

*Corresponding author: lsq@lzb.ac.cn

Wang 2013), the active layer has become much thicker, and the landscape over the TP has changed (Xue et al. 2009, Wu et al. 2012). From model simulations in the Coupled Model Intercomparison Project Phase 5 (CMIP5), the TP appears more sensitive and susceptible to climate change than other Chinese areas in future emission scenarios (You et al. 2014). A climate model has also demonstrated that air temperatures will rise, leading to a reduction in ice and frost over the TP in the next 35 yr (Zhu et al. 2013). This deterioration of frozen ground may influence the energy and hydrological cycles of the TP, ultimately leading to further climate change (Chen et al. 2014).

A noteworthy feature of the TP frozen ground is that its temperature has continually risen over the past several decades. Based on monitoring data from 27 sites collected from 2006–2010, the mean rate of temperature increase of the permafrost at a depth of 6.0 m was 0.2°C decade^{-1} (Wu et al. 2012). According to remote sensing data, surface soil temperatures have increased at an average rate of 0.6°C decade^{-1} (Xue et al. 2009), and from borehole investigations, the temperature of the supra permafrost water has increased by 0.5–0.7°C between 1980 and 2005 in the Yellow River source region (Jin et al. 2009). Increases in soil temperature as a result of a warmer climate will have profound effects on surface and subsurface hydrologic conditions, thawing permafrost in high latitudes and at high altitudes, accelerating the decomposition of organic carbon in the soil, increasing the release of CO_2 from the soil to the atmosphere and altering terrestrial ecosystems, thereby creating a positive feedback loop leading to further climate change (Nelson 2003, Davidson & Janssens 2006, Kurylyk et al. 2014). Therefore, analysing the variability and trends in long-term soil temperatures at various depths may contribute to our understanding of the consequences of a warmer climate on surface energy processes and regional environmental and climatic conditions (Hu & Feng 2003, Qian et al. 2011, Alamusa et al. 2014, Bai et al. 2014, Yeşilırmak 2014).

The Three River Source Region (TRSR), located in the northeastern TP, is an area in which China's 2 longest rivers (the Yangtze and Yellow) and a transnational river (the Mekong River; called Lancang in China) originate. This region has been given the name 'Chinese water tower', and is a mosaic transition zone of seasonal frozen ground and areas of discontinuous and continuous permafrost. In recent years, the trends in many climate parameters, including temperature, rainfall, evaporation, relative humidity and wind speed have been studied in the

TRSR because this region is particularly sensitive to the impacts of climate change (You et al. 2008, 2014, 2015, Yi et al. 2011, Guo & Wang 2012, Xu et al. 2012, Liang et al. 2013). Meteorological data has demonstrated that air temperatures in the TRSR have been rising at an average rate of 0.32°C decade^{-1} for the past half century (Yi et al. 2011, Liang et al. 2013). The annual number of warm days and nights increased from 1961–2005 (You et al. 2008), and these warming trends have caused an increase in the number of permafrost thawing days and a thickening of the permafrost active layer (Xue et al. 2009). Data from boreholes indicates that the lower limit of the permafrost has risen by 50–80 m, and that the average maximum depth of frost penetration has decreased by 0.1–0.2 m in the Yellow River source region since 1980 (Jin et al. 2009). Some permafrost has already disappeared, which has resulted in the disappearance of, or a lowering of the level of groundwater in the seasonal thawing layer. According to data collected at the Maduo weather station, seasonal frozen ground depths have decreased by 0.2 m since the 1980s (Jin et al. 2009).

Taken together, the research mentioned above has enhanced our understanding of the local frozen ground response to climate change in the TRSR. However, there are limitations to these studies, in that they focus only on local effects and use short-term datasets. In order to achieve a more complete understanding, it is necessary to examine temporal and spatial changes in frozen ground relative to present conditions in the TRSR. In this paper, we analysed trends in the seasonal and annual means of soil temperatures from 1980–2014, using data collected at 9 meteorological stations across the TRSR at depths of 0, 5, 10, 20, 40, 80, 160 and 320 cm, and examined their relationships to air temperature, precipitation, maximum depth of frozen ground and maximum snow depth.

2. STUDY REGION AND DATA

The TRSR, located in the interior of the TP, China, contains the headwaters of the Yellow, Yangtze and Mekong Rivers (Fig. 1). It covers an area of 3.258×10^5 km^2 with an average elevation of 4483 m above mean sea level (AMSL; range: 3217–6575 m AMSL). It is located in the sub-frigid zone of the TP, with annual air temperatures ranging between 4.2 and −5.4°C (Zheng et al. 2010, Xu et al. 2012). Climate regionalization within the TP is characterized by humid, sub-humid and semi-arid zones (Zheng et al.

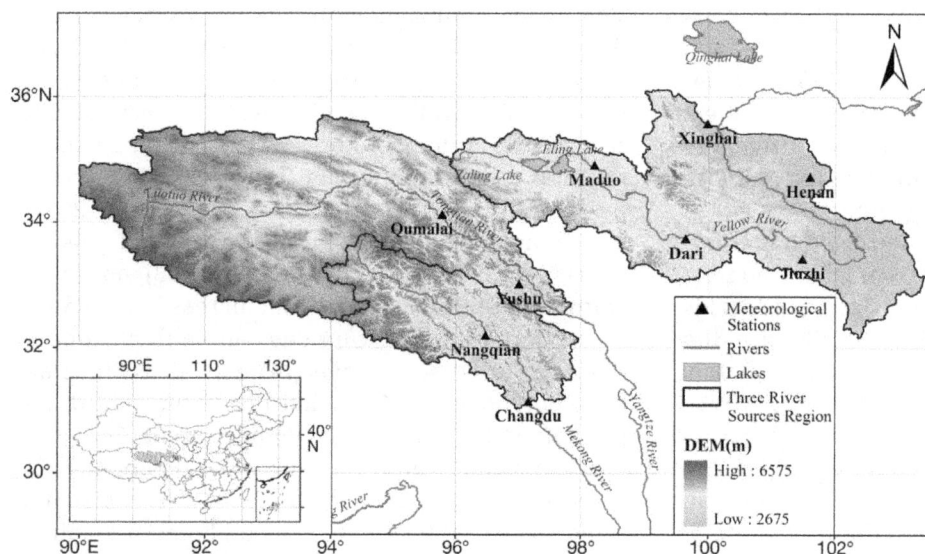

Fig. 1. Study area and locations of the meteorological stations (▲) used to collect climate data from 1980–2014 in the Three River Source Region (TRSR) of the Tibetan Plateau, China (inset)

2010). Precipitation varies, with annual accumulations ranging from 649 mm in the eastern part of the region (Ruoergai meteorological station) to 299 mm in the western precipitation occurs during the rainy season (May–September).

The data analysed in this study included surface (0 cm), shallow layer (depths of 5, 10, 15 and 20 cm) and deep layer (depths of 40, 80, 160 and 320 cm) soil temperatures collected from 9 meteorological stations operated by the China Meteorological Data Sharing Service System. In addition, data on air temperature, precipitation, maximum depth of frozen ground and maximum depth of snow were also included in this study. Soil temperature monitoring was conducted each day by trained professional meteorological technicians at all stations. The temperatures at depths of 0, 5, 10, 15, 20 and 40 cm were measured 4 times d^{-1} (02:00, 08:00, 14:00, and 20:00 h Beijing Time) and averaged as a daily mean; temperatures at depths of 80, 160 and 320 cm were measured once d^{-1} (14:00 h Beijing Time) (CMA 2003). Surface and shallow layer soil temperatures were measured in bare soil with no overlying vegetation, whereas deep layer soil temperatures were measured in natural vegetation (CMA 2003). The depth of the frozen ground was observed once d^{-1} (08:00 h Beijing Time) by frozen soil apparatus when ground surface temperature was below 0°C (CMA 2003). All data were collected monthly at each station for 35 yr (from 1980–2014); the monthly data were then used to create seasonal and annual time series. Seasons were defined as follows: spring = March, April and May; summer = June, July and August; autumn = September, October and November; and winter = December, January and February. A list of the stations and their associated information is provided in Table 1; their locations are shown on Fig. 1.

Table 1. Details of the meteorological stations used to collect climate data from 1980–2014; height: m above mean sea level; ST: soil temperature

Station name	Station number	Latitude (°N)	Longitude (°E)	Height (m)	Data period for 0 cm ST	Data period for shallow ST	Data period for deep ST
Yellow River							
Xinghai	52 943	35°35′	99°59′	3323	1980–2014	1980–2014	none
Henan	56 065	34°44′	101°36′	3500	1980–2014	1980–2014	1981–2014
Jiuzhi	56 067	33°26′	101°29′	3629	1980–2014	1980–2014	1980–2014
Dari	56 046	33°45′	99°39′	3968	1980–2014	1980–2014	1980–2014
Maduo	56 033	34°55′	98°13′	4272	1980–2014	1980–2014	1980–2014
Yangtze River							
Qumalai	56 021	34°08′	95°47′	4175	1980–2014	1980–2014	1982–2014
Yushu	56 029	33°01′	97°01′	3681	1980–2014	1980–2014	none
Mekong River							
Nangqian	56 125	32°12′	96°29′	3644	1980–2014	1980–2014	none
Changdu	56 137	31°09′	97°10′	3306	1980–2014	1980–2014	1980–2014

3. METHODS

For this study, 2 non-parametric methods, the modified Mann-Kendall trend (MMK) test (Mann 1945, Kendall 1955, Hamed & Rao 1998) and Sen's slope estimator (Sen 1968), were used in Matlab (MathWorks) to detect trends in soil temperatures and other meteorological variables. These methods used have been widely used to quantify the significance of trends in hydro-meteorological time series (Gocic & Trajkovic 2013). Correlation analysis was used to identify relationships between soil temperature variability and other climate variables at the same locations, which is a commonly used method of statistical diagnosis in modern climatic analysis studies (Wei 2007).

3.1 Modified Mann-Kendall trend test

The MMK (Mann 1945, Kendall 1955, Hamed & Rao 1998) test statistic (S) was calculated as:

$$S = \sum_{i=1}^{n-1} \sum_{j=i+1}^{n} \mathrm{sgn}(x_j - x_i) \tag{1}$$

where n is the number of data points, x_i and x_j are the data values in time series i and j ($j > i$), respectively and $\mathrm{sgn}(x_j - x_i)$ is:

$$\mathrm{sgn}(x_j - x_i) = \begin{cases} +1, \text{if } x_j - x_i > 0 \\ 0, \text{ if } x_j - x_i = 0 \\ -1, \text{if } x_j - x_i < 0 \end{cases} \tag{2}$$

The variance was computed as:

$$\mathrm{Var}(S) = \frac{n(n-1)(2n+5) - \sum_{i=1}^{m} t_i(t_i - 1)(2t_i + 5)}{18} \tag{3}$$

where n is the number of data points, m is the number of tied groups and t_i denotes the number of ties of extent i. A tied group is a set of sample data having the same value. The null hypothesis in the Mann-Kendall test is that the data are independent and randomly ordered. However, the existence of positive autocorrelation in the data increases the probability of detecting trends when none actually exist, and vice versa (Hamed & Rao 1998). In order to remove the effects of autocorrelation for the data, we amended the modified variance, $\mathrm{Var}^*(S)$, as follows:

$$\beta = \mathrm{median}\frac{x_j - x_i}{j - i} \quad (1 \le i < j \le n) \tag{4}$$

where β is the trend estimator based on the sequence rank. The stable sequence corresponding

with the original, which eliminated the trend $\{y_i\}_{i=1}^{n}$, is represented as:

$$y_i = x_i - \beta \times i \tag{5}$$

$$r(i) = \frac{\sum_{k=1}^{n-i} (R_k - R)(R_{k+i} - R)}{\sum_{k=1}^{n} (R_k - R)^2} \tag{6}$$

where $r(i)$ is the autocorrelation function of the ranks of the observations, R_i is the rank of y_i, and R is the mean rank order of y_i. Using:

$$\eta = 1 + \frac{2}{n(n-1)(n-2)} \times \sum_{i=1}^{n-1} (n-i)(n-i-1)(n-i-2)r(i) \tag{7}$$

where η represents a correction due to the autocorrelation in the data, we can calculate:

$$\mathrm{Var}^*(S) = \eta \times \mathrm{Var}(S) \tag{8}$$

In cases where the sample size $n > 10$, the standard normal test statistic, Z, is computed using Eq. (4):

$$Z = \begin{cases} \frac{S-1}{\sqrt{\mathrm{Var}^*(S)}}, & \text{if } S > 0 \\ 0, & \text{if } S = 0 \\ \frac{S+1}{\sqrt{\mathrm{Var}^*(S)}}, & \text{if } S < 0 \end{cases} \tag{9}$$

Positive values of Z indicate increasing trends while negative values show decreasing trends.

Testing trends is done at a specific significance level. When $|Z| > Z_{1-\alpha/2}$, the null hypothesis is rejected and a significant trend exists in the time series. $Z_{1-\alpha/2}$ is obtained from the standard normal distribution table. In this study, significance levels of $\alpha = 0.01$ and 0.05 were used. At the 5% significance level, the null hypothesis of no trend was rejected if $|Z| > 1.96$, and at the 1% significance level if $|Z| > 2.576$.

3.2 Sen's slope estimator

Sen (Sen 1968) developed the following non-parametric procedure for estimating the slope of the trend (Q_i) in a sample of N pairs of data:

$$Q_i = \frac{x_j - x_k}{j - k} \quad for \ i = 1, ..., N, \tag{10}$$

where x_j and x_k are data values at times j and k ($j > k$), respectively.

If there is only one datum in each time period, then $N = \frac{n(n-1)}{2}$, where n is the number of time periods. If there are multiple observations in one or more time periods, then $N < \frac{n(n-1)}{2}$, where n is the total number of observations.

The N values of Q_i are ranked from smallest to largest and the median of the slope, or Sen's slope estimator (Q_{med}), is computed as:

$$Q_{med} = \begin{cases} Q_{[(N+1)/2]}, & \text{if } N \text{ is odd} \\ \dfrac{Q_{[N/2]} + Q_{[(N+2)/2]}}{2}, & \text{if } N \text{ is even} \end{cases} \quad (11)$$

The Q_{med} sign reflects data trend, while its value indicates the steepness of the trend. To determine whether the median slope is statistically different than zero, the confidence interval of Q_{med} at specific probability can be obtained. The confidence interval about the time slope (Gilbert 1987, Hollander et al. 1999) is computed as follows:

$$C_a = Z_{1-\alpha/2}\sqrt{Var^*(S)} \quad (12)$$

where $Var^*(S)$ is defined in Eq. (8), and $Z_{1-\alpha/2}$ is obtained from the standard normal distribution table. In this study, the confidence interval was computed at 2 significance levels ($\alpha = 0.01$ and 0.05).

Then, $M_1 = \dfrac{N - C_\alpha}{2}$ and $M_2 = \dfrac{N + C_\alpha}{2}$ are computed. The lower and upper limits of the confidence interval, Q_{min} and Q_{max}, are the M_1^{th} largest and the $(M_2 + 1)^{th}$ largest of the N ordered slope estimates (Gilbert 1987).

The slope of Q_{med} is statistically different than zero if the 2 limits (Q_{min} and Q_{max}) have similar sign.

4. SOIL TEMPERATURE CLIMATOLOGY

Changes in seasonal soil temperature cycles from 0–320 cm depth recorded at the 9 meteorological stations between 1980 and 2012 are shown in Fig. 2. The annual mean soil temperature during that period was 4.693°C . Regarding spatial distribution throughout the study region, the Maduo station recorded the lowest soil temperature (1.629°C), while the Changdu station had the highest (10.832°C). Soil temperatures are usually affected by solar radiation, air tem-

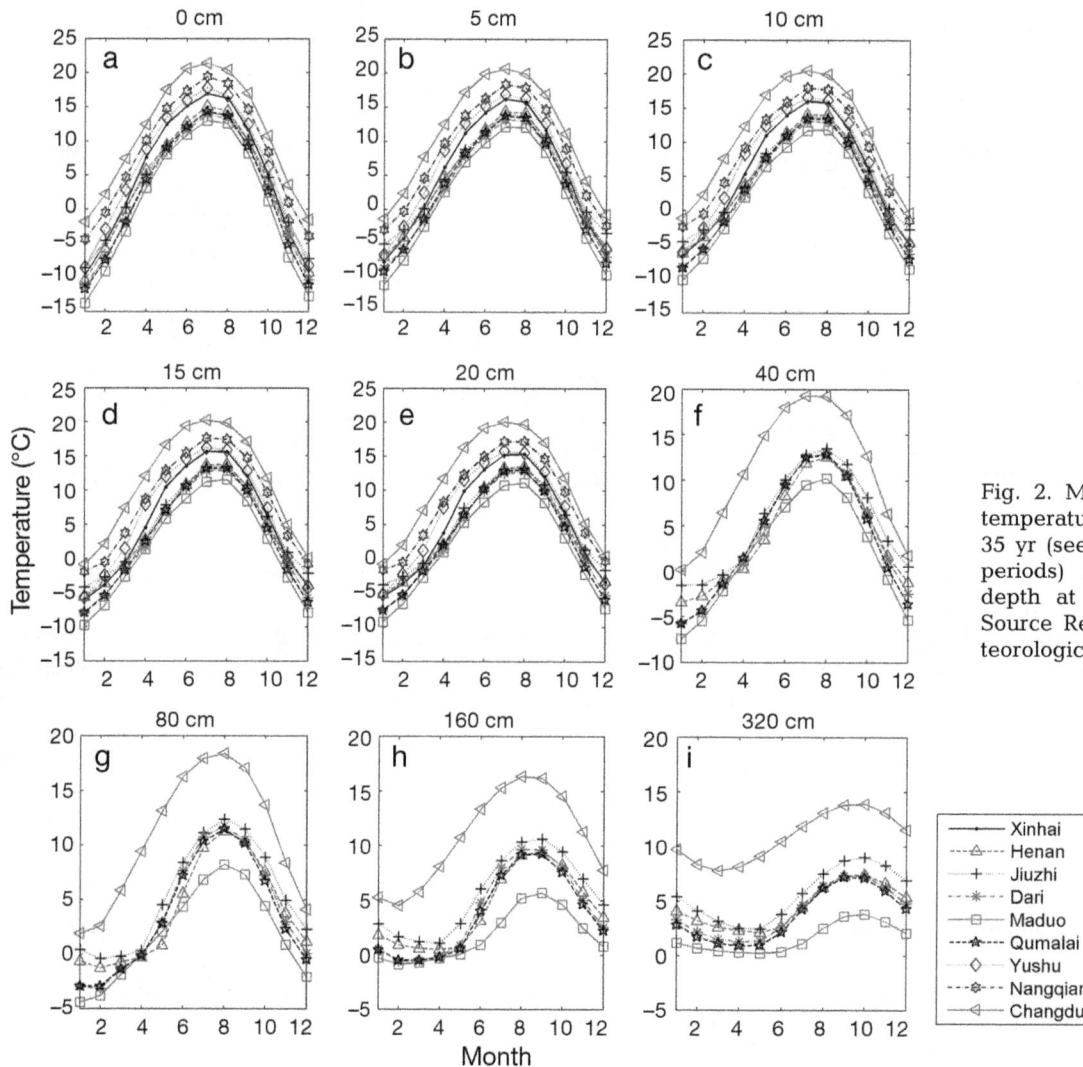

Fig. 2. Monthly mean soil temperature cycles across 35 yr (see Table 1 for data periods) from 0–320 cm depth at the Three River Source Region (TRSR) meteorological stations (see Fig. 1)

Table 2. Trends in seasonal and annual mean soil temperatures (°C decade^{-1}) across 35 yr (see Table 1 for data periods) at the 9 Three River Source Region (TRSR) meteorological stations (Fig. 1) for 3 different layers (surface, shallow and deep). Z: modified Mann-Kendall test statistic; slope: Sen's slope estimator; significance levels: $^*\alpha = 0.05$, $^{**}\alpha = 0.01$

	—— Spring ——		—— Summer ——		——Autumn ——		——Winter ——		——Annual——	
	Z	Slope	Z	Slope	Z	Slope	Z	Slope	Z	Slope
Surface (0 cm)										
Xinghai	4.460**	0.606**	2.798**	0.500**	4.162**	0.667**	3.198**	0.571**	5.568**	0.583**
Henan	4.460**	0.988**	5.015**	0.950**	4.447**	0.754**	4.276**	0.912**	5.340**	0.946**
Jiuzhi	4.192**	0.773**	4.533**	0.636**	4.035**	0.583**	4.716**	1.161**	5.368**	0.768**
Dari	3.324**	0.507**	2.330*	0.321*	2.884**	0.410**	2.969**	0.652**	3.891**	0.492**
Maduo	3.963**	0.756**	3.908**	0.584**	4.035**	0.667**	3.765**	0.778**	4.773**	0.688**
Qumalai	4.235**	0.650**	3.410**	0.500**	4.275**	0.800**	5.541**	1.222**	5.624**	0.820**
Yushu	3.352**	0.500**	2.785**	0.476**	5.085**	0.624**	4.503**	1.455**	5.724**	0.768**
Nangqian	3.750**	0.758**	3.623**	0.689**	4.221**	0.604**	5.285**	0.939**	4.830**	0.722**
Changdu	3.838**	0.521**	1.492	0.391	2.787**	0.424**	4.517**	0.902**	4.290**	0.569**
Average	3.908**	0.669**	3.282**	0.561**	3.965**	0.615**	4.255**	0.955**	5.011**	0.706**
Shallow (average of 5, 10, 15, 20 cm)										
Xinghai	2.970**	0.414**	2.694**	0.403**	3.768**	0.383**	3.419**	0.361**	4.740**	0.394**
Henan	4.115**	0.615**	4.801**	0.721**	4.029**	0.442**	2.886**	0.326**	5.177**	0.553**
Jiuzhi	4.949**	0.632**	5.321**	0.651**	4.420**	0.499**	4.514**	0.450**	5.931**	0.542**
Dari	3.657**	0.408**	3.827**	0.441**	2.581**	0.246**	3.369**	0.389**	4.457**	0.364**
Maduo	4.171**	0.525**	4.851**	0.710**	3.830**	0.449**	1.776	0.306	5.423**	0.495**
Qumalai	4.950**	0.610**	4.760**	0.612**	3.681**	0.401**	4.101**	0.557**	5.876**	0.578**
Yushu	0.910	0.119	3.745**	0.459**	4.404**	0.469**	4.753**	0.927**	5.486**	0.495**
Nangqian	3.716**	0.476**	3.954**	0.592**	3.206**	0.391**	5.323**	0.393**	4.865**	0.453**
Changdu	3.600**	0.370**	1.659	0.342	2.150*	0.332*	4.355**	0.675**	3.661**	0.423**
Average	3.671**	0.463**	3.957**	0.548**	3.563**	0.401**	3.833**	0.487**	5.068**	0.477**
Deep (average of 40, 80, 160, 320 cm)										
Henan	4.809**	0.321**	4.932**	0.793**	4.846**	0.461**	4.425**	0.277**	5.446**	0.469**
Jiuzhi	4.453**	0.424**	5.720**	0.746**	3.473**	0.314**	1.146	0.077	5.238**	0.395**
Dari	3.647**	0.310**	4.677**	0.514**	3.747**	0.312**	3.866**	0.257**	4.880**	0.346**
Maduo	6.243**	0.477**	6.031**	1.243**	5.483**	0.638**	0.261	-0.137	6.431**	0.549**
Qumalai	3.563**	0.294**	4.628**	0.593**	2.929**	0.248**	1.302	0.110	4.038**	0.320**
Changdu	2.305*	0.364*	1.898	0.444	2.824**	0.465**	3.944**	0.471**	2.706**	0.422**
Average	4.170**	0.365**	4.647**	0.722**	3.883**	0.406**	2.491*	0.176*	4.790**	0.417**

perature and terrain variables. Maduo is a northerly station, located at the maximum height AMSL of all the stations, while Changdu is an extreme southern station, at the minimum height (Table 1).

For surface (0 cm) and 5 cm soils, the minimum recorded temperature occurred in January and the maximum in July at all stations. Below the surface, the minimum and maximum temperatures occurred later than their corresponding values at the surface, with time lag increasing with depth. The minimum soil temperatures occurred in January, and the maximum in July or August at depths of 10, 15, 20 and 40 cm. The minimum temperature at 80 cm depth was recorded in January (Maduo and Changdu) or February (Henan, Jiuzhi, Dari and Qumalai), and the maximum in August. The minimum temperature at 160 cm occurred in February (Dari, Maduo and Changdu), March (Qumalai) or April (Henan and Jiuzhi), while the maximum occurred in August (Dari and Changdu) or September (Henan, Jiuzhi, Maduo

and Qumalai). The minimum temperature at 320 cm occurred in March (Changdu), April (Dari and Qumalai) or May (Henan, Jiuzhi and Maduo), and the maximum occurred in September (Dari and Qumalai) or October (Henan, Jiuzhi, Maduo and Changdu). The maximum seasonal amplitude of soil temperature (25.161°C) occurred at 0 cm, from 22.822°C (Jiuzhi) to 26.852°C (Maduo); this value decreased with increasing depth. The soil temperature seasonal amplitudes for the 4 shallow layers (5–20 cm) averaged 22.094, 21.278, 20.561 and 19.903°C, respectively, which were higher than that of the 4 deep layers (40–320 cm), at 17.548, 13.955, 9.456 and 5.633°C, respectively.

5. SOIL TEMPERATURE TRENDS

Trends in annual, seasonal and monthly mean soil temperatures at the 9 TRSR stations collected during

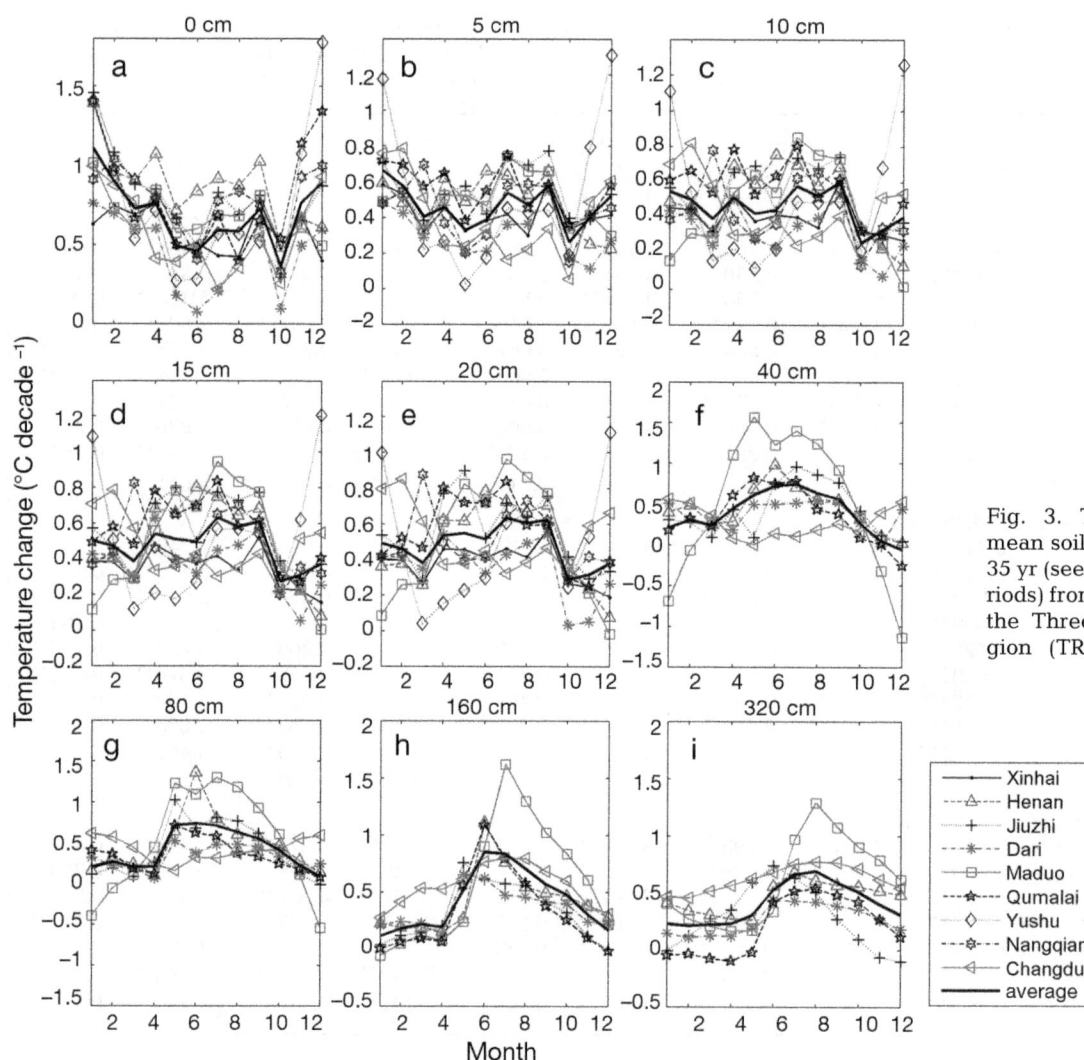

Fig. 3. Trends in monthly mean soil temperature across 35 yr (see Table 1 for data periods) from 0–320 cm depth at the Three River Source Region (TRSR) meteorological stations

1980–2014 are shown in Table 2 and Fig. 3. The spatial distributions of each layer with increasing, decreasing or no trends for the seasonal and annual data series are also presented in Figs. S1 & S2 in the Supplement at www.int-res.com/articles/suppl/c067p241_supp.pdf.

5.1 Surface soil temperature trends

Seasonal and annual mean surface soil temperatures for the entire TRSR increased significantly between 1980 and 2014 (Table 2). On an annual time scale, increasing trends were detected at a 1% significance level at all stations, with an average rate of 0.706°C decade^{-1}, varying between 0.946°C decade^{-1} at Henan station and 0.492°C decade^{-1} at Dari station.

Regarding monthly distributions (Fig. 3a), larger warming trends of surface soil temperatures were

observed in January, November and December, while smaller trends occurred in June and October. These trends also exhibited significant seasonal variation. As shown in Table 2, the maximum warming trends occurred in winter for 7 stations, with the exception of Xinghai and Henan. The average trend in winter exhibited an increase of 0.955°C decade^{-1}, which was more than 70% that of the summer rate (0.561°C decade^{-1}). The values in spring and autumn were 0.669°C decade^{-1} and 0.615°C decade^{-1}, respectively. These results indicate that the rate of increase in surface soil temperatures is greater in winter than during the other seasons.

5.2 Shallow soil temperature trends

Annual mean soil temperatures in the 4 shallow layers warmed significantly (at $\alpha = 0.01$) at all 9 stations across the TRSR over the 35 yr analysed, with an average rate increase of 0.477°C decade^{-1} (Table 2). In terms of vertical distribution (Fig. S1), there was no significant increase or decrease in warming trend with soil depth. Regarding spatial distribution across the TRSR, the trends of the shallow layers were similar to that at the surface, with greater warming occurring at Henan (0.553°C decade^{-1}) and Qumalai stations (0.578°C decade^{-1}), respectively, and a less of a warming trend seen at Dari station (0.364°C decade^{-1}).

The monthly trends in shallow soil temperatures differed considerably across the TRSR (Fig. 3b–e). The greatest warming trends in the 4 shallow depths were recorded in July, August and September at most stations. Significant warming trends were also found in January and December in stations such as Henan and Changdu—a similar pattern to that observed in surface soil temperatures. There was a common characteristic, in that the warming trends were relatively small in October at all depths and all stations. One of the major reasons for this occurrence is that the freezing process is a buffer to seasonal changes in soil temperatures (discussed in more detail in Section 6.2). On a seasonal scale, the percentage of stations characterized by significant increasing trends (at $\alpha = 0.05$) in shallow layer soil temperatures was 88.89% for spring, 94.44% for summer, 97.22% for autumn and 91.67% for winter. As shown in Fig. S1, the strongest warming trends occurred in winter for 5 stations at 5 cm soil layers, while the strongest warming trends occurred in summer at 10, 15 and 20 cm soil layers for most stations (5, 6 and 6 stations, respectively). As shown in Table 2, the largest warming trends in all shallow layers occurred in the summer at 6 stations (Henan, Jiuzhi, Dari, Maduo, Qumalai and Nangqian), in winter at 2 stations (Yushu and Changdu), and in spring at 1 station (Xinghai). This also suggest that the increasing rate in summer (0.548°C decade^{-1}) is greater than that of other seasons (0.463°C decade^{-1}, 0.401°C decade^{-1} and 0.487°C decade^{-1}, in spring, autumn and winter, respectively) for all shallow layer soils.

5.3 Deep soil temperature trends

The soil temperatures of the 4 deep layers increased significantly (at $\alpha = 0.01$) at 6 stations across the TRSR over 35 yr, with an average increasing rate of 0.417°C decade^{-1}. Warming that occurs in the upper layers could take some time to reach the deeper layers. Other recent studies have produced similar results (Qian et al. 2011). With respect to spatial distribution, the greatest warming trend for the deep layer soils occurred at the Maduo station (0.549°C decade^{-1}), while the lowest trend occurred at the Qumalai station (0.320°C decade^{-1}).

The trends in temperatures of the deep layer soils also exhibited significant seasonal variation. Significant soil warming occurred from April or May until September or October at depths of 40 and 80 cm for 5 of the 6 stations (except Changdu; Fig. 3f,g). Significant soil warming from May or June until October or November was found at depths of 160 and 320 cm at all stations (Fig. 3h,i). The strongest warming trends in summer (average rate: 0.722°C decade^{-1}) was nearly 4 times that in winter (0.176°C decade^{-1}) in the TRSR. Soil temperature warming trends in spring and autumn were 0.365°C decade^{-1} and 0.406°C decade^{-1}, respectively. The percentage of stations characterized by significant at ($\alpha = 0.05$) increasing trends was 91.67% for spring, summer and autumn, and 62.50% for winter. As shown in Fig. S2, there were only 2 significant changes in decreasing direction (at $\alpha = 0.05$): during winter at Maduo station at 40 and 80 cm depths, at rates of −0.667 and −0.353°C decade^{-1}, respectively. The strongest trends occurred in the summer for 5 stations at the 40 and 80 cm depths, (again except for Changdu), for which the strongest trend occurred in winter. For the 160 and 320 cm soil layers, the strongest trends were also observed in summer at all 6 stations, except for the 320 cm soil layer at Maduo station, for which the strongest warming occurred in autumn. This implies that increases in the summer contributed more to increases in the winter for all deep layer soil in the TRSR.

6. DISCUSSION

6.1 Trends in other climate variables

Annual means and trends in air temperature, total precipitation, maximum depth of frozen ground and maximum snow depth are provided in Table 3 and in Figs. 4 & Fig. S3 in the Supplement at www.int-res.com/articles/suppl/c067p241_supp.pdf. As shown in Table 3, over the last 35 yr the mean annual air temperature in the TRSR was 1.502°C. Half of the meteorological stations measured air temperatures below 0°C. Maduo station recorded the lowest air temperature at −3.250°C, while Changdu station had the highest air temperature at 7.834°C, which was similar to the soil temperatures in the study region. All

Table 3. Annual means in air temperature, total precipitation, maximum depth of frozen ground and maximum depth of snow recorded at the Three River Source Region (TRSR) meteorological stations across 35 yr (see Table 1)

	Air temperature (°C)	Total precipitation (mm)	Max. frozen depth (cm)	Max. snow depth (cm)
Xinghai	1.726	374.457	156.971	4.629
Henan	−0.137	564.526	122.114	8.800
Jiuzhi	1.112	737.423	90.114	11.400
Dari	−0.454	565.937	196.559	9.943
Maduo	−3.250	335.383	213.059	7.229
Qumalai	−1.660	430.266	202.121	7.171
Yushu	3.706	492.849	89.706	6.857
Nangqian	4.637	549.251	66.029	6.229
Changdu	7.834	487.286	48.429	4.514
Average	1.502	504.156	131.678	7.419

meteorological stations measured mean annual air temperatures that were lower than the mean annual soil temperatures. These results indicate that 91.67 % of stations were characterized by significant (at α = 0.05) increasing trends in air temperature across all 4 seasons, with an average rate of 0.458°C decade^{-1} over the last 35 yr. This rate was greater than the rate of 0.32°C decade^{-1} that occurred during the past 50 yr in this region (Liang et al. 2013). On a monthly time scale, the greatest warming trends in air temperatures occurred in January, February, July, August and September, while smaller warming trends were observed in March, June and October (Fig. 4). For seasonal distributions, greater warming trends in air temperatures were recorded in winter for 7 of the 9 meteorological stations, with the exception of Xinghai and Henan. The average trends in winter (0.582°C decade^{-1}) and summer (0.490°C de-

cade^{-1}), were larger than in autumn (0.380°C decade^{-1}) and spring (0.366°C decade^{-1}). The greater rate of soil temperature warming in winter was likely caused by the vegetation growing over the TP in summer, which would attenuate surface warming (Shen et al. 2015). As described in Section 5, surface and 5 cm depth soil warming occurred mainly during the winter, with soil temperature increases in the summer more evident in the other layers (10–320 cm) at most stations. These results indicate that only surface soil temperatures are consistent with air temperatures. The mean annual increasing trends for surface soil temperatures (0 cm) were greater than those for air temperature at all stations. For all shallow layer soils, the mean annual increasing trends were greater than those for air temperature at 5 stations (Xinghai, Henan, Jiuzhi, Yushu and Changdu). For all deep layer soils, 2 stations (Maduo and Changdu) showed mean annual increasing trends for soil temperature that were greater than those for air temperature. Similar results (i.e. increasing trends in soil temperature being stronger than that for air temperature) have been found elsewhere in the TP and in Russia (Zhang et al. 2001, García-Suárez & Butler 2006, Du et al. 2007).

As shown in Table 3, mean annual precipitation for the 9 stations across the TRSR measured 504.156 mm, with the least amount (335.383 mm) recorded in Maduo and the highest (737.423 mm) in Jiuzhi. A positive trend in seasonal and annual precipitation was detected at about two-thirds of the stations (Fig. S3). There were 3 significant changes in increasing direction (at α = 0.05): spring in Yushu,

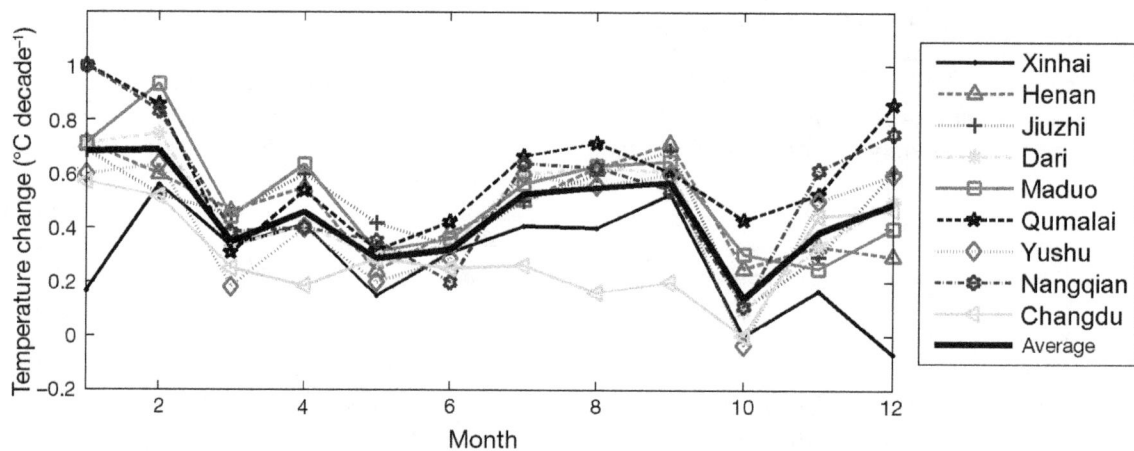

Fig. 4. Trends in monthly mean air temperature across 35 yr (see Table 1 for data periods) at the Three River Source Region (TRSR) meteorological stations (see Fig. 1)

Table 4. Correlation coefficients for seasonal and annual air temperature and soil temperature across 35 yr (see Table 1) for the 9 TRSR stations (Fig. 1). Significance levels: *α = 0.05; **α = 0.01

	0 cm	5 cm	10 cm	15 cm	20 cm	40 cm	80 cm	160 cm	320 cm
Spring									
Xinghai	0.562**	0.662**	0.684**	0.676**	0.652**	none	none	none	none
Henan	0.681**	0.733**	0.732**	0.728**	0.731**	0.767**	0.781**	0.820**	0.776**
Jiuzhi	0.727**	0.719**	0.690**	0.698**	0.676**	0.625**	0.611**	0.562**	0.630**
Dari	0.644**	0.645**	0.671**	0.670**	0.627**	0.610**	0.623**	0.628**	0.528**
Maduo	0.654**	0.556**	0.600**	0.609**	0.641**	0.637**	0.651**	0.667**	0.653**
Qumalai	0.585**	0.588**	0.599**	0.590**	0.584**	0.607**	0.656**	0.652**	0.526**
Yushu	0.505**	0.610**	0.601**	0.602**	0.574**	none	none	none	none
Nangqian	0.649**	0.636**	0.633**	0.637**	0.634**	none	none	none	none
Changdu	0.566**	0.559**	0.553**	0.526**	0.539**	0.483**	0.495**	0.490**	0.434**
Summer									
Xinghai	0.676**	0.742**	0.713**	0.722**	0.714**	none	none	none	none
Henan	0.794**	0.796**	0.815**	0.824**	0.820**	0.740**	0.775**	0.770**	0.780**
Jiuzhi	0.697**	0.828**	0.829**	0.810**	0.784**	0.838**	0.855**	0.856**	0.828**
Dari	0.742**	0.795**	0.795**	0.811**	0.795**	0.761**	0.723**	0.763**	0.733**
Maduo	0.676**	0.686**	0.702**	0.658**	0.694**	0.753**	0.765**	0.729**	0.691**
Qumalai	0.748**	0.771**	0.742**	0.763**	0.737**	0.683**	0.727**	0.723**	0.576**
Yushu	0.653**	0.604**	0.622**	0.626**	0.617**	none	none	none	none
Nangqian	0.830**	0.865**	0.854**	0.854**	0.842**	none	none	none	none
Changdu	0.718**	0.724**	0.746**	0.757**	0.751**	0.801**	0.783**	0.741**	0.689**
Autumn									
Xinghai	0.437**	0.476**	0.444**	0.446**	0.431**	none	none	none	none
Henan	0.565**	0.587**	0.600**	0.583**	0.587**	0.444**	0.524**	0.488**	0.541**
Jiuzhi	0.541**	0.609**	0.659**	0.656**	0.652**	0.638**	0.645**	0.604**	0.567**
Dari	0.567**	0.677**	0.652**	0.657**	0.678**	0.630**	0.576**	0.554	0.421
Maduo	0.523**	0.543**	0.528**	0.531**	0.507**	0.527**	0.479**	0.416	0.439
Qumalai	0.602**	0.601**	0.600**	0.606**	0.606**	0.509**	0.447**	0.374*	0.388*
Yushu	0.543**	0.554**	0.560**	0.539**	0.549**	none	none	none	none
Nangqian	0.654**	0.664**	0.647**	0.654**	0.637**	none	none	none	none
Changdu	0.685**	0.691**	0.705**	0.703**	0.685**	0.652**	0.623**	0.551**	0.446**
Winter									
Xinghai	0.590**	0.586**	0.651**	0.616**	0.623**	none	none	none	none
Henan	0.748**	0.732**	0.739**	0.746**	0.751**	0.650**	0.714**	0.704**	0.674**
Jiuzhi	0.745**	0.732**	0.737**	0.725**	0.702**	0.659**	0.644**	0.610**	0.657**
Dari	0.818**	0.813**	0.792**	0.775**	0.760**	0.646**	0.674**	0.627**	0.573**
Maduo	0.755**	0.728**	0.743**	0.702**	0.673**	0.722**	0.686**	0.641**	0.607**
Qumalai	0.790**	0.772**	0.773**	0.782**	0.780**	0.768**	0.778**	0.761**	0.679**
Yushu	0.599**	0.591**	0.595**	0.594**	0.580**	none	none	none	none
Nangqian	0.784**	0.758**	0.769**	0.780**	0.769**	none	none	none	none
Changdu	0.617**	0.652**	0.628**	0.615**	0.610**	0.467**	0.519**	0.544**	0.564**
Annual									
Xinghai	0.571**	0.924**	0.932**	0.916**	0.898**	none	none	none	none
Henan	0.894**	0.909**	0.921**	0.921**	0.924**	0.823**	0.897**	0.889**	0.880**
Jiuzhi	0.861**	0.920**	0.924**	0.903**	0.781**	0.876**	0.866**	0.819**	0.840**
Dari	0.889**	0.669**	0.926**	0.923**	0.914**	0.826**	0.451**	0.791**	0.693**
Maduo	0.885**	0.843**	0.870**	0.835**	0.825**	0.880**	0.854**	0.780**	0.781**
Qumalai	0.888**	0.692**	0.702**	0.735**	0.729**	0.836**	0.838**	0.794**	0.696**
Yushu	0.742**	0.590**	0.619**	0.617**	0.637**	none	none	none	none
Nangqian	0.918**	0.913**	0.914**	0.923**	0.908**	none	none	none	none
Changdu	0.907**	0.921**	0.922**	0.913**	0.906**	0.820**	0.832**	0.803**	0.739**

spring in Nangqian, and summer in Maduo, at rates of 10.773, 11.412 and 16.000 mm decade^{-1}, respectively.

The average maximum depth of frozen ground at the 9 stations was 131.678 cm, with the lowest depth (48.429 cm) at Changdu and the highest depth (213.059 cm) at Maduo (Table 3). This depth gradually decreased for all stations over time, with an average decreasing trend over 35 yr of 10.161 cm decade^{-1}. Statistically significant trends (at α = 0.01) at

Table 5. Correlation coefficients for seasonal and annual precipitation and soil temperature across 35 yr (see Table) for the 9 TRSR stations (Fig. 1). Significance levels: $^*\alpha = 0.05$; $^{**}\alpha = 0.01$

	0 cm	5 cm	10 cm	15 cm	20 cm	40 cm	80 cm	160 cm	320 cm
Spring									
Xinghai	−0.170	−0.267	−0.166	−0.217	−0.194	none	none	none	none
Henan	−0.190	−0.256	−0.227	−0.213	−0.198	−0.126	−0.090	−0.149	−0.132
Jiuzhi	0.041	0.140	0.135	0.143	0.125	0.222	0.246	0.275	0.260
Dari	0.175	0.215	0.224	0.237	0.240	0.210	0.217	0.251	0.273
Maduo	−0.339*	−0.194	−0.198	−0.180	−0.150	−0.166	−0.118	−0.105	−0.082
Qumalai	0.243	0.222	0.210	0.238	0.234	0.193	0.161	0.168	0.186
Yushu	0.452**	0.353*	0.364*	0.363*	0.381*	none	none	none	none
Nangqian	0.287	0.282	0.323	0.333*	0.308	none	none	none	none
Changdu	0.096	0.045	0.082	0.074	0.084	−0.009	0.044	0.071	0.116
Summer									
Xinghai	0.187	0.167	0.183	0.143	0.132	none	none	none	none
Henan	−0.065	−0.237	−0.210	−0.194	−0.178	0.024	0.026	−0.069	−0.060
Jiuzhi	0.163	0.118	0.131	0.073	0.072	0.193	0.183	0.291	0.194
Dari	0.374*	0.258	0.278	0.323	0.269	0.315	0.458**	0.400*	0.411*
Maduo	0.376*	0.442**	0.416*	0.478**	0.464**	0.430*	0.479**	0.447**	0.441**
Qumalai	0.314	0.240	0.273	0.289	0.300	0.384*	0.347*	0.314	0.338
Yushu	−0.184	−0.095	−0.091	−0.072	−0.098	none	none	none	none
Nangqian	−0.044	−0.113	−0.166	−0.159	−0.162	none	none	none	none
Changdu	−0.166	−0.227	−0.216	−0.216	−0.187	−0.238	−0.188	−0.134	−0.114
Autumn									
Xinghai	0.146	0.236	0.232	0.157	0.145	none	none	none	none
Henan	0.247	0.033	0.055	0.073	0.097	−0.008	0.110	0.056	0.170
Jiuzhi	−0.113	−0.180	−0.106	−0.157	−0.116	−0.010	−0.006	0.092	0.026
Dari	−0.081	0.001	−0.030	−0.086	−0.100	−0.068	−0.056	−0.105	−0.009
Maduo	0.027	−0.019	0.016	0.047	0.082	0.078	0.163	0.183	0.179
Qumalai	0.190	0.107	0.086	0.123	0.094	0.193	0.171	0.166	0.199
Yushu	−0.179	−0.087	−0.093	−0.062	−0.068	none	none	none	none
Nangqian	0.067	0.067	0.125	0.146	0.190	none	none	none	none
Changdu	−0.282	−0.195	−0.216	−0.274	−0.272	−0.383*	−0.349*	−0.323	−0.261
Winter									
Xinghai	−0.072	−0.128	−0.185	−0.083	−0.080	none	none	none	none
Henan	−0.151	−0.063	−0.072	−0.094	−0.114	−0.042	−0.140	−0.129	−0.142
Jiuzhi	0.040	0.142	0.154	0.168	0.130	0.146	0.172	0.232	0.187
Dari	0.035	0.041	0.074	0.089	0.105	0.276	0.232	0.254	0.266
Maduo	−0.219	−0.083	−0.133	−0.112	−0.129	−0.238	−0.229	−0.221	−0.210
Qumalai	0.146	0.104	0.113	0.135	0.138	0.202	0.140	0.121	0.086
Yushu	0.016	0.001	0.010	−0.036	0.051	none	none	none	none
Nangqian	−0.089	0.030	0.079	0.088	0.072	none	none	none	none
Changdu	−0.058	−0.012	−0.042	−0.051	−0.083	0.124	0.063	0.002	−0.010
Annual									
Xinghai	0.128	0.115	0.160	0.083	0.079	none	none	none	none
Henan	0.008	−0.232	−0.194	−0.172	−0.147	−0.022	0.037	−0.073	−0.085
Jiuzhi	0.070	0.054	0.092	0.038	0.046	0.201	0.207	0.331	0.232
Dari	0.338*	0.297	0.305	0.322	0.273	0.334*	0.456**	0.400*	0.464**
Maduo	0.163	0.252	0.243	0.309	0.323	0.281	0.371*	0.360*	0.363*
Qumalai	0.385*	0.290	0.306	0.340*	0.338*	0.422*	0.374*	0.348*	0.380*
Yushu	−0.098	−0.021	−0.017	0.005	−0.007	none	none	none	none
Nangqian	0.062	0.001	−0.014	0.005	0.011	none	none	none	none
Changdu	−0.218	−0.241	−0.233	−0.259	−0.234	−0.333	−0.270	−0.214	−0.163

8 of the meteorological stations (except Maduo) were all negative.

Average maximum snow depth at the 9 stations measured 7.419 cm, with the thinnest layer (4.5 cm) at Changdu and the deepest (11.4 cm) at Jiuzhi. The maximum depth of snow at 4 stations gradually decreased, but the trend was not obvious. There was only 1 location (Changdu) that showed a significant (at $\alpha = 0.01$) decreasing trend, with a rate of −1.200 cm decade^{-1}.

Table 6. Correlation coefficients for maximum depth of frozen ground and snow with soil temperature across 35 yr (see Table 1) for the 9 TRSR stations (Fig. 1). Significance levels: $^*\alpha = 0.05$; $^{**}\alpha = 0.01$

	0 cm	5 cm	10 cm	15 cm	20 cm	40 cm	80 cm	160 cm	320 cm
Max. depth of frozen ground									
Xinghai	−0.679**	−0.722**	−0.758**	−0.783**	−0.804**	none	none	none	none
Henan	−0.464**	−0.528**	−0.575**	−0.617**	−0.653**	−0.727**	−0.860**	−0.896**	−0.755**
Jiuzhi	−0.592**	−0.649**	−0.688**	−0.722**	−0.748**	−0.817**	−0.871**	−0.867**	−0.619**
Dari	−0.447**	−0.502**	−0.533**	−0.560**	−0.583**	−0.659**	−0.622**	−0.807**	−0.700**
Maduo	−0.173	−0.231	−0.266	−0.293	−0.319	−0.388*	−0.555**	−0.798**	−0.789**
Qumalai	−0.264	−0.316	−0.348*	−0.373*	−0.400*	−0.537**	−0.622**	−0.674**	−0.507**
Yushu	−0.636**	−0.645**	−0.655**	−0.662**	−0.670**	none	none	none	none
Nangqian	−0.824**	−0.836**	−0.845**	−0.853**	−0.858**	none	none	none	none
Changdu	−0.790**	−0.797**	−0.800**	−0.801**	−0.804**	−0.799**	−0.780**	−0.664**	−0.286
Max. depth of snow									
Xinghai	−0.429*	−0.443**	−0.404*	−0.356*	−0.352*	none	none	none	none
Henan	−0.495**	−0.439**	−0.446**	−0.468**	−0.487**	−0.453**	−0.450**	−0.420*	−0.394*
Jiuzhi	−0.084	−0.047	−0.122	−0.146	−0.144	−0.100	−0.089	−0.020	−0.064
Dari	−0.018	0.014	0.037	0.019	−0.026	−0.061	0.034	0.034	0.069
Maduo	−0.075	−0.051	−0.032	−0.048	−0.008	−0.072	−0.088	−0.048	−0.055
Qumalai	−0.059	−0.050	−0.069	−0.064	−0.075	0.000	0.035	0.052	0.088
Yushu	0.232	0.149	0.175	0.177	0.227	none	none	none	none
Nangqian	−0.280	−0.200	−0.165	−0.154	−0.126	none	none	none	none
Changdu	−0.148	−0.077	−0.092	−0.114	−0.141	−0.073	−0.149	−0.261	−0.333

6.2 Comparing soil temperature to other climate characteristics

Relationships between air temperature, precipitation, maximum depth of frozen ground and maximum depth of snow with soil temperature are given in Tables 4–6. As expected, the correlations between seasonal and annual soil and air temperatures were significant and positive (at $\alpha = 0.01$) at all stations and at all depths (Table 4). This relationship was stronger in the summer and winter than in spring and autumn. There were no evident decreasing correlation coefficients of air temperature on soil temperatures from the upper to deeper layers.

We expect that soil temperature and some climatic variables are interrelated, but there are many factors that influence soil temperatures at a local or site-specific scale. As shown in Table 5, there was no significant correlation between soil temperature and seasonal or annual precipitation at most stations in the TRSR region. Except at Maduo station, snow cover typically disappears by summer in this area of the TP. As shown in Table 6, there was no obvious correlation between soil temperature and maximum depth of snow, except at Xinghai and Henan stations, where there were negative correlations between these 2 factors. Previous studies have shown that more precipitation or snow on the ground normally results in lower soil temperatures

in spring, summer and autumn, but higher soil temperatures in winter due to the thermal insulation effect of snow cover (Qian et al. 2011, Ye ılırmak 2014). A similar situation occurred in spring at the Yushu station and in winter at Maduo station. In spring, significant increases in precipitation at Yushu station (Fig. S3a2) resulted in a reduced rate of soil temperature increase, with the average rise in shallow soil temperature of only 0.119°C decade^{-1} (Table 2). During winter, the soil temperature increase was relatively small (even negative), with an average rate of 0.306°C decade^{-1} in the shallow layers and −0.137°C decade^{-1} in the deep layers (Table 2) as a result of less winter snow cover at Maduo station (Fig. S3d2). Precipitation or soil moisture plays an important role in soil temperature trends. An increase in soil moisture causes an increase in evaporation rates, which results in energy being absorbed from the surrounding soil, creating a soil temperature decrease despite air temperature increases; an effect known as the soil moisture feedback mechanism (Zhang et al. 2001). A similar situation occurred in summer at Henan station. In terms of the means of the trend magnitudes, soil temperatures at all layers increased faster (Table 2, Fig. S3b1) than the air temperature in summer, possibly due to slightly decreasing precipitation. An opposite tendency was observed in summer at the Maduo station. Soil temperatures at all layers, especially in the deep layers, increased faster (0.710°C decade⁻¹ for shallow and 1.243°C decade⁻¹ for deep layers) than the air temperature (0.526°C decade⁻¹) with significant increasing precipitation (16.000 mm decade⁻¹) in summer. Significant positive correlations (at $\alpha = 0.05$) between soil temperatures and summer precipitation were observed at the Maduo station (Table 5). As shown in Table 3, the maximum depth of frozen ground was more than 2 m deep, and it occurred in spring or summer at Maduo. An increase in precipitation or soil moisture resulted in an increase in thawing rate, which released energy to the surrounding soil because those temperatures were higher than that of the frozen soil, creating a soil temperature increase. This resulted in the significantly positive correlations observed in summer. These results indicate that more precipitation or soil moisture on the ground also results in higher soil temperatures in summer during soil freezing season. Recent studies have shown that summer precipitation in the eastern region of the TP has increased over the past 50 yr (Zhang et al. 2009, Hu & Liang 2013). Increasing trends of frozen ground soil temperature in summer might be enhanced by the increased summer precipitation in the eastern TP.

The maximum depth of frozen ground had a clear and significant effect on soil temperatures at all depths, with negative correlations between these variables. As seen in Table 6, the largest negative correlations were associated with the maximum depth of frozen ground for all stations. These results indicate that frozen ground has some inhibitory effect on increasing soil temperatures because of absorbing phase change energy. As described in Section 5, the strongest trends at the 0 and 80 cm soil layers occurred in winter for the Changdu station. These results also indicate that the warming trends in soil temperature are more affected by air temperature when depth of frozen ground is shallower. Moreover, the changes that occur during the freeze/thaw cycle could affect the increasing trends of soil temperature. From Section 5, smaller warming trends in October were evident at depths of 0–20 cm at all stations. Similarly, smaller warming trends were observed in March, February and January at the above-mentioned depths. Similar results regarding the buffering effect of the freezing and thawing process for seasonal changes in soil temperatures have been found in a simulation experiment conducted by Chen et al. (2014). As reported in Section 5, there were larger warming trends recorded in September at depths of 0–20 cm at all

stations. One possible reason is that the date of the start of soil freezing is occurring later as a result of climate change (Gao et al. 2003). Slower cooling results in greater soil temperature warming trends observed in September. With the earth warming and the maximum depth of frozen ground gradually decreasing, the warming trends in soil temperature will continue due to the weakened inhibitory effect of frozen soil. This will contribute to a positive feedback loop which will continue to hasten soil and air warming.

7. CONCLUSIONS

Significant soil warming was detected in the TRSR of the TP between 1980 and 2014. Rates of increase were 0.706°C decade⁻¹ for surface soils (0 cm), 0.477°C decade⁻¹ for shallow layer soils (0–20 cm) and 0.417°C decade⁻¹ for deep layer soils (40–320 cm). At most of the meteorological stations, surface and 5 cm depth layers of the soil were affected by warming during the winter, whereas soil temperature increases in the summer were more evident in the deeper layers (10–320 cm) than in the upper layers (0–5 cm).

Air temperatures also rose significantly in the TRSR during the 35 yr analysed, with an annual average increase of 0.458°C decade⁻¹. There was a clear effect of air temperature on soil temperatures, as seen from the significant and concurrently increasing trends in both air and soil temperatures, and from the strong correlation between these temperatures. This relationship was stronger in summer and winter than that in spring and autumn.

Precipitation amounts at most stations increased over the study period, but not obviously so. The relationship between precipitation and soil temperatures is very complicated, with precipitation feedback mechanisms occurring in spring and thermal insulating effect of snow occurring in winter. More precipitation on the ground also results in higher soil temperatures during the summer in the frozen soil. The freezing and thawing process also has some inhibitory effects on increasing soil temperatures. The warming trends of soil temperature will continue with the degradation of frozen soil in the TRSR.

Acknowledgements. This work was supported by the National Natural Science Foundation of China (No. 41375077 41130961,and 91537104). We are grateful to the China Meteorological Data Sharing Service System for supplying the data for the TRSR. Special thanks are given to the anonymous reviewers and the editor for very constructive comments.

LITERATURE CITED

Alamusa, Niu C, Zong Q (2014) Temporal and spatial changes of freeze-thaw cycles in Ulan'aodu region of Horqin Sandy Land, northern China in a changing climate. Soil Sci Soc Am J 78:89–96

► Bai Y, Scott TA, Min Q (2014) Climate change implications of soil temperature in the Mojave Desert, USA. Front Earth Sci 8:302–308

► Chen B, Luo S, Lü S, Zhang Y, Ma D (2014) Effects of the soil freeze-thaw process on the regional climate of the Qinghai-Tibet Plateau. Clim Res 59:243–257

Cheng G (1998) Glaciology and geocryology of China in the past 40 years: progress and prospect. J Glaciol Geocryol 20:213–216 (in Chinese with English abstract)

► Cheng G, Jin H (2013) Permafrost and groundwater on the Qinghai-Tibet Plateau and in northeast China. Hydrogeol J 21:5–23

CMA (China Meteorological Administration) (2003) Specifications for surface meteorological observation. China Meteorological Press, Beijing (in Chinese)

► Davidson EA, Janssens IA (2006) Temperature sensitivity of soil carbon decomposition and feedbacks to climate change. Nature 440:165–173

Du J, Li C, Liao J, Lhak P, Lu H (2007) Responses of climatic change on soil temperature at shallow layers in Lhasa from 1961 to 2005. Meteorol Monogr 33:61–67 (in Chinese with English abstract)

Gao R, Wei Z, Dong W (2003) Interannual variation of the beginning date and the ending date of soil freezing in the Tibetan Plateau. J Glaciol Geocryol 25:49–54 (in Chinese with English abstract)

► García-Suárez AM, Butler CJ (2006) Soil temperatures at Armagh Observatory, Northern Ireland, from 1904 to 2002. Int J Climatol 26:1075–1089

Gilbert RO (1987) Statistical methods for environmental pollution monitoring. Wiley, New York, NY

► Gocic M, Trajkovic S (2013) Analysis of changes in meteorological variables using Mann-Kendall and Sen's slope estimator statistical tests in Serbia. Global Planet Change 100:172–182

► Guo D, Wang H (2012) The significant climate warming in the northern Tibetan Plateau and its possible causes. Int J Climatol 32:1775–1781

► Guo D, Wang H (2013) Simulation of permafrost and seasonally frozen ground conditions on the Tibetan Plateau, 1981–2010. J Geophys Res 118:5216–5230

► Hamed KH, Rao AR (1998) A modified Mann-Kendall trend test for autocorrelated data. J Hydrol 204:182–196

Hollander M, Wolfe DA, Chicken E (2013) Nonparametric statistical methods, 3rd edn. John Wiley & Sons, New York, NY

► Hu Q, Feng S (2003) A daily soil temperature dataset and soil temperature climatology of the contiguous United States. J Appl Meteorol 42:1139–1156

Hu H, Liang L (2013) Temporal and spatial variations of rainfall at the east of Qinghai-Tibet Plateau in last 50 years. Plateau Mountain Meteorol Res 33:1–7 (in Chinese with English abstract)

► Jin H, He R, Cheng G, Wu Q, Wang S, Lü L, Chang X (2009) Changes in frozen ground in the source area of the Yellow River on the Qinghai-Tibet Plateau, China, and their eco-environmental impacts. Environ Res Lett 4:045206, doi:10.1088/1748-9326/4/4/045206

Kendall MG (1955) Rank correlation methods. Charles Griffin & Co, London

► Kurylyk BL, MacQuarrie KTB, McKenzie JM (2014) Climate change impacts on groundwater and soil temperatures in cold and temperate regions: implications, mathematical theory, and emerging simulation tools. Earth Sci Rev 138: 313–334

► Lawrence DM, Slater AG, Swenson SC (2012) Simulation of present-day and future permafrost and seasonally frozen ground conditions in CCSM4. J Clim 25:2207–2225

► Liang L, Li L, Liu C, Cuo L (2013) Climate change in the Tibetan Plateau Three Rivers Source Region: 1960-2009. Int J Climatol 33:2900–2916

► Liu X, Chen B (2000) Climatic warming in the Tibetan Plateau during recent decades. Int J Climatol 20: 1729–1742

► Mann HB (1945) Nonparametric tests against trend. Econometrica 13:245–259

► Nelson FE (2003) (Un)frozen in time. Science 299:1673–1675

► Qian B, Gregorich EG, Gameda S, Hopkins DW, Wang XL (2011) Observed soil temperature trends associated with climate change in Canada. J Geophys Res 116:D02106, doi:10.1029/2010JD015012

► Ran Y, Li X, Cheng G, Zhang T, Wu Q, Jin H, Jin R (2012) Distribution of permafrost in China: an overview of existing permafrost maps. Permafrost Periglac Process 23: 322–333

► Sen PK (1968) Estimates of the regression coefficient based on Kendall's Tau. J Am Stat Assoc 63:1379–1389

► Shen M, Piao S, Jeong SJ, Zhou L and others (2015) Evaporative cooling over the Tibetan Plateau induced by vegetation growth. Proc Natl Acad Sci USA 112: 9299–9304

Wei F (2007) Modern climatic statistical diagnosis and prediction technology; 2nd edn. China Meteorological Press, Beijing (in Chinese)

► Wu Q, Zhang T, Liu Y (2012) Thermal state of the active layer and permafrost along the Qinghai-Xizang (Tibet) railway from 2006 to 2010. The Cryosphere 6:607–612

Xu W, Gu S, Su W, Jiang S, Xiao R, Xiao J, Zhang J (2012) Spatial pattern and its variations of aridity/humidity during 1971-2010 in Three River Source Region on the Qinghai Tibet Plateau. Arid Land Geogr 35:46–55 (in Chinese with English abstract)

► Xue X, Guo J, Han B, Sun Q, Liu L (2009) The effect of climate warming and permafrost thaw on desertification in the Qinghai-Tibetan Plateau. Geomorphology 108: 182–190

Yeşilırmak E (2014) Soil temperature trends in Büyük Menderes Basin, Turkey. Meteorol Appl 21:859–866

Yi X, Yin Y, Li G, Peng J (2011) Temperature variation in recent 50 years in the Three-River headwaters region of Qinghai Province. Acta Geogr Sin 66:1451–1465 (in Chinese with English abstract)

You Q, Kang S, Li C, Yan Y, Yan S (2008) Change in extreme temperature over San Jiang Yuan region in the period from 1961 to 2005. Resour Environ Yangtze Basin 17: 232–236 (in Chinese with English abstract)

► You Q, Min J, Fraedrich K, Zhang W, Kang S, Zhang L, Meng X (2014) Projected trends in mean, maximum, and minimum surface temperature in China from simulations. Global Planet Change 112:53–63

► You Q, Min J, Lin H, Pepin N, Sillanpää M, Kang S (2015) Observed climatology and trend in relative humidity in the central and eastern Tibetan Plateau. J Geophys Res 120:3610–3621

Zhang T, Barry RG, Gilichinsky D, Bykhovets SS, Soro-kovikov VA, Ye J (2001) An amplified signal of climatic change in soil temperatures during the last century at Irkutsk, Russia. Clim Change 49:41–76

Zhang W, Li S, Pang Q (2009) Changes of precipitation spatial-temporal over the Qinghai-Tibet Plateau during last 40 years. Ad Water Sci 20:168–176 (in Chinese with English Abstract)

Zheng J, Yin Y, Li B (2010) A new scheme for climate region-alization in China. Acta Geogr Sin 65:3–12 (in Chinese with English abstract)

▶ Zhu X, Wang W, Fraedrich K (2013) Future climate in the Tibetan Plateau from a statistical regional climate model. J Clim 26:10125–10138

11

Long-term seasonality of rainfall in the southwest Florida Gulf coastal zone

Margaret Gitau[1],*

[1]Agricultural and Biological Engineering, Purdue University, 225 South University Street, West Lafayette, IN 47907, USA

ABSTRACT: In addition to supporting fisheries and wildlife, coastal water resources provide a variety of ecosystem services including water purification and storm surge protection, and the intangible such as educational, spiritual, and inspirational benefits. The integrity of these resources is, however, threatened by the inherent impacts of climate variability and change. Seasonality of climate is particularly important as it influences ecosystem diversity and other sensitive ecosystem components that are important to water resource integrity. For example, rainfall seasonality affects water availability, timing of inputs, seasonal water balance, ecological responses, and inter-annual responses of water resource systems. This study examines long-term seasonality of rainfall in the coastal zone with particular focus on the southwest Florida Gulf coastal zone. Analyses show marked inter-annual variations in rainfall seasonality, although trends are not significant. Decadal patterns show a primarily seasonal regime ($0.6 \leq$ Decadal Seasonality Index ≤ 0.79) with one markedly wet season and 2 drier periods. A cyclic pattern in seasonality is discernible on a regional basis, although there seems to have been a shift in cycle spans from 20 yr in the earlier decades to 30 yr in more recent decades, which could render water resource systems more vulnerable to climate change effects. The analyses showed a tendency towards long drier periods along with increases in rainy season (June–September) rainfall and progressive decreases in October–December rainfall. Results provide information useful for management decision-making and a basis for further assessments in the region. Approaches and methodologies are applicable to other coastal areas.

KEY WORDS: Water resources · Coastal zone · Climate · Seasonality · Rainfall

1. INTRODUCTION

In addition to oceans and upwelling areas, coastal water resources comprise a complex combination of estuaries, coastal wetlands, seagrass meadows, bays, gulfs, lagoons and/or sounds, among other features. Estuaries are particularly important in that they provide a mixing zone for freshwater emanating from upstream areas and saltwater tidal influxes. This mixing creates a transitional zone between land and sea that is a diverse and rich ecosystem comprising a unique composition of flora and fauna (US Environmental Protection Agency 2012, Wilson & Farber undated). Coastal water resources provide a variety of ecosystem services including water purification and storm surge protection, and the impalpable such as spiritual and inspirational benefits (US Commission on Ocean Policy 2004, US Environmental Protection Agency 2012, Wilson & Farber undated), in addition to supporting fisheries and wildlife. These resources are currently threatened by a variety of factors including: population growth, rapid urbanization, hydromodification, loss of water to upstream (primarily) agricultural abstraction, water quality impairment, and energy development (US Commission on Ocean Policy 2004, Gill 2005, US Environmental Protection Agency 2012, Wong et al. 2014, Wilson & Farber undated). These impacts are further exacerbated by the inherent impacts of climate variability and change.

*Corresponding author: mgitau@purdue.edu

Climate is in itself an important factor in working towards sustainable development and management of coastal water resources. Seasonality of climate is particularly important as it influences ecosystem diversity and other sensitive ecosystem components (Feng et al. 2013). In particular, rainfall seasonality affects water availability, timing of (both direct and indirect) inputs, seasonal water balance, ecological responses, and inter-annual responses of water resource systems (Tedesco et al. 2008, Feng et al. 2013, Berghuijs et al. 2014). Rainfall seasonality has also been found to play a very important role in watershed classifications (Coopersmith et al. 2012).

Walsh & Lawler (1981) designed an index to assess rainfall seasonality. The index was designed to be a simple and yet effective way to get around the limitation of assessments based on basic analysis of monthly rainfall distributions. Coopersmith et al. (2012) formulated an index derived from the Walsh & Lawler (1981) index. This adaptation was designed to accommodate daily precipitation and was used as part of a suite of measures aimed at characterizing hydrologic similarity of watersheds. Feng et al. (2013) presented an alternate index relating normalized mean monthly

rainfall with rainfall entropy, with the intent being to formulate a global measure of seasonality that captured both magnitude and concentration of rainy seasons. Both the Walsh & Lawler (1981) and Coopersmith et al. (2012) indices have associated interpretive codes for use with the indices, with the former being more detailed and descriptive. The Walsh & Lawler (1981) index is the more commonly used index, with applications primarily relating to spatial comparisons and analysis of long-term variability and trends in precipitation (e.g. Sumner et al. 2001, Kanellopoulou 2002, Hu et al. 2003, Livada & Asimakopoulos 2005, Celleri et al. 2007, Guhathakurta & Saji 2013). This index can also be applied to other climatological and hydrologic components (Walsh & Lawler 1981).

The present study examines rainfall seasonality in the coastal zone with particular focus on the southwest Florida Gulf coastal zone. Specifically, this study determined long-term seasonality of rainfall with a view to identifying existing trends for various stations and across the region. Results provide information that is useful for management decision-making while also providing a basis for further assessments in the region. Approaches and methodologies are applicable to other coastal areas.

2. MATERIALS AND METHODS

2.1. General description

As defined for this study, the southwest Florida Gulf coastal region (Fig. 1) comprises the region between the Florida Keys and Tampa Bay, covering the region west of Lake Okeechobee. In just over 100 years, this once expansive wetland has been transformed into a region characterized by agriculture, industry, and residential development (Charlotte Harbor National Estuary Program 2010), this being largely associated with population growth in the region (Obeysekera et al. 1999). Rainfall occurs throughout the year, although the region experiences distinct wetter and drier periods. The mean annual precipitation in the region ranges between 1320 mm (52 in) and 1420 mm (56 in) based on 1981–2010 climate normals for Arcadia, Fort Myers, and Naples (Fig. 1) as obtained from the Florida Climate Center (http://climatecenter.fsu. edu/products-services/data). Also characteristic of the region are tropical storms and hurricanes, which typically occur about once a year and once in 10 yr, respectively, and usually in August through October (Obeysekera et al. 1999). Notably, the region was impacted either directly or indirectly by major hurricanes

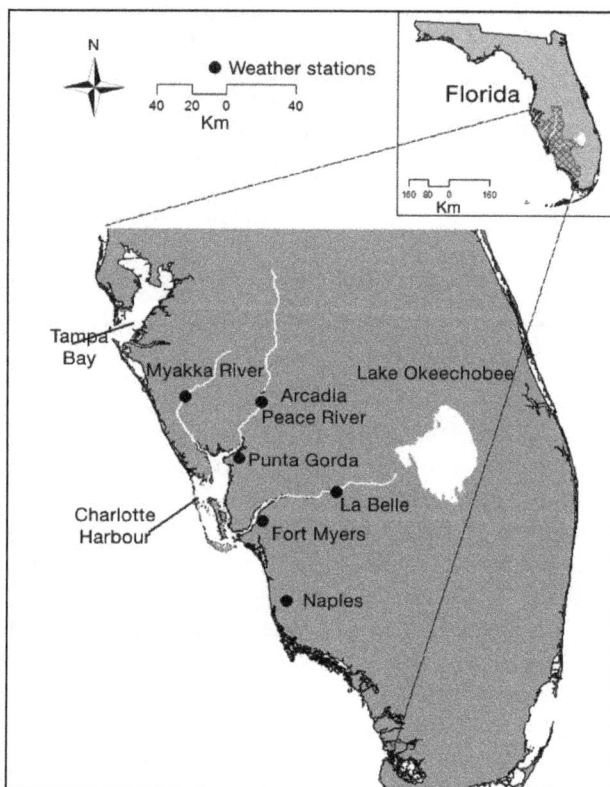

Fig. 1. Southwest Florida region showing weather station locations. Cross-hatching: southwest Florida Gulf coastal region. Rivers from northwest to southeast: Mykka River, Peace River, and Caloosahatchee River

in 1926 (Great Miami), 1928 (San Fellpe–Okeechobee), 1935 (FL Keys Labor Day), 1960 (Donna), 1992 (Andrew), 1998 (Mitch – as tropical storm), 2004 (Charley, Frances, Ivan, Jeanne – as tropical storm), and 2005 (Katrina – as tropical storm, Wilma) (NOAA undated). Temperatures are generally mild, with mean daily temperatures ranging between 15.6°C (60°F, January) and 28.9°C (84°F, August), and maximum temperatures ranging between 22.8°C (73°F, January) and 33.9°C (93°F, July). Hence, seasons in the region are generally related to precipitation rather than temperatures (Obeysekera et al. 1999). The region has long periods of precipitation data ranging from 50 yr at Punta Gorda to 123 yr at Fort Myers (Fig. 1). These data were used in the analyses as described in ensuing subsections.

2.2. Rainfall data preprocessing

Raw rainfall data were obtained through the Florida Climate Center using the center's Downloadable Data Tool (http://climatecenter.fsu.edu/climate-data-access-tools/downloadable-data). This tool allows users to access data from National Weather Service first order and cooperative stations as well as Federal Aviation Administration stations. For each of the stations, all available daily precipitation data were preprocessed to prepare them for analyses. The data were assigned decadal designations (1900s, 1950s, 2000s, etc.) based on year. The decades were further assigned the numbers 0 (1890s) to 12 (2010s) for ease of analyses and cross-reference among datasets. The 2010s were not included in the main analyses; rather, the data were used for comparison purposes and to provide indications of current directions relative to historical trends.

As a starting point, total precipitation for each station was computed on a monthly and annual basis. For analyses, months with missing data were excluded, consistent with Lana et al. (2004). For example, monthly totals were only preserved for months that had no missing values and annual totals were only preserved for years in which all 12 months had a full set of data. The remaining data constituted a sizeable dataset with monthly data ranging from 50 yr at Punta Gorda to 113 yr at Fort Myers, and annual data from 26 yr at La Belle to 84 yr at Fort Myers. Data preserved for decadal analysis were determined considering the extent of completeness in the monthly datasets; because there are 10 yr in each decade, a complete set for any one month in the decade would have 10 entries (10 Jan, 10 Feb, etc.), these being years for

which the specific month had complete data (30 or 31 d as appropriate, 28 or 29 for February). Generally, there were at least 7 entries for each month in each decade (that is, available data for the specific month were complete in at least 7 of the years within the decade) with the exception of La Belle, which had very few data for the 1920s and the 1990s and beyond. Thus, for this station, most of the analyses were carried out for the 1930s to 1980s. Both Arcadia and Fort Myers had some data from the late 1890s; however, these data were largely incomplete, and this decade (1890s) was thus excluded from all analyses. All other available data were used in the analyses.

As part of the preprocessing, initial analysis was conducted on annual rainfall to establish trends in the data on both an annual and a decadal basis. This was done to provide a comparison with subsequent analyses of seasonal trends.

2.3. Rainfall seasonality analyses

For seasonality analyses, this study used the Walsh & Lawler (1981) seasonality index:

$$\overline{\mathrm{SI}} = \frac{1}{\overline{P}} \sum_{n=1}^{12} \left| \overline{P}_n - \frac{\overline{P}}{12} \right| \qquad (1)$$

where $\overline{P}$ is the average annual precipitation for the study period and $\overline{P}_n$ is the mean monthly precipitation for month n over the same period. This index was selected as it is simple and it has been used successfully and reliably in several previous applications (e.g. Sumner et al. 2001, Kanellopoulou 2002, Hu et al. 2003, Celleri et al. 2007, Guhathakurta & Saji 2013).

For this study, an individual year seasonality index (SI_a) was first calculated for each year of data for each station:

$$\mathrm{SI}_a = \frac{1}{P_a} \sum_{n=1}^{12} \left| P_n - \frac{P_a}{12} \right| \qquad (2)$$

where P_a is the total annual precipitation for each year and P_n is the total monthly precipitation for month n. SI_a is only computed for years with complete datasets. The resulting seasonality values were then averaged across the decades (Eq. 3) to provide an indication of decadal seasonality:

$$\overline{\mathrm{SI}}_a = \frac{1}{N} \sum_{i=1}^{N} (\mathrm{SI}_a)_i \qquad (3)$$

where N is the total number of years with complete (12 month) datasets in the decade. This was done so as not to lose important variability information through smoothing of the data (Sumner et al. 2001, as would be the case if the index were computed directly on decadal data.

Resulting SI_a and $\overline{SI}_a$ values were then examined to determine variations in seasonality both on an annual and a decadal basis. Annual seasonality values were also plotted against annual rainfall amounts to determine relationships, if any, between seasonality and rainfall. Further, SI_a values were plotted and analyzed for trends using the non-parametric Kendall test. Finally, $\overline{SI}_a$ values were combined for all stations and plotted on a decadal basis to provide an indication of the time evolution of seasonality across the region.

2.4. Rainfall seasonal trends

Although both SI_a and $\overline{SI}_a$ provide a good indication of rainfall regimes, these indices do not give detailed indications of the distribution of rainfall over the year or of associated amounts and trends, if any, occurring during individual seasons. For example, an SI_a value of 1.1 (most rain falling in 3 months or less; Walsh & Lawler 1981) would be obtained regardless of whether the rainfall occurred in consecutive months or at different points during the year, and similarly for years and/or areas with different amounts of rainfall, as long as most of the rainfall occurred in 3 months or less.

To distinguish between wetter and drier periods, long-term monthly medians were first computed for each station and then plotted on a monthly basis. Seasonal distinctions were then made based on a comparison of individual monthly rainfall to the average of the data. Total rainfall was then computed on an annual basis for each season as identified, and then analyzed for trends. Further, data were analyzed for seasonal distribution (% rain falling in any one season) on both an annual and a long-term basis. Finally, seasonal medians were determined on a decadal basis for each station. These data were then combined and plotted on a decadal basis to provide an indication of the time evolution of seasonal precipitation across the region.

3. RESULTS

3.1. Long-term and decadal trends in annual rainfall

Fig. 2 shows scatter plots of annual rainfall for the 6 stations along with associated long-term trends. Fig. 3 shows decadal median annual rainfall and trends for the same stations. Based on Fig. 2, a downward trend is visible for Arcadia and Punta Gorda rainfall and upwards trends are visible for all other stations. Based on Fig. 3, downward trends were visible in decadal rainfall at Arcadia, Myakka River, and Punta Gorda, while upward trends were visible at Fort Myers, La Belle, and Naples. For the period for which data for Punta Gorda were available (recent but shorter period), a downward trend was visible at La Belle, and the downward trend observed at Myakka River was more pronounced. Although visible, however, none of the trends were significant on either an annual or a decadal basis based on the trends statistics ($p > 0.05$).

3.2. Seasonality analysis

Table 1 shows maximum and minimum individual rainfall seasonality index values for the 6 climate stations in the region while Table 2 shows the seasonality values on a decadal basis. Based on Table 1, the seasons in 1958 were generally equable based on SI_a limits provided in Walsh & Lawler (1981) ($0.20 \leq SI_a \leq 0.39$), although there remained a definite wetter season. A long drier season was experienced at 3 of the stations in 1974 and at one in 1971. Based on seasonality calculations (Table 1), Myakka River seems to be the only station that experienced a long drier season in recent years. On the decadal time scale, the region experienced a primarily seasonal rainfall regime ($0.6 \leq SI_a \leq 0.79$, based on Walsh & Lawler 1981), with the exception of Punta Gorda and Fort Myers, which experienced long drier periods in the 2000s and Myakka River, which experienced long drier seasons in the 1940s and short drier seasons in the 1950s.

Fig. 4 shows a comparison of monthly rainfall patterns for years with high (long drier periods) and low (marked wetter periods) seasonality index values with recent (2010) patterns. Periods in which rainfall was more or less equable over the year are visible in the charts as well as those in which there were shorter wetter periods. Seasonality index values were not necessarily related to total annual precipitation (Fig. 5), thus, years with definite wetter seasons or short drier periods are not necessarily ones with the highest amount of rainfall and those with longer drier seasons will not necessarily have the lowest annual rainfall. Correlations ranged from −0.0449 (La Belle) to −0.3593 (Naples), with the only significant relationship being that for Naples ($p = 0.0247$). Other p-values ranged from 0.0616 (Punta Gorda) to 0.8276 (La Belle).

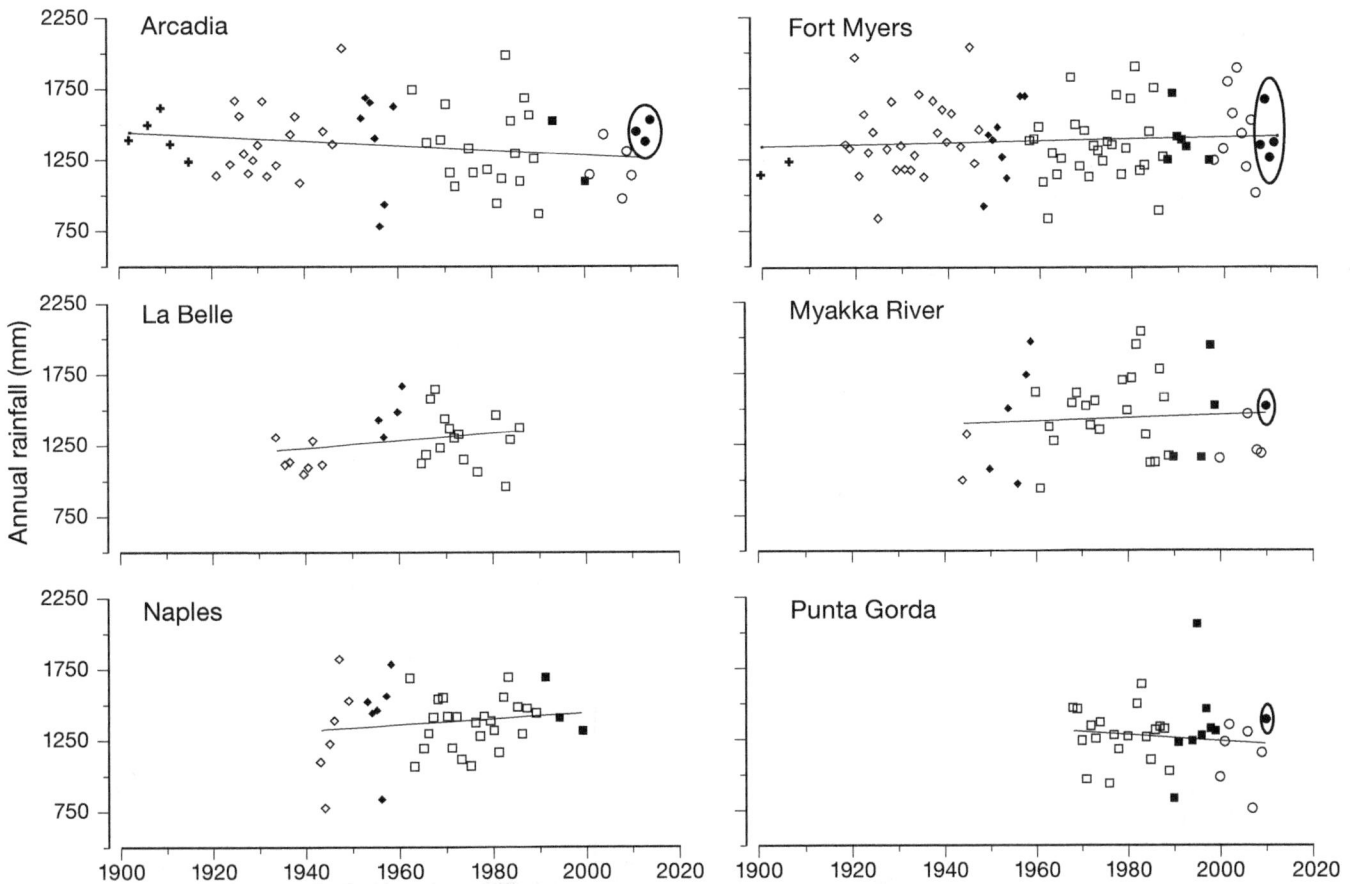

Fig. 2. Trends in annual rainfall for the regional stations. Ellipses show current data (2010–2014) as available for each of the stations. Different markers represent the different decades

Based on Fig. 6, seasonality varied from year to year at all stations. Slight downward trends (tendency towards short drier periods) were observed at Arcadia and Naples, while upward trends (tendency towards long drier periods) were observed at Fort Myers, La Belle, and Punta Gorda. However, none of the trends were significant. A cyclic pattern (marked variations in SI_a followed by periods with less variation) is, however, observable particularly for stations with longer periods of record (Arcadia, Fort Myers, and Myakka River).

Fig. 7 shows a time evolution of $\overline{SI}_a$ across the region. A cyclic pattern in $\overline{SI}_a$ can be observed, with values increasing to a high point and then decreasing to a low point before increasing again. This is not unlike the cyclic patterns generally observed with rainfall over a long period of time. Based on the data shown in Fig. 7, however, there seems to have been a shift in the cyclic periods from 20 yr in the earlier decades up to the 1940s to 30 yr in more recent decades (1950s onwards). Values for the 2010s show a drop from the 2000s $\overline{SI}_a$ levels, which is expected

based on the pattern. However, this (2010s) level is much higher than the previous low (1980s) and is even above the overall mean line ($\overline{SI}_a = 0.7$). As only 5 yr of data are available for the current decade at this point, it is possible that the value could become lower when the decade can be considered in its entirety. It is also worthwhile noting that the high value for the 2000s is higher than other high values in past cycles and that the minimum values in these high-value decades (marked with boxes) show progressive increases across cycles from the 1940s. Higher values of $\overline{SI}_a$ are associated with longer drier seasons, which could indicate a seasonal shift in that direction.

3.3. Analysis of seasonal rainfall

Fig. 8 shows long-term median monthly precipitation at the 6 stations in the region. While seasonality is highly variable across years and decades, as observed in prior analyses, when data are aggregated over the

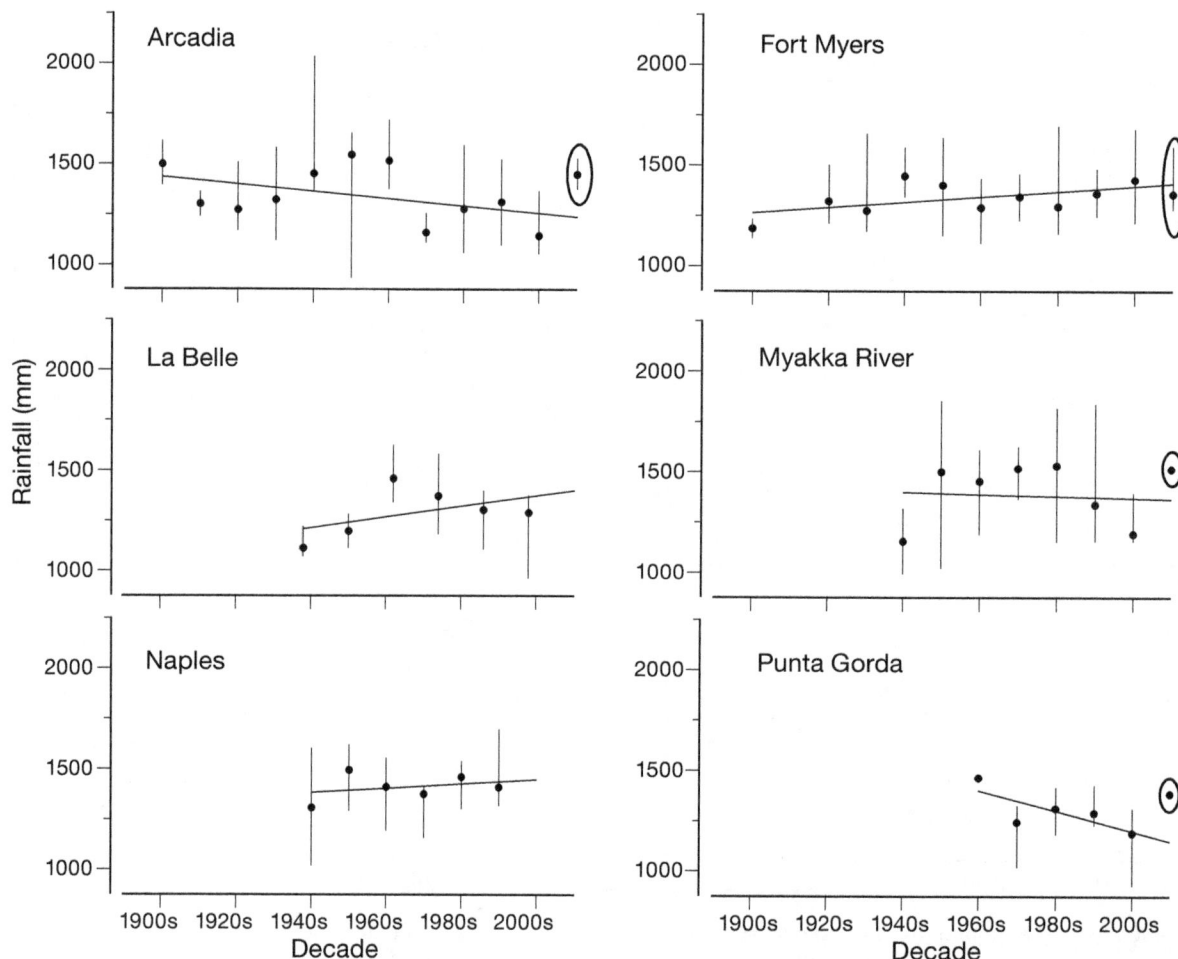

Fig. 3. Trends in decadal median annual rainfall for the regional stations. Error bars: interquartile ranges. Ellipses show current data (2010s) as available for each of the stations

Table 1. Individual seasonality index (SI_a) values for the 6 climate stations in the region. **Bold**: years that were common among the stations with respect to minimum or maximum SI_a

Station	SI_a				
	Minimum		Maximum		
	Value (decade)	Interpretation[a]		Value (decade)	Interpretation
Arcadia	0.38 (**1958**)	Equable but with definite wetter season		0.96 (**1974**)	Markedly seasonal with long drier season
Fort Myers	0.49 (1931, **1958**)	Rather seasonal with short drier season		1.10 (**1974**)	Most rain in ≤3 mo
LaBelle	0.38 (**1958**)	Equable but with definite wetter season		0.90 (1981)	Markedly seasonal with long drier season
Myakka River	0.33 (**1958**)	Equable but with definite wetter season		0.97 (2006)	Markedly seasonal with long drier season
Naples	0.44 (1983)	Rather seasonal with short drier season		0.99 (1971)	Markedly seasonal with long drier season
Punta Gorda	0.45 (1983)	Rather seasonal with short drier season		1.06 (**1974**)	Most rain in ≤3 mo

[a]Source Walsh & Lawler (1981). Larger SI_a values are associated with long drier seasons

long term, a distinct wetter period (June–September) and drier periods (January–May and October–December) are observable. This is consistent with information from the Charlotte Harbor National Estuary Program (2010) and Obeysekera et al. (1999), both documenting a wetter period running from June to September and a drier period lasting from October of one year through May of the following year. For the present analysis, this drier period is split into 2 periods (January–May and October–December) with the wetter period in between, consistent with Fig. 8. While these distinctions provide a generalized picture of the

Table 2. Decadal seasonality index ($\overline{SI}_a$) values for the 6 climate stations in the region

| Station | $\overline{SI}_a$ | | | |
| | Minimum | | Maximum | |
	Value (decade)	Interpretation[a]	Value (decade)	Interpretation
Arcadia	0.60 (1930s)	Seasonal	0.74 (1900s, 2000s)	Seasonal
Fort Myers	0.69 (1930s)	Seasonal	0.82 (2000s)	Markedly Seasonal with a long drier season
LaBelle[b]	0.59 (1950s)	Rather seasonal with a short drier season	0.76 (1980s)	Seasonal
Myakka River	0.58 (1950s)	Rather seasonal with a short drier season	0.85 (1940s)	Markedly Seasonal with a long drier season
Naples[c]	0.65 (1950s)	Seasonal	0.79 (1940s)	Seasonal
Punta Gorda	0.60 (1980s)	Seasonal	0.88 (2000s)	Markedly Seasonal with a long drier season

[a]Source Walsh and Lawler (1981). Larger SIa values are associated with longer drier seasons
[b]Complete datasets not available for 1990s and 2000s
[c]Complete datasets not available for 2000s

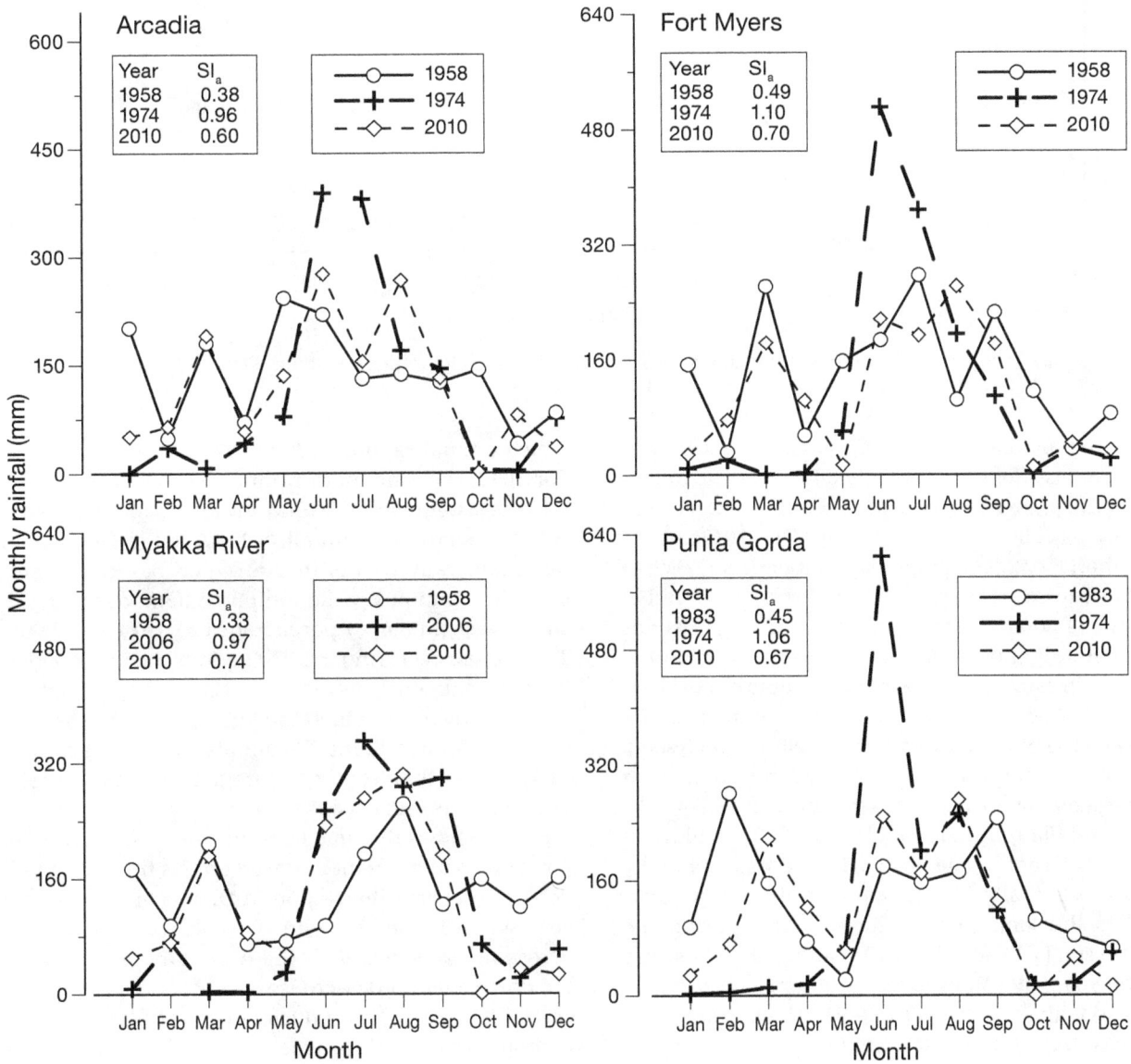

Fig. 4. Monthly rainfall at selected stations for years with high (long drier periods) and low (marked wetter periods) individual seasonality index (SI$_a$) values in comparison to a recent year (2010)

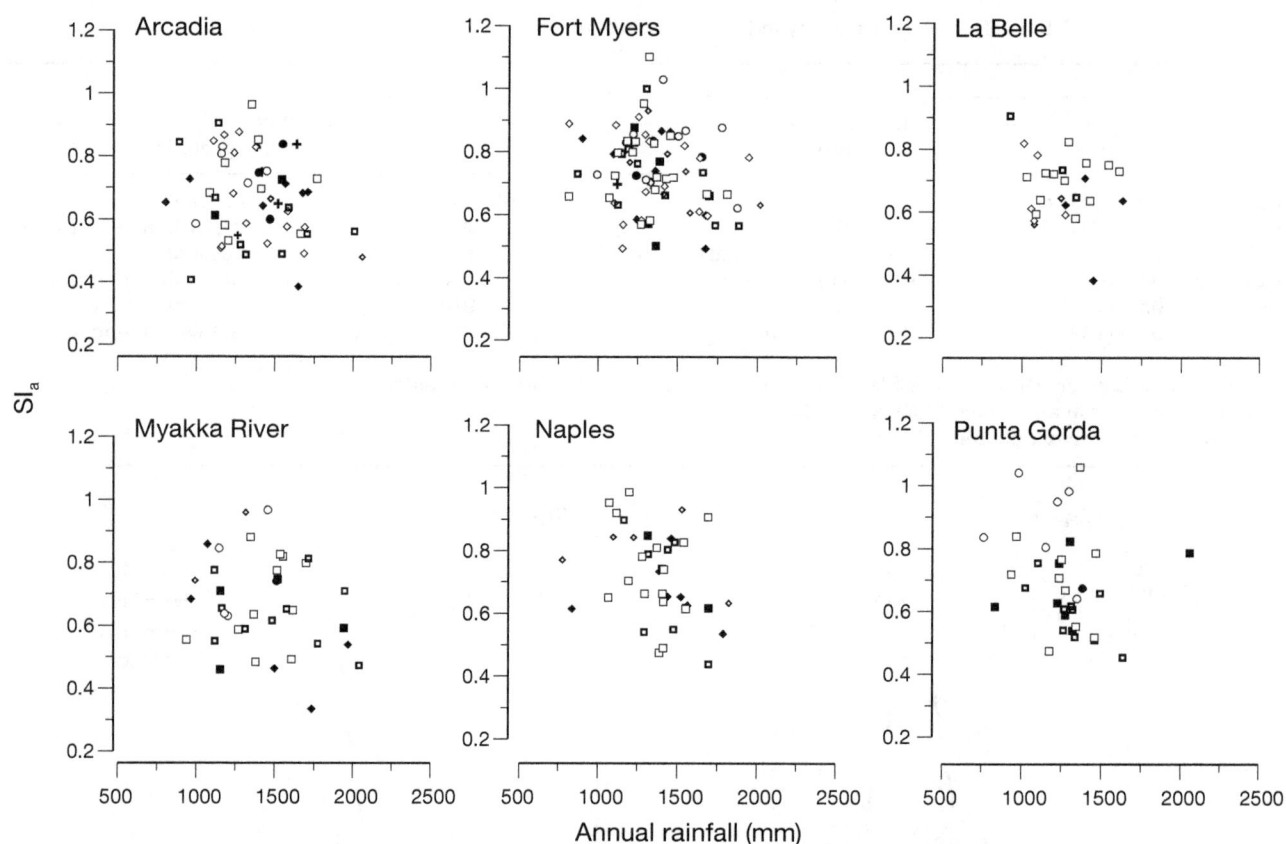

Fig. 5. Relationship between annual seasonality index (SI$_a$) and annual rainfall for regional stations. Different markers represent the different decades

rainfall regime, based on the previous analyses, they provide a basis for separating data for seasonal analyses. For the present study, seasonal designations were assigned as follows: January–May: Dry 1; October–December: Dry 2; and June–September: Rainy. While variations in seasonal precipitation were observed at the various stations across the years (Fig. 9), trends were generally not significant, with the exception of increases in rainy season rainfall amounts at Fort Myers, which were significant on both an annual and a decadal basis (p = 0.0207 and p = 0.0286, respectively). Over the long term, rainfall distribution was more or less the same for all stations, with approximately 63 % of rainfall falling in Rainy and 24 and 13 % falling in Dry 1 and Dry 2, respectively. The region generally had a substantially higher amount of rainfall (compared to the long-term average) falling during the rainy season (>80 %) in both 1974 and 2006. These are also the years that showed longer drier seasons based on seasonality analyses (Table 1). These years also generally had the lowest amount of rainfall experienced across the stations for the years considered, with annual rainfall being at about the climate normal. This is with the exception of Punta Gorda, for

which annual rainfall (1298 mm) was below normal. For 2010, the amount of rainfall falling during Rainy was consistent with the long-term average. However, substantially more rainfall (>30 %) fell in Dry 1 than what generally occurs in the region based on long-term distributions, while the percentage occurring in Dry 2 was similar to percentages in 1974 and 2006. The rainfall occurring in 2010 generally corresponded to regional normals (1320–1420 mm) except at Myakka River, for which the annual rainfall was substantially higher. Rainfall amounts were also substantially higher than regional normal in 1983, especially at Fort Myers and Myakka River.

Fig. 10 shows the time evolution of seasonal rainfall for stations in the region. Based on this figure, rainfall occurring across the region in Dry 1 (January–May) showed variations but did not change substantially across the decades, with the exception of the 2000s, for which a marked decrease was observed. However, rainfall falling in Dry 2 (October–December) showed a progressive decline from about the 1950s to the present decade, while Rainy (June–September) rainfall showed higher levels from the 1990s through the current decade.

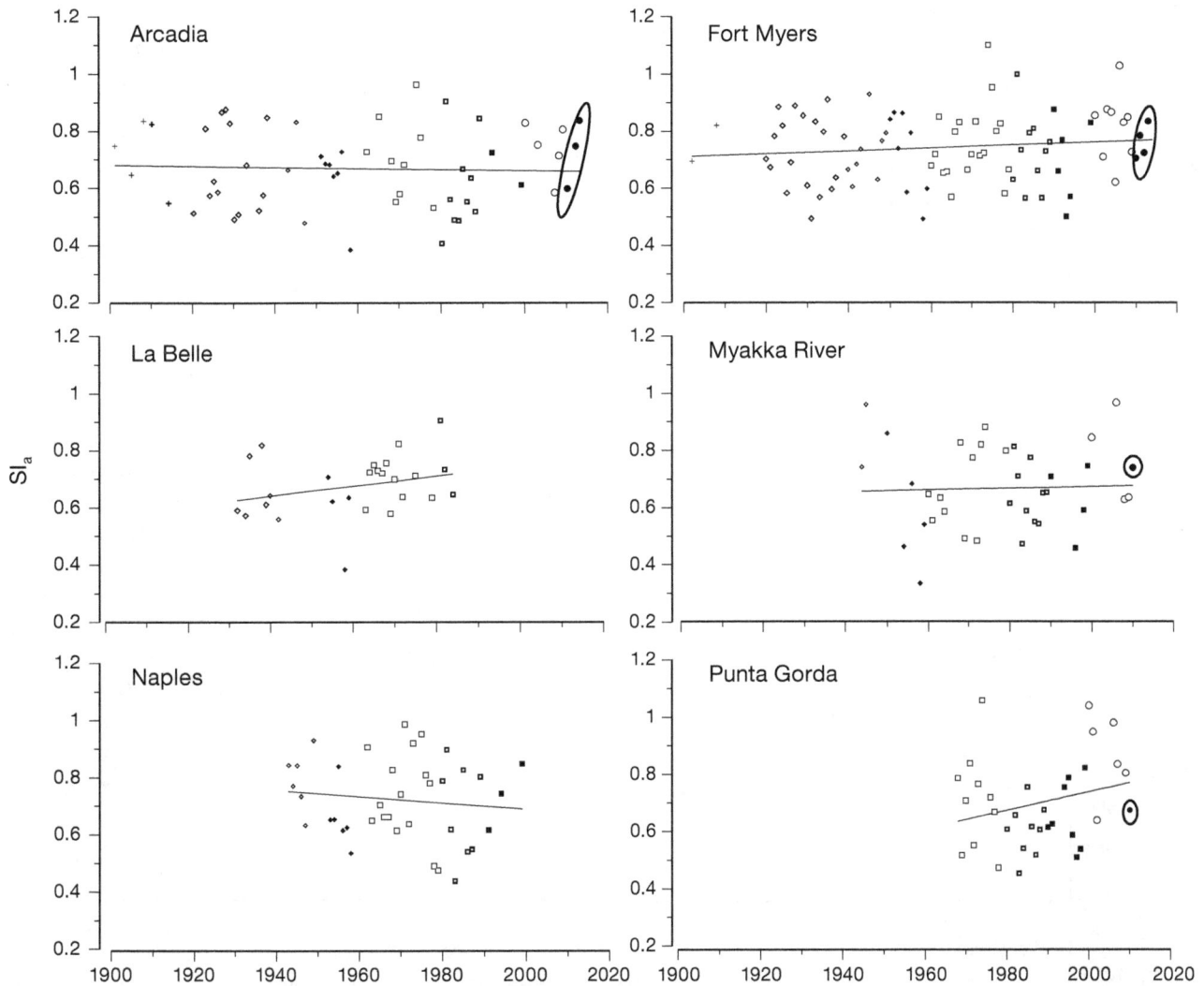

Fig. 6. Trends in annual seasonality index (SI$_a$) for the regional stations. Ellipses show current data (2010–2014) as available for each of the stations. Different markers represent the different decades

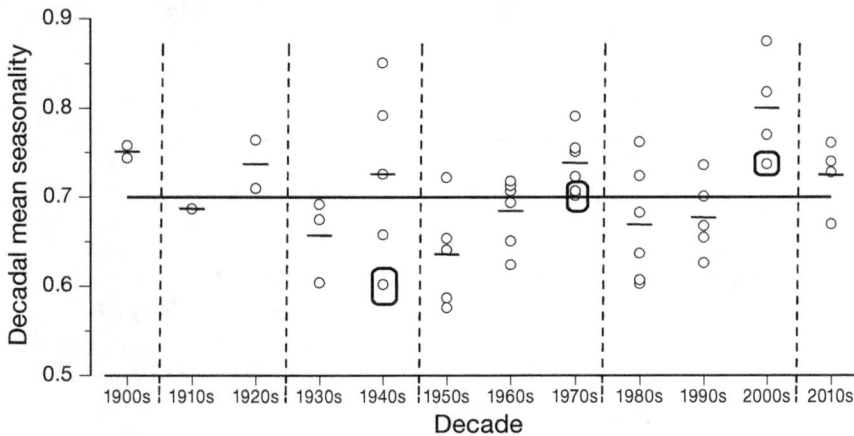

Fig. 7. Time evolution of $\overline{SI}_a$ showing decadal variations across the region. Short horizontal lines: averages for the respective decades. Solid line represents the mean of the data. Dotted vertical lines demarcate periods in which mean seasonality increases progressively from a low point to a high point before dropping back to a low point. Rounded rectangles highlight evolution of lowest values for decades in which highest means occur

4. DISCUSSION

4.1 Seasonal variability and change

Based on the analyses, the region experiences marked variations in rainfall, seasonality, and seasonal rainfall. Within the region, the amount of rainfall falling during a year is affected by climate phenomena such as the El Niño–Southern Oscillation (ENSO), which primarily affects winter rainfall (NOAA 2015), and the Atlantic Multidecadal Oscillation, which affects total rainfall amounts and the occurrence of hurricanes (NOAA 2005). Winters in the region are generally cool and wet during El Niño years and warm and

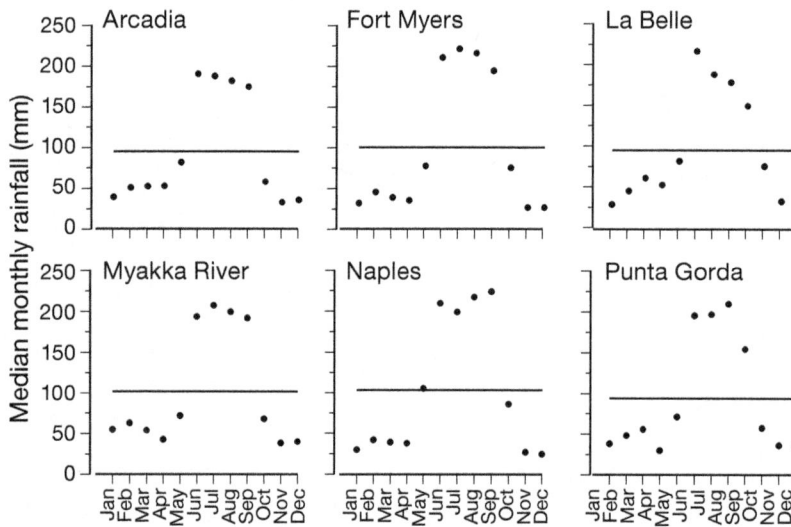

Fig. 8. Long-term median monthly precipitation at the 6 stations in the region. Solid lines: mean of the data

dry during La Niña conditions. More rainfall is experienced in the region when the Atlantic is in its warm cycle. Warm cycles were experienced in 1860–1880 and 1940–1960, while cool cycles were experienced in 1905–1925 and 1970–1990 (Enfield et al. 2001). The ocean was in a warm cycle in 2005 based on NOAA data (NOAA 2005). Similarly, effects on rainfall amounts and seasonality are noted by Poveda et al. (2001) due to ENSO, and by Luo and Zhang (2015) and Arias et al. (2012) due to different climate modulators. These phenomena can also experience shifts, thereby also impacting seasonality (e.g. Luo & Zhang 2015). Thus, there is the need to determine the extent to which these phenomena, and any associated shifts, impact seasonal-

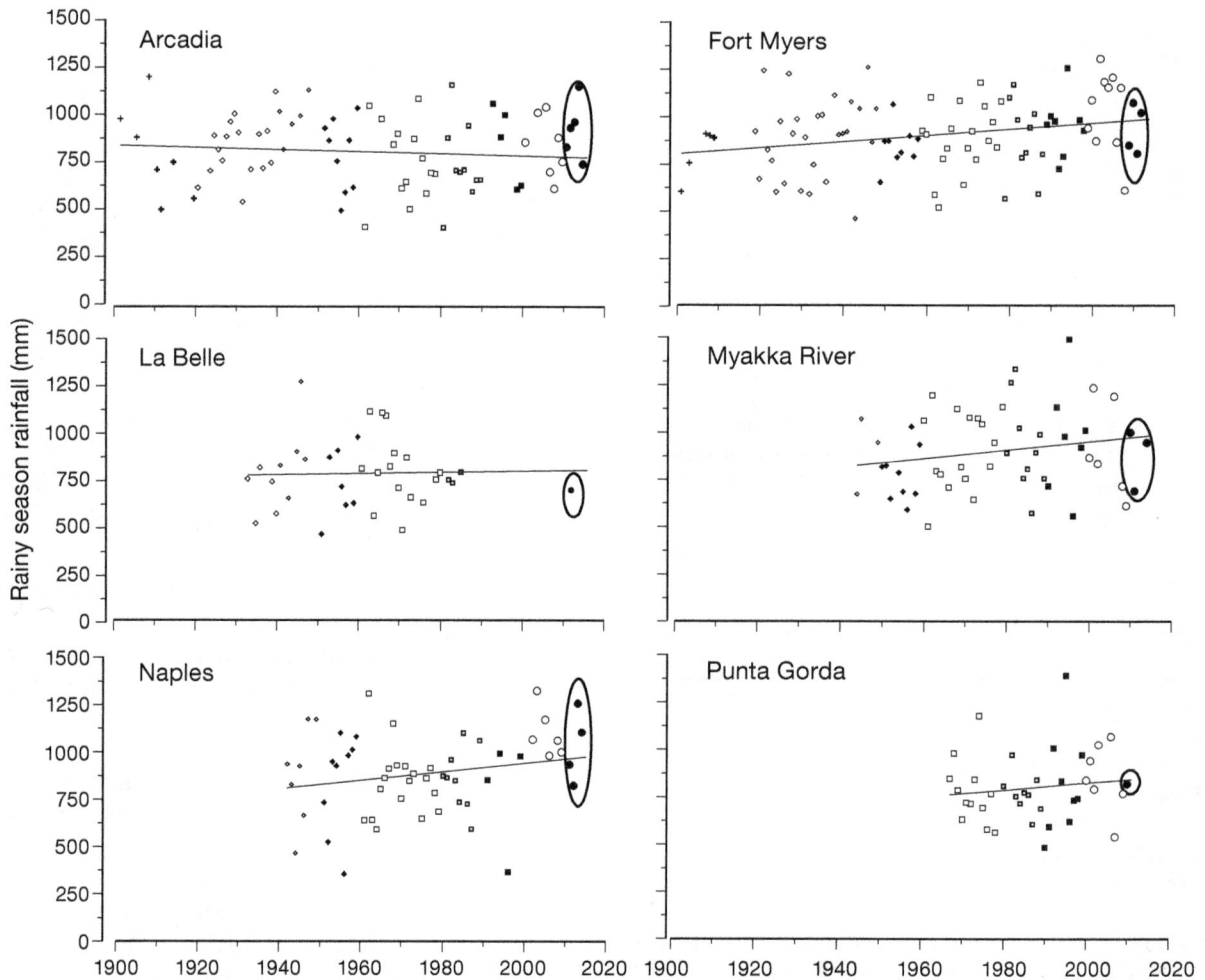

Fig. 9. Trends in rainy season rainfall at region stations. Ellipses show current data (2010–2014) as available for each of the stations. Different markers represent the different decades

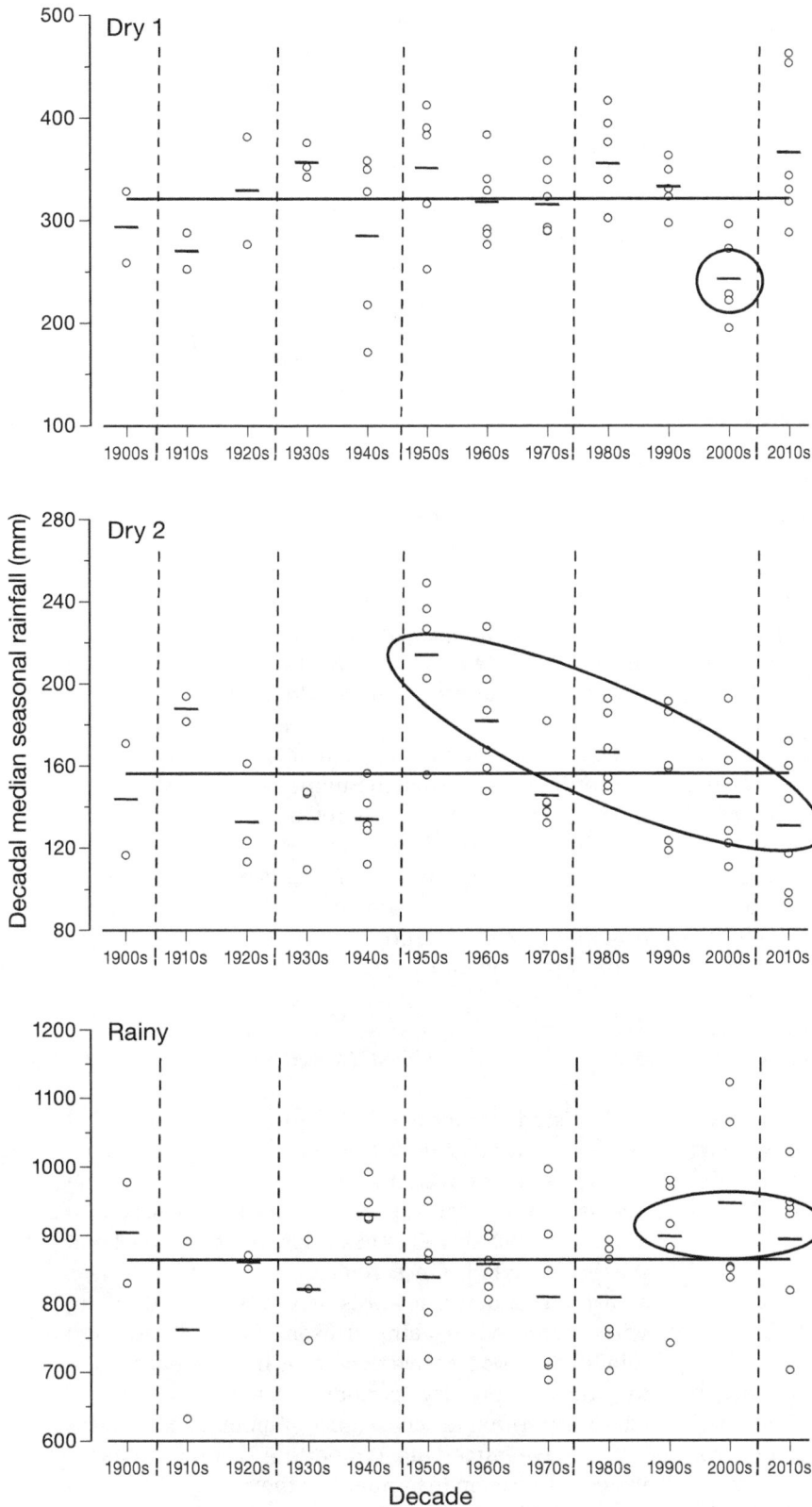

Fig. 10. Time evolution of seasonal rainfall showing decadal variations across the region. Ellipses: occurrences and patterns that warrant attention. See Section 3.3. Short horizontal lines: averages for the respective decades. Solid line: mean of the data

ity in the region. However, such analysis is beyond the scope of this study.

From the analyses, significant trends were not observed in long-term annual rainfall when data were considered on either an annual or a decadal basis. This was consistent with observations by Charlotte Harbor National Estuary Program (2010), although these authors observed statistical significance when data were considered on a moving (10 yr) average basis. The lack of statistically significant trends in annual rainfall amounts could, in part, reflect the effects of the cyclic climate phenomena, as also noted by Pal et al. (2013). However, lack of statistical significance at the annual level does not necessarily imply a static rainfall regime, as variabilities occurring within the year could change over time (for example, as documented by Pryor & Schoof 2008). In the present study, analyses showed variations in seasonality with a tendency towards long drier periods. Variations were also observed in rainfall distribution, with rainfall being distributed more or less evenly in some of the years and occurring primarily during the rainy periods in others. Further, variations were observed in seasonal rainfall, with increases in rainy season rainfall and progressive decreases in one of the dry periods. This is consistent with conservative climate predictions for the region (wetter wet seasons and drier dry seasons; Charlotte Harbor National Estuary Program 2010). Thus an evaluation of only the annual rainfall data would lead to the erroneous conclusion that climate was unchanging in the region.

Changes in rainfall occurrence, amounts, and distribution have immediate and pertinent implications on water resources management in the region, for example, on cycles of water releases and retention in Lake Okeechobee (Enfield et al. 2001), which, in turn, affect freshwater and saltwater influxes in the estuarine region and ultimately impact ecosystem integrity in the region. Long drier periods accompanied by decreases in dry period rainfall constitute drought conditions, thus also posing water management challenges in the region including providing water for agriculture and ensuring adequate supplies for rapidly growing urban populations. Drought conditions also bring about increased risk of wildfire damage and dry-outs of coastal wetlands, which can make it more difficult to adapt to other climate change effects such as sea level rise. Increase in rainy season precipitation also presents challenges, including the need for flood control while also considering the need to protect estuarine areas from excessive freshwater influxes. Fraser (1997) observed that species diversity was impacted by freshwater inflows, with richer diversity being attributed to lower inflows and associated higher salinity but less so during wetter periods.

Among the climate change predictions for the study region are the occurrence of more variable rainfall with longer dry spells and increases in hurricane severity (Stanton & Ackerman 2007). An overall decrease in hurricane and tropical storm activity is also predicted (Bender et al. 2010, Christensen et al. 2013), although increases are predicted in the frequencies of category 4 and 5 hurricanes (Bender et al. 2010) and the nature of anticipated changes in extreme events remains uncertain (Christensen et al. 2013). The tendency towards long drier spells and decreasing rainfall in one of the drier periods is seen in this study. However, changes in seasonality associated with extreme events are difficult to discern from this analysis. None of the maximum or minimum SI_a values were associated with years in which the region was impacted by major hurricanes. Rust et al. (2009) proposed a method of evaluating seasonality in relation to extreme events based on a seasonally varying generalized extreme value model, while Dhakal et al. (2015) used non-parametric approaches to assess temporal changes in the same. Similar analyses would provide insights into changes, if any, occurring in extreme events and their associated seasonality in the study region.

4.2. Seasonality analysis

The seasonality index does not generally account for temporal distribution of rainfall or rainfall amounts, although this can be discerned for SI_a values of zero or near zero, which indicate that rainfall is distributed equally across all months of the year (Walsh & Lawler 1981, Sumner et al. 2001). For all other values, however, an in-depth look at the monthly rainfall is needed to obtain a perspective on when and how much rainfall occurs during the year. An alternative would be to use principal component analysis to study the distribution of rainfall in each year, as well as associated changes and possible trends. This methodology would require datasets without missing values, but could also employ a regional dataset developed by aggregating normalized data from the different regional stations.

The seasonality index, however, has value in that it provides an overall picture of rainfall variability that is easy to compute and understand, and that can be used to assess changes in rainfall regimes over time and across regions. Guhathakurta & Saji (2013) and Sumner et al. (2001), for example, both found increases in seasonality (tendency towards long drier periods) in their respective study regions as well as differences in rainfall regimes across the regions.

A completely equable distribution would mean that each month contributes 8.33% of the total annual rainfall. For the study region, this would translate to 42% of the rainfall occurring in Dry 1, 25% occurring in Dry 2, and 33% occurring in Rainy 1. This scenario is seen in an approximate sense in 1958 at Myakka River but is otherwise not typical of the region. The relative contribution of each month to SI_a can be determined as detailed in Sumner et al. (2001), allowing the evaluation of changes in the significance of contributions from any given month. An analysis of daily values would provide insights on within-season variations and also provide a means of getting around the lack of trend resulting from the impacts of natural phenomena.

5. CONCLUSIONS

This study determined long-term seasonality of rainfall in the coastal zone with particular focus on the southwest Florida Gulf coastal zone. Based on the analyses, the region experiences marked variations in annual rainfall as well as in seasonality and associated seasonal rainfall. On a regional basis, the analyses showed a tendency towards long drier periods along with increases in rainy season (June–September) rainfall and progressive decreases in seasonal rainfall in one of the dry periods (October–December). Changes in rainfall occurrence, amounts, and distribution have immediate and pertinent implications for water resource management in the region. Thus, results provide information that is useful for management decision-making while also providing a basis for further assessments in the region. Approaches and methodologies are applicable to other coastal areas.

Acknowledgements. Many thanks to Sara Raissa Brito Bezerra for her work in organizing and preprocessing the climate data.

LITERATURE CITED

Arias P, Fu R, Mo K (2012) Decadal variation of rainfall seasonality in the North American monsoon region and its potential causes. J Clim 25:4258–4274

Bender MA, Knutson TR, Tuleya RE, Sirutis JJ, Vecchi GA, Garner ST, Held IM (2010) Modeled impact of anthropogenic warming on the frequency of intense Atlantic hurricanes. Science 327:454–458

Berghuijs W, Sivapalan M, Woods R, Savenije H (2014) Patterns of similarity of seasonal water balances: a window into streamflow variability over a range of time scales. Water Resour Res 50:5638–5661

Celleri R, Willems P, Buytaert W, Feyen J (2007) Space–time rainfall variability in the Paute Basin, Ecuadorian Andes. Hydrol Processes 21:3316–3327

Charlotte Harbor National Estuary Program (2010) Charlotte Harbor regional climate change vulnerability assessment. Available at www.chnep.org/GrantsReceived/CRE/Vulner abilityAssessment2-19-10.pdf (accessed 12/17/2015)

Christensen JH, Krishna Kumar K, Aldrian E, An SI and others (2013) Climate phenomena and their relevance for future regional climate change. In: Stocker TF, Qin D, Plattner G-K, Tignor M and others (eds) Climate change 2013: the physical science basis. Contribution of Working Group I to the Fifth Assessment Report of the Intergovernmental Panel on Climate Change. Cambridge University Press, Cambridge

Coopersmith E, Yaeger M, Ye S, Cheng L, Sivapalan M (2012) Exploring the physical controls of regional patterns of flow duration curves. 3. A catchment classification system based on regime curve indicators. Hydrol Earth Syst Sci 16:4467–4482

Dhakal N, Jain S, Gray A, Dandy M, Stancioff E (2015) Nonstationarity in seasonality of extreme precipitation: a nonparametric circular statistical approach and its application. Water Resour Res 51:4499–4515

Enfield D, Mestas-Nunez A, Trimble P (2001) The Atlantic multidecadal oscillation and its relation to rainfall and river flows in the continental US. Geophys Res Lett 28: 2077–2080

Feng X, Porporato A, Rodriguez-Iturbe I (2013) Changes in rainfall seasonality in the tropics. Nat Clim Change 3: 811–815

Fraser TH (1997) Abundance, seasonality, community indices, trends and relationships with physicochemical factors of trawled fish in upper Charlotte Harbor, Florida. Bull Mar Sci 60:739–763

Gill A (2005) Offshore renewable energy: ecological implications of generating electricity in the coastal zone. J Appl Ecol 42:605–615

Guhathakurta P, Saji E (2013) Detecting changes in rainfall pattern and seasonality index vis-a-vis increasing water scarcity in Maharashtra. J Earth Syst Sci 122:639–649

Hu Z, Yang S, Wu R (2003) Long-term climate variations in China and global warming signals. J Geophys Res D Atmos 108:4614

Kanellopoulou EA (2002) Spatial distribution of rainfall seasonality in Greece. Weather 57:215–219

Lana X, Martinez M, Serra C, Burgueno A (2004) Spatial and temporal variability of the daily rainfall regime in Catalonia (northeastern Spain), 1950-2000. Int J Climatol 24:613–641

Livada I, Asimakopoulos DN (2005) Individual seasonality index of rainfall regimes in Greece. Clim Res 28:155–161

Luo X, Zhang Y (2015) Interdecadal change in the seasonality of rainfall variation in South China. Theor Appl Climatol 119:1–11

NOAA (National Oceanic and Atmospheric Administration) (2005) Frequently asked questions about the Atlantic Multidecadal Oscillation. www.aoml.noaa.gov/phod/amo_faq.php (accessed 12/10/2015). NOAA, Silver Spring, MD

NOAA (National Oceanic and Atmospheric Administration) (2015) South Florida climate page. www.srh.noaa.gov/mfl/ ?n=winteroutlookforsouthflorida (accessed 12/10/2015). NOAA, Silver Spring, MD

NOAA (National Oceanic and Atmospheric Administration) (undated) Hurricanes in history. NOAA, Silver Spring, MD. https://nhc.noaa.gov/outreach/history/#top (accessed 12/17/2015)

Obeysekera J, Browder J, Hornung L, Harwell MA (1999) The natural South Florida system. I. Climate, geology, and hydrology. Urban Ecosyst 3:223–244

Pal I, Anderson B, Salvucci G, Gianotti D (2013) Shifting seasonality and increasing frequency of precipitation in wet and dry seasons across the US. Geophys Res Lett 40: 4030–4035

Poveda G, Jaramillo A, Gil M, Quiceno N, Mantilla R (2001) Seasonality in ENSO-related precipitation, river discharges, soil moisture, and vegetation index in Colombia. Water Resour Res 37:2169–2178

Pryor S, Schoof J (2008) Changes in the seasonality of precipitation over the contiguous USA. J Geophys Res D Atmos 113:D21108, doi:10.1029/2008JD010251

Rust H, Maraun D, Osborn T (2009) Modelling seasonality in extreme precipitation. Eur Phys J Spec Top 174:99–111

Stanton EA, Ackerman F (2007) Florida and climate change: the costs of inaction. Tufts University, Global Development and Environment Institute and Stockholm Environment Institute-US Center

Sumner G, Homar V, Ramis C (2001) Precipitation seasonality in eastern and southern coastal Spain. Int J Climatol 21:219–247

Tedesco PA, Hugueny B, Oberdorff T, Durr HH, Merigoux S, de Merona B (2008) River hydrological seasonality influences life history strategies of tropical riverine fishes. Oecologia 156:691–702

US Commission on Ocean Policy (2004) An ocean blueprint for the 21st century. Final Report. US Commission on Ocean Policy, Washington, DC. http://govinfo.library.unt.edu/ oceancommission/documents/full_color_rpt/000_ocean_ full_report.pdf

US Environmental Protection Agency (2012) National Coastal Condition Report IV. US EPA, Washington, DC

Walsh P, Lawler D (1981) Rainfall seasonality: description, spatial patterns and change through time. Weather 36: 201–208

Wilson MA, Farber S (undated) Accounting for ecosystem goods and services in coastal estuaries. In: Pendleton LH (ed) The economic and market value of coasts and estuaries: What's at stake? Restore America's Estuaries, Arlington, VA, p 14–36. www.habitat.noaa.gov/pdf/economic_ and_market_valueofcoasts_and_estuaries.pdf (accessed June 2016)

Wong PP, Losada IJ, Gattuso JP, Hinkel J and others (2014) Climate change 2014: impacts, adaptation, and vulnerability. Part A. Global and sectoral aspects. Contribution of Working Group II to the Fifth Assessment Report of the Intergovernmental Panel on Climate Change. Cambridge University Press, Cambridge

12

Bridge over troubled water-valuing Russia's scientific landscape

Ulf Büntgen[1,2,3,*]

[1]Swiss Federal Research Institute WSL, 8903 Birmensdorf, Switzerland
[2]Oeschger Centre for Climate Change Research, 3012 Bern, Switzerland
[3]Global Change Research Centre AS CR, 61300 Brno, Czech Republic

ABSTRACT: Environmental and climate change not only implies many research needs, but also offers a wide arena for (re)activating collaborations between Russia and the international scientific community. Despite a variety of economic and logistic challenges, as well as political and administrative caveats, I advocate how to help mitigate a deterioration of the Russian Academy of Sciences, to reduce the brain drain from the world's largest country, and to facilitate access to, and the exploring of, unique paleo-archives.

KEY WORDS: Environmental change · Brain drain · Interdisciplinary research · Paleo-archives · Russia · Scientific collaboration

1. MOTIVATION AND BACKGROUND

The negative effects of a dwindling global oil price on the Russian rouble and the country's overall economy, together with increasing political isolation, cast a long shadow on Russia's international reputation, with severe consequences for its scientific community. The situation is further exacerbated by the ongoing 'scientific reform'. For example, the Russian Academy of Sciences is currently experiencing damage to its established structure that is unprecedented since its foundation in 1724 CE by Peter the Great (Gelfand 2013, Schiermeier 2013, 2015a, Yablokov 2014, Stone 2016). Moreover, the philanthropist Dmitry Zimin has recently fallen out of favour with the Kremlin (Schiermeier 2015b), and was thus forced by the Ministry of Justice to close his Dynasty Foundation (Kondrashov et al. 2015) — the nation's first private science-funding organization that was finally labelled a 'foreign agent' by the Russian government (Pokrovsky 2015).

These situations are symptomatic of the present instability in the world's largest country, with wide-ranging consequences. Inflating bureaucracy, politi-cal conflicts of interest and a marked tendency towards 'climate-change scepticism', in tandem with meaningless formalism as well as an increase in pseudoscience and pseudoscientific project mongering, are triggering a new wave of scientific emigration from Russia (Schiermeier 2014), mainly among the youngest and most talented scholars. If this continues, this trend will further reduce the quantity and quality of internationally peer-reviewed publications with Russian contributions — last year Russia published approximately the same number of scholarly papers as Iran (Stone 2016).

It cannot be denied that Russia has made positive commitments, with regard to the 'Paris Agreement on Climate Change' in December 2015, to keep the global temperature rise at <2°C until 2100, and has also recently improved the country's grant system via the Russian Foundation for Basic Research, the new Russian Science Foundation, and via the most recent approval of the next round (no. 5) of 'mega grants' by the Director of the Department of Science and Technologies of the Ministry of Education and Science of the Russian Federation (11 March 2016). Neverthe-

*Corresponding author: buentgen@wsl.ch

less, it is alarming that the President's Scientific Council just announced a further reduction of the number of research institutions (often via combining several institutions) from around 1000 to approximately 150 (Stone 2016).

Although less severe and difficult to compare, the current situation in Russia is somehow similar to the turmoil that occurred just after the transformation of the Soviet Union, when funding was exceptionally low. At that time, Russian scholars had to either leave the country or increase their collaborations with western partners. A prime example of successful international partnership was the first upsurge of Russian tree-ring research in the 1990s (Briffa et al. 1998a, 1998b, Vaganov et al. 1999). Russia's current political and economic instability once more calls for global community support and transnational funding, superimposed on a solid background of trust. More than coincidentally,

the global oil price was at its record low when the Soviet Union collapsed, and is now, after a long-term high, again dramatically dwindling (Fig. 1).

Despite potential caveats, I advocate a twofold action plan: first, to launch and enhance international and interdisciplinary research collaborations with Russia; second, to establish new and maintain existing educational programs and scientific infrastructure in the world's largest country.

2. RESEARCH FOCI

Environmental changes not only generate a wide range of scientific questions and associated tasks, but also offer ample research opportunities between Russia and the international community. Joining forces appears particularly timely in light of Russia's

Fig. 1. Permafrost thawing along the Yana River in northeastern Siberia (70° N, 135° E) emphasizes the severity of recent global warming, and also stresses the urgent necessity for prompt scientific endeavour. Safeguarding unique plant and animal macrofossils released from thawing ice wedges requires state-of-the-art research skills and infrastructure, together with adequate funding schemes, which can only be achieved by strong international alliances. Inset reveals long-term fluctuations of the global oil price, with record lows in the 1980s, before the collapse of the Soviet Union, and again in 2016

ongoing economic crisis and political upheavals (Stone 2016), as well as the forecast rate of climate change (IPCC 2014). Understanding the effects of past, present and predicted temperature and precipitation variability on natural and human systems requires long-term and cross-disciplinary scientific investigations (Büntgen & Di Cosmo 2016, Büntgen et al. 2016). Apart from historical and future time scales, the recent warming trend across the high-northern latitudes seems unprecedented (IPCC 2014). Rapidly rising temperatures not only result in sea-ice reduction, Arctic greening and permafrost thawing (Fig. 1), but also influence river runoff, vegetation dynamics, species composition and unexpected disease revivals, such as the most recent anthrax outbreak in northwestern Siberia. These changes have implications not only for ecosystem services and the global carbon cycle (Gauthier et al. 2015), but also for human and animal health, for instance. At the same time, depopulation and reforestation across large fractions of the agriculturally valuable parts of western Russia cause social and biological challenges, respectively, with hydroclimatic regime shifts further affecting ecosystem functioning and productivity.

Geographical foci, in addition to the boreal taiga (Hellmann et al. 2016), sub-Arctic tundra and Arctic ecosystems, are mountainous biodiversity hotspots such as the Altai Mountains and the Caucasus region that have both experienced glacier retreats in recent decades (Solomina et al. 2016). The Baikal region and the Kamchatka peninsula, as well as the Lena delta, represent ideal locations for cross-disciplinary evaluations of various terrestrial, as well as aquatic and/or marine processes across a wide range of spatiotemporal scales (Büntgen et al. 2014, Meyer et al. 2015). Large parts of Siberia comprise exclusive permafrost sites, where remains of ancient humans and animals can be found, together with plant macrofossils (Pitulko et al. 2016b). Eastern Siberia still offers ample unexplored landscapes (Fig. 1), with Beringia being particularly important for testing the 'standstill hypothesis' (Hoffecker et al. 2014, 2016), as well as for researching conditions of the ancient 'mammoth flora and fauna' (Pitulko et al. 2016a), and Lake El'-gygytgyn representing a unique paleo-archive (Melles et al. 2012). Moreover, the vast expanse of the Russian landscape allows standardized research protocols to be conducted along extensive latitudinal and longitudinal gradients. Thus any serious assessment at the scale of the northern hemisphere or even globally depends on free data access from Russia, be it meteorological measurements or any other kind of environmental recording.

3. STRUCTURAL REQUIREMENTS

Outreach-oriented endeavours, as well as the establishment of a sustainable research infrastructure, should supplement my scientific recommendations. Encouraging interdisciplinary approaches to address timely subjects will generate high-impact publications and/or help obtain lucrative grants. Contributions from eastern and western partners should be carefully balanced with respect to funding, scientific content, and logistical and administrative obligations. International summer schools and field weeks in Russia should be organized and attended by the best lecturers and students from all over the world. Bilateral educational programs for the next generation of Russian scientists should involve PhD and post doc exchange programs, as well as open-access data portals and information forums. Infrastructure investments should include the establishment of long-term monitoring programs, and support for state-of-the-art laboratories and field stations. Nevertheless, most, if not all, of the suggested mid- to large-scale initiatives will require some official approval, and possibly even direct involvement of high-level officials, which might not always be easy to achieve (Nature 2015). Moreover, adequate financing of the necessary programs appears to be difficult (Kondrashov et al. 2015), because many international funding agencies require matching contributions from the Russian side, and the scientific budgets of the Russian Foundation for Basic Research and the Russian Science Foundation are steadily decreasing. Another non-trivial task is the introduction of western scientists to the conventions of Russia's political system. This and related aspects require respectful handling at eye level to best achieve common goals. While scientific interests should always be of the upmost importance, any kind of political and/or economic manipulation must be prohibited.

4. CONCLUSIONS

Like a bridge over troubled water, climate and environmental changes not only open up a wide range of research questions and tasks, but also call for new scientific alliances between Russia and the international community. To mitigate the dramatic deterioration of the Russian Academy of Sciences, to stop the ongoing brain drain and to facilitate access to excellent paleo-archives, global partnerships are urgently needed.

Acknowledgements. Alexander Kirdyanov was instrumental in stimulating this discussion. Olga Solomina, from the Department of Geography within the Russian Academy of Sciences in Moscow, Quirin Schiermeier, German Correspondent of Nature in Munich, as well as 2 anonymous referees kindly provided feedback on earlier versions of this manuscript. I received funding from the Ministry of Education, Youth and Sports of the Czech Republic within the National Sustainability Program I (NPU I), grant number LO1415. This study partly resulted from the interdisciplinary and international framework of the PAGES Euro-Med 2k project.

LITERATURE CITED

▶ Briffa KR, Schweingruber FH, Jones PD, Osborn TJ, Shiyatov SG, Vaganov EA (1998a) Reduced sensitivity of recent tree-growth to temperature at high northern latitudes. Nature 391:678–682

▶ Briffa KR, Jones PD, Schweingruber FH, Osborn TJ (1998b) Influence of volcanic eruptions on Northern Hemisphere summer temperature over the past 600 years. Nature 393:450–455

▶ Büntgen U, Di Cosmo N (2016) Climatic and environmental aspects of the Mongol withdrawal from Hungary in 1242 CE. Sci Rep 6:25606

▶ Büntgen U, Kirdyanov AV, Hellmann L, Nikolayev A, Tegel W (2014) Cruising an archive: on the palaeoclimatic value of the Lena Delta. Holocene 24:627–630

▶ Büntgen U, Myglan VS, Charpentier Ljungqvist F, McCormick M and others (2016) Cooling and societal change during the Late Antique Little Ice Age from 536 to around 660 AD. Nat Geosci 9:231–236

▶ Gauthier S, Bernier P, Kuuluvainen T, Shvidenko AZ, Schepaschenko DG (2015) Boreal forest health and global change. Science 349:819–822

▶ Gelfand M (2013) What is to be done about Russian science? Nature 500:379

Hellmann L, Agafonov L, Charpentier Ljungqvist F, Churakova (Sidorova) O, and others (2016) Diverse growth trends and climate responses across Eurasia's boreal forest. Environ Res Lett 11:074021

▶ Hoffecker JF, Elias SA, O'Rourke DH (2014) Out of Beringia? Science 343:979–980

▶ Hoffecker JF, Elias SA, O'Rourke DH, Scott GR, Bigelow NH (2016) Beringia and the global dispersal of modern humans. Evol Anthropol 25:64–78

IPCC (2014) Climate Change 2014: synthesis report. Contribution of Working Groups I, II and III to the Fifth Assessment Report of the Intergovernmental Panel on Climate Change. IPCC, Geneva, Switzerland, 151 pp.

▶ Kondrashov FA, Kondrashov AS, Gelfand MS (2015) Russian science loses to politics. Nature 522:419

Melles M, Brigham-Grette J, Minyuk PS, Nowaczyk NR and others (2012) 2.8 million years of arctic climate change from Lake El'gygytgyn, NE Russia. Science 337:315–320

▶ Meyer H, Opel T, Laepple T, Dereviagin AY, Hoffmann K, Werner M (2015) Long-term winter warming trend in the Siberian Arctic during the mid- to late Holocene. Nat Geosci 8:122–125

Nature (2015) Russian roulette. Nature 526:475

▶ Pitulko VV, Pavlova EY, Basilyan AE (2016a) Mass accumulations of mammoth (mammoth 'graveyards') with indications of past human activity in the northern Yana-Indighirka lowland, Arctic Siberia. Quat Int 406:202–217

▶ Pitulko VV, Tikhonov AN, Pavlova EY, Nikolskiy PA, Kuper KE, Polozov RN (2016b) Early human presence in the Arctic: evidence from 45 000-year-old mammoth remains. Science 351:260–263

▶ Pokrovsky V (2015) Russia targets Western ties. Science 349: 224–225

▶ Schiermeier Q (2013) Russian academy awaits new head. Nature 497:420–421

▶ Schiermeier Q (2014) Putin's Russia divides scientists. Nature 516:298–299

Schiermeier Q (2015a) Russian science minister explains radical restructure. Nature, doi:10.1038/nature.2015.16776

▶ Schiermeier Q (2015b) Russia turns screw on science foundation. Nature 521:273

▶ Solomina O, Bushueva I, Dolgova E, Jomelli V, Alexandrin M, Mikhalenko V, Matskovsky V (2016) Glacier variations in the Northern Caucasus compared to climatic reconstructions over the past millennium. Global Planet Change 140:28–58

▶ Stone R (2016) Only the strong survive. Science 352:134–138

▶ Vaganov EA, Hughes MK, Kirdyanov AV, Schweingruber FH, Silkin PP (1999) Influence of snowfall and melt timing on tree growth in subarctic Eurasia. Nature 400:149–151

▶ Yablokov A (2014) Academy 'reform' is stifling Russian science. Nature 511:7

13

Projection of drought-inducing climate conditions in the Czech Republic according to Euro-CORDEX models

Petr Štěpánek[1,2,*], Pavel Zahradníček[1,2], Aleš Farda[1], Petr Skalák[1,2], Miroslav Trnka[1,3], Jan Meitner[1], Kamil Rajdl[1]

[1]Global Change Research Institute CAS, Belidla 986/4a, Brno 60300, Czech Republic
[2]Czech Hydrometeorological Institute, Na Šabatce 2050/17, Praha 14306, Czech Republic
[3]Institute of Agriculture Systems and Bioclimatology, Mendel University in Brno, Zemedelska 1, 613 00 Brno, Czech Republic

ABSTRACT: The end of the 20th century and the beginning of the 21st century in the Czech Republic were characterized by frequent extreme water cycle fluctuations, i.e. the occurrence of increased incidences of flood and drought events. Drought occurs irregularly in the Czech Republic during periods with low precipitation amounts. The most noteworthy droughts with significant impact, especially to agriculture, occurred in the years 2000, 2003, 2007, 2009, 2012, 2014 and 2015. A significant increase in frequency and length of drought periods was detected in future climate projections based on the latest model outputs, such as from the Euro-CORDEX 0.11° resolutions for the European area. For these model experiments, the following greenhouse gas emissions scenarios were used: Representative Concentration Pathway (RCP) 4.5 (milder scenario) and RCP8.5 (pessimistic scenario). Since the climate models suffer from potentially severe biases, it is necessary to statistically correct their outputs. For this purpose, a suitable reference dataset was prepared, based on quality-controlled, homogenized and gap-filled station time series. The correction method applied was based on variable correction using individual percentiles. From the corrected model outputs, selected extreme indexes with respect to drought analysis were calculated. From the results, it follows that we can expect both an increase in air temperature and in precipitation (with increased amounts per event), as well as an increase in other extremes with the capability of inducing drought (number of tropical days, heat waves, etc.).

KEY WORDS: Euro-CORDEX simulations · Model bias correction · Climate change · Drought indices · Czech Republic

1. INTRODUCTION

As is shown in several studies in this CR Special, drought constitutes a potential threat to people's livelihoods and socioeconomic development, including in the Central European region (e.g. Brázdil et al. 2016, this issue). Compared to hazards such as floods, drought tends to occur less frequently. However, when it does occur, it generally affects a broad region for seasons or years at a time (UNISDR 2009). Drought originates from a deficiency of precipitation over an extended period of time, usually a season or more. This deficiency can result in water shortage for some activities, groups or environmental sectors. Drought is different from other hazards in that it develops slowly, sometimes over years, and its onset can be masked by a number of factors. Drought is an issue concerning all European Union (EU) countries. According to Spinioni et al. (2016) the drought episodes affected, on average, 15 % of the EU territory and 17 % of the EU population from 2006 to 2010. This caused considerable damages and economic losses that were esti-

*Corresponding author: stepanek.p@czechglobe.cz

mated at over €100 billion (e.g. Spinioni et al. 2016). Water shortage in Europe is a serious problem in many regions (Vogt & Somma 2013), and estimates for the 21st century show an increasing chance of drought across most of the European continent (Stocker et al. 2013). However, the character of these changes is important for adaptation options. For instance, for a number of catchments in Central Europe, the projected average annual changes in water level are relatively small, despite considerable changes in seasonal distribution, even for high-impact scenarios (e.g. Hanel et al. 2012, 2013). This suggests that the development of accumulation capacities might be an adaptation option. However, if the periods of drought cover multiple years, this adaptation option may no longer be efficient. As another example, the timing of drought onset and the maximum deficit is important for the impacts on agricultural production. In the studied region, more crops and larger areas are vulnerable to drought in the first half of the growing season (especially April–June) than in July–September, with the notable exception of maize, potatoes and sugar beet (Hlavinka et al. 2009). The year-to-year variability of drought can be related to circulation patterns, as shown e.g. by Trnka et al. (2009) and Kingston et al. (2015). There is evidence that the trends in soil moisture anomalies in Central Europe are indeed linked to the occurrence of atmospheric circulation patterns that are conducive to drought. It also appears that long-term trends in the frequency of drought-conducive circulation patterns have contributed to a change in the duration and intensity of drought episodes (Trnka et al. 2009). This phenomenon is particularly pronounced during the early vegetation period (April–June), which is crucial both for the productivity of managed ecosystems (e.g. rain-fed field crops) as well as for the net primary production of central European ecosystems as a whole. Recent studies have introduced evidence of decreasing soil moisture content since 1961 (Trnka et al. 2015), and attributed it to increasing CO_2 levels related to anthropogenic forcing (Brázdil et al. 2015).

High-resolution information about future climate is needed for proper adaptation and mitigation of the impacts of climate change and variability. Driven by a suite of IPCC assessment reports and accompanied by increasing public awareness of ongoing climate change, the past decades have seen rapid development in the corresponding methods for climate scenario generation (Kotlarski et al. 2014). The primary tools used for this task are climate models. Unfortunately, high-resolution climate simulations are still not computationally affordable with global climate models (GCMs). A coarse resolution also precludes global models from providing an accurate description of extreme events, which are of fundamental importance to users of climate information with respect to the regional and local impacts of climate variability and change (Giorgi et al. 2009). To obtain climate change information at the regional to local scale, different downscaling techniques are applied on GCMs' outputs. Dynamical downscaling (Giorgi & Mearns 1991) using a regional climate model (RCM) is an example of such a technique.

RCMs (Giorgi & Mearns 1999, Wang et al. 2004) are widely used tools for providing regional climate information over limited areas. The availability and reliability of RCM simulations for Europe has increased rapidly in recent years, thanks to projects such as PRUDENCE (Christensen & Christensen 2007), STARDEX (Goodess et al. 2012), ENSEMBLES (van der Linden & Mitchell 2009), and recently, NAR-CCAP (Mearns et al. 2012) and CORDEX (Giorgi et al. 2009). However, RCMs feature considerable systematic errors (see e.g. Frei et al. 2003, Suklitsch et al. 2010), which hampers easy application of RCM results in climate change impact research.

Since model outputs suffer from systematic errors (due to a necessary simplification of complex real-world processes: coarser spatial resolution, parameterizations, etc.), it is necessary to correct them to obtain meaningful results on the simulated properties of the climate system. Generally, when dealing with mean values of meteorological elements (e.g. seasonal and annual values), the changes given by the models can be treated as they are, without any modifications. The problem occurs in the analysis of daily data and extreme values (such as temperature maxima and minima, precipitation values over given thresholds, etc.), where incorrect statistical distribution simulated by a model (in particular withregards to its tail parts) for a given meteorological element may lead to incorrect conclusions. To cope with distorted statistical moments of different order, different model correcting techniques are applied (a list of these is given in e.g. Themessl et al. 2012). In the present work, the model outputs were corrected using our own correction method (distribution adjusting by percentiles, or DAP) that is based on the quantile mapping (QM) approach of Déqué (2007) (see details in Section 2.3).

Our previous analyses of climate change for the Czech Republic (Štěpánek et al. 2012, Brázdil & Trnka 2015) were based only on the Special Report on Emissions Scenarios (SRES) emissions scenario A1B and 2 models, ALADIN-Climate/CZ, either in 25 or 10 km resolution (Farda et al. 2010), or RegCM (Pal et al. 2007). Here we present new results based on an ensemble of 11 simulations of Euro-CORDEX RCMs (described in Section 2.2). Euro-CORDEX simulations are based on Representative Concentration Pathway (RCP) scenarios (Moss et al. 2008). These scenarios take radiative forcing (W m^{-2}) as the characteristic driving variable, instead of the concentration of the equivalent CO_2 (ppm). The RCPs are consistent with a wide range of possible changes in future anthropogenic greenhouse gas emissions. RCP2.6 assumes that global annual greenhouse gas emissions will peak around 2010–2020, with emissions declining thereafter. Emissions in RCP4.5 are expected to peak around 2040, and then decline. In RCP6 (not used in the present study), emissions peak around 2080; in RCP8.5, emissions continue to rise throughout the 21st century (Meinshausen et al. 2011). The differences between the older results (based on SRES scenarios) and these new results are discussed in Section 4.

Since the volume of obtained results is enormous and there is not enough space in this article to show them all, only selected features of a combination of all 11 experiments for the whole Czech Republic (spatial aggregates) will be presented here (in Section 3). Nonetheless, the obtained material will serve for further analysis and be published on the web portal designated for the exchange of information on climate change impacts, vulnerability and adaptation measures on the territory of the Czech Republic (www.klimatickazmena.cz/en/).

2. METHODS

2.1. Station data

For proper validation of the RCM outputs and their later correction, station data of the highest quality has to be used. First, the underlying raw station data should be subject to thorough quality control. Data quality control applicable to large datasets was developed by Štěpánek et al. (2009). Automation of the process (preserving a good ratio of true and false alarms) was achieved through a combination of several methods of temporal and spatial analysis.

Once the erroneous data are removed from the series during quality control, the series are the subject of homogenization, applying several statistical tests for the detection of inhomogeneities, and found discontinuities are corrected in the daily scale (again, several methods are applied to decrease the uncertainty of the correction estimates). Further details on the homogenization can be found in e.g. Štěpánek et al. (2013) or in the documentation of the software (Štěpánek 2010). Quality control and correction of inhomogeneities were performed on a daily (subdaily) basis for all key meteorological variables (air temperature, precipitation, sunshine duration, relative humidity, wind speed) over the territory of the Czech Republic since 1961 (as well as for neighboring countries, such as the Slovak Republic and Austria, within international projects).

After quality control and homogenization, missing values were filled in. Calculation of the 'new' values was based on geostatistical interpolation methods, improved by standardization of neighboring stations' values to the altitude of a given location by means of regional regression analysis (Štěpánek et al. 2011). Parameter settings of the calculation differ for each meteorological element, and the optimal settings were found by means of cross-validation.

Data quality control, homogenization and the filling in of missing values led to the creation of a so-called 'technical' series for mean, maximum and minimum temperatures, precipitation totals, sums of sunshine duration, relative humidity (mean water vapor pressure) and wind speed. These were calculated for 268 climatological and 787 rain-gauge stations of the Czech Hydrometeorological Institute (CHMI) network in the 1961–2009 period, and actual values are continually being added. Despite the fact that a smaller number of stations was available for some of the studied characteristics (e.g. for sunshine duration or water vapor pressure), the 'technical' series were completely calculated (both for arbitrary station location and regular gridded network). In this way, we have a complex set of meteorological variables for each position of a climatological station, which could easily be applied in any climate analysis or impact study in this territory.

2.2. Model simulations

Our analysis of future climate conditions is based on RCM simulations prepared within the European part of the global Coordinated Regional Climate Downscaling Experiment (CORDEX, www.cordex.org). CORDEX is an international effort supported by the World Climate Research Programme (WRCP), aimed at producing a set of climate change projections covering individual world regions with multiple RCMs and several emissions scenarios. Thus, the climate research community gets more reliable information on future climate parameters, including information on related uncertainty. To account for greenhouse gases and aerosol forcing, RCP scenarios are used (van Vuuren et al. 2011). The GCM output from the Coupled Model Intercomparison Project Phase 5 (CMIP5; Taylor et al. 2012) has been utilized as the source of driving data for the RCMs. Generally, CORDEX models follow a unified model setup, including control, historic (hindcast) and future climate projection experiments.

The European domain of CORDEX is covered within the frame of the Euro-CORDEX sub-project (www.euro-cordex.net). Model experiments are performed here in 2 spatial resolutions: 0.44° and 0.11°. In total, 10 different RCMs and 13 driving GCMs have been employed. In our paper, we focus only on the 0.11° resolution experiments forced by the RCP2.6 (van Vuuren et al. 2007), RCP4.5 (Clarke et al. 2007) and RCP8.5 (Riahi et al. 2007) scenarios, respectively. The following RCMs were used in our study: ALADIN53, CCLM4-8-17, HIRHAM5, RACMO22E and RCA4. Two of the 5 RCMs were driven by >1 GCM. The selection of the experiments was based on their availability in July 2015, and is summarized in Table 1. Also included in Table 1 is the identification code of the particular simulation (e.g. r1i1p1) taken from the CMIP5 GCM ensemble to drive an RCM.

2.3. RCM bias correction

The climate simulated by numerical models shows systematic deviations from reality (true observed climate), which limits their applicability for impact models. Therefore, climate model outputs have to be post-processed to match the observed climate (Christensen et al. 2008, Maraun 2013). One common way of dealing with model errors in climate change impact studies is the delta change approach (Räty et al. 2014). Besides the delta approach, more sophisticated RCM post-processing methods have been proposed and evaluated, and their list is given in e.g. Themessl et al. (2012). These approaches belong to the family of model output statistics (MOS), a concept developed in weather forecasting and now commonly used in climate science (Maraun et al. 2010).

In a comprehensive intercomparison study of 7 empirical-statistical downscaling and error correction methods (DECMs) for daily precipitation from a 10 km resolved RCM, Themessl et al. (2011) conclude that QM outperforms all other investigated DECMs. In ad-

Table 1. Selected Euro-CORDEX experiments of regional climate models (RCMs) and their driving global climate models (GCM)

RCM	Driving GCM	GCM ensemble member	Scenarios
ALADIN53	CNRM-CM5	r1i1p1	RCP4.5, RCP8.5
CCLM4-8-17	CNRM-CM5	r1i1p1	RCP4.5, RCP8.5
	EC-EARTH	r12i1p1	RCP4.5, RCP8.5
	MPI-ESM-LR	r1i1p1	RCP4.5, RCP8.5
HIRHAM5	EC-EARTH	r3i1p1	RCP4.5, RCP8.5
RACMO22E	EC-EARTH	r1i1p1	RCP4.5, RCP8.5
RCA4	CNRM-CM5	r1i1p1	RCP4.5, RCP8.5
	EC-EARTH	r12i1p1	RCP2.6, RCP4.5, RCP8.5
	HadGEM2-ES	r1i1p1	RCP4.5, RCP8.5
	IPSL-CM5A-MR	r1i1p1	RCP4.5, RCP8.5
	MPI-ESM-LR	r1i1p1	RCP4.5, RCP8.5

dition, they also show — at least for daily precipitation linear regression approaches, although optimized by predictor transformation and randomization — that RCM error characteristics are not systematically reduced by these methods. The distribution mapping method was recommended as the best-performing correction method by Teutschbein & Seibert (2013), where various bias correction techniques were compared (delta change correction, linear transformation, local intensity scaling [LOCI], power transformation, variance scaling, distribution mapping), finding that QM was best able to cope with non-stationary conditions.

Based on the those results, QM was chosen for the bias correction purposes in the present work. Our method originates from an approach described in e.g. Déqué (2007). It is applied as parameter-free (using empirical cumulative density distributions, rather than theoretical cumulative distribution functions). An empirical method is recommended over the parametric one, since the latter is not robust enough, given the limited length of the time period (Gutjahr & Heinemann 2013), and also, using theoretical distribution, QM becomes less flexible in its application to different parameters and regions as *a priori* information about the shape of the probability density functions is needed (Themessl et al. 2012).

Based on validation of the QM method within model control runs, we further adopted some settings that best suit the purpose of bias correction of various meteorological elements (including precipitation, which is difficult to handle on both distribution tails). For example, the final corrections (obtained for individual percentiles) were smoothed with a low-pass Gaussian filter (over 20 percentiles) to reduce noise in the individual percentile values. Each month was treated separately and a time window including the previous and following month was applied: thus,

Table 2. Model bias for air temperature (°C) as difference between original (uncorrected) model and reality, areal averages for different altitudes

Altitude (m)	CNRM-CM5_ ALADIN	EC-EARTH_R ACMO	EC-EARTH_ RCA	HadGEM2-ES_ RCA	MPI-ESM-LR_ CCLM
0–300	−2	−2.46	−1.81	−0.3	−0.7
300–600	−1.92	−2.21	−1.76	−0.24	−0.66
600–900	−1.39	−1.92	−1.6	−0.06	−0.43
900–1200	−0.51	−1.42	−1.19	0.34	0.18
Above 1200	0.37	−0.4	−0.32	1.23	1.23
Whole Czech Republic	−1.83	−2.21	−1.74	−0.21	−0.62

we get rid of the steps between the individual months and at the same time comply with different bias sizes in different parts of a year. To preserve reasonable extrapolated values (in the tails of the distribution), changes between individual values of the highest (or lowest) percentiles (likely to be very noisy) were limited to certain values (such as a coefficient of 1.5 for maximal extrapolated value compared to the last percentile, and a ratio of 3.0 as a change between the last percentiles values).

The QM method was applied on a daily basis and for each grid cell/location separately. To be suitable for impact studies where station data are preferred (and because these data are also available for the current climate), the correction/localization was done by finding the nearest grid points for a given location (station) and applying the correction several times, 5 times in the case of precipitation to 10 times in the case of other elements. In practice, the first (nearest) neighbor was applied as the final correction, but the other results were used to evaluate uncertainty coming from the correction process.

We call this correction method DAP (distribution adjusting by percentiles), simply to distinguish it from other QM methods, since it differs by the above-mentioned parameters settings. For the data processing, the software packages AnClim (Štěpánek 2008), LoadData and ProClimDB (Štěpánek 2010) were created. They offer a complex solution, from tools for handling databases, through data quality control, to homogenization of time series, as well as time series analyses, extreme value evaluation and model output verification and correction. The software is available on the webpage www.climahom.eu.

3. RESULTS

3.1. Bias in model data

Over the Czech Republic, we found bias patterns similar to those discussed in Kotlarski et al. (2014) and briefly described in Section 4 (the present study). In this subsection we summarize our main findings for some of the meteorological variables influencing evapotranspiration and drought occurrence.

Biases between projection and reality were analyzed, in detail, mainly for 5 selected experiments. A control run was compared with the real meteorological data. For spatial comparison, individual maps with values interpolated into 500 m resolution were obtained for each data source (stations or model grid points).

Air temperature is underestimated by uncorrected models (Table 2). The greatest differences were observed for the experiment EC-EARTH_RACMO for which the average annual temperature is about 2.2°C lower than reality. In spring, it is underestimated by about 4°C. The lowest biases were achieved by HadGEM2-ES_RCA, where the difference from reality is only −0.2°C. Overall, for all 5 selected experiments, the worst results were found in the spring season (Fig. 1).

A bias analysis was also performed in regard to different altitudes (since model orography differs from actual orography). We chose 5 levels: up to 300 m, 301–600 m, 601–900 m, 901–1200 m and >1200 m. The results are surprising. The highest model biases are observed within the lower altitudes (up to 300 m); in contrast, the model simulations for mountain regions are relatively non-biased. Two of the experiments are different, HadGEM2-ES_RCA and MPI-ESM-LR_CCLM, which have quite accurate results. These 2 experiments did, however, overestimate the temperature in the highest mountains (Fig. 2).

For a selected experiment (EC-EARTH_RACMO22, whose values are, after bias correction, in the middle of a value spread of other models), we tested whether the bias is constant or changes over time. Spatial biases for different decades of the control run are shown in Fig. S1 (in the Supplement at www.int-res.com/articles/suppl/c070p179_supp.pdf). The biggest underestimation is observed in the case of older values (first decades of the control run period). A bias of about

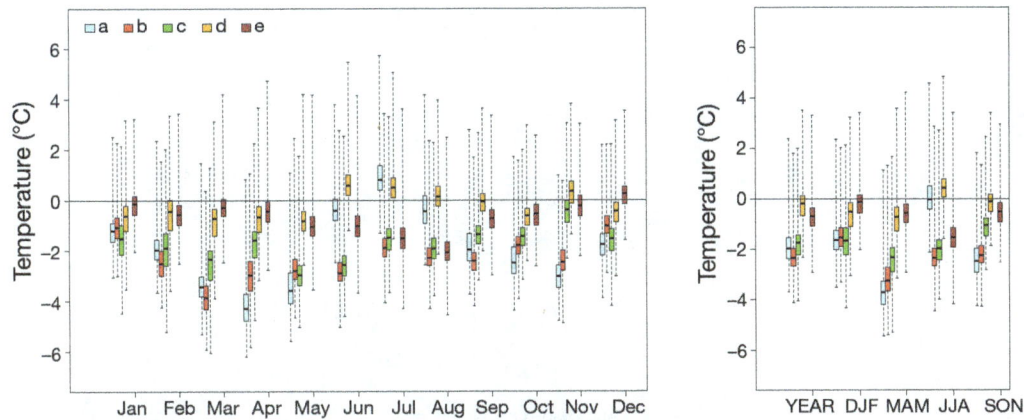

Fig. 1. Temperature bias, difference between original (uncorrected) model and reality, for 5 experiments, by month (left panel) and by season (right panel): (a) CNRM-CM5_ALADIN (1961–2005), (b) EC-EARTH_RACMO (1961–2005), (c) EC-EARTH_RCA (1970–2005), (d) HadGEM2-ES_RCA (1970–2005), (e) MPI-ESM-LR_CCLM (1961–2005). Boxplots—central line: median; box: interquartile range (IQR); whiskers: outlier limits (1.5 × the IQR)

−2.5°C is found for the period 1961–1970 (Table 3), while a bias of only −2°C is found in the last years of the control run (1991–2005). This means that the modeled air temperature increase in the current climate is more rapid than it is in reality.

Precipitation sums are overestimated by the uncorrected model outputs (Fig. 3). From the 5 selected experiments, the wettest conditions are modeled by MPI-ESM-LR_CCLM; its average daily precipitation is higher by about 0.65 mm (Table 4). In contrast, almost bias-free precipitations are simulated by the EC-EARTH_RACMO experiment. The remaining 3 models overestimated the precipitation by about 0.35 mm d^{-1}. Spring is wetter compared to the other seasons.

More precipitation is simulated for the Bohemia (west) region than for Moravia (east) (Fig. 4). Spatial differences of biases by altitude are not as evident as in the case of the air temperature. The precipitation sums in lowlands are overestimated, especially by

the CNRM-CM5_ALADIN and MPI-ESM-LR_CCLM experiments (Table 4). Mountain regions are modeled with a higher amount of precipitation in the case of the EC-EARTH_RCA and HadGEM2-ES_RCA experiments. In contrast, the EC-EARTH_RACMO experiment predicts lower precipitation sums than the reality for altitudes above 600 m.

Table 3. Model bias for air temperature (°C) as difference between original (uncorrected) EC-EARTH_RACMO22 and reality, areal averages for the Czech Republic

Period	Average	Minimum	Maximum
1961–1970	−2.46	−4.18	0.38
1971–1980	−2.27	−4.11	0.66
1981–1990	−2.1	−3.89	0.84
1991–2000	−2.09	−3.8	0.97
2001–2005	−1.9	−3.69	1.28

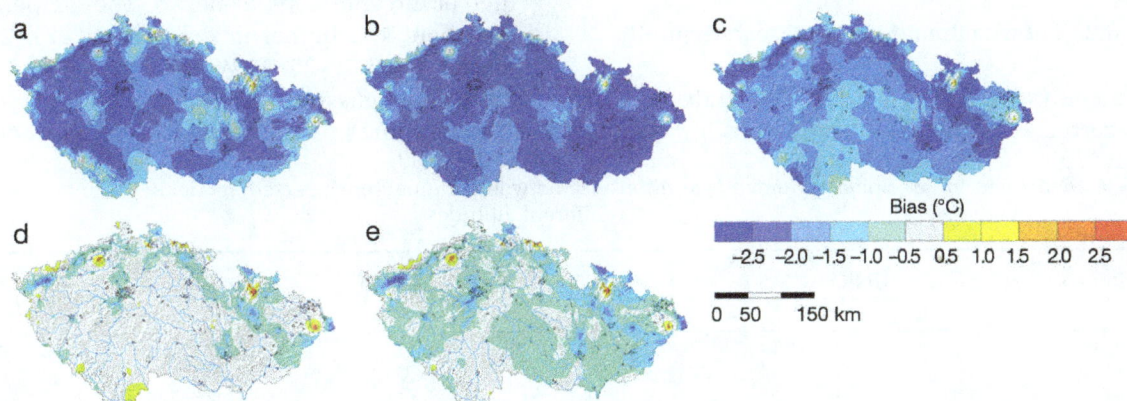

Fig. 2. Temperature spatial bias for 5 experiments across the Czech Republic: (a) CNRM-CM5_ALADIN (1961–2005), (b) EC-EARTH_RACMO (1961–2005), (c) EC-EARTH_RCA (1970–2005), (d) HadGEM2-ES_RCA (1970–2005), (e) MPI-ESM-LR_CCLM (1961–2005)

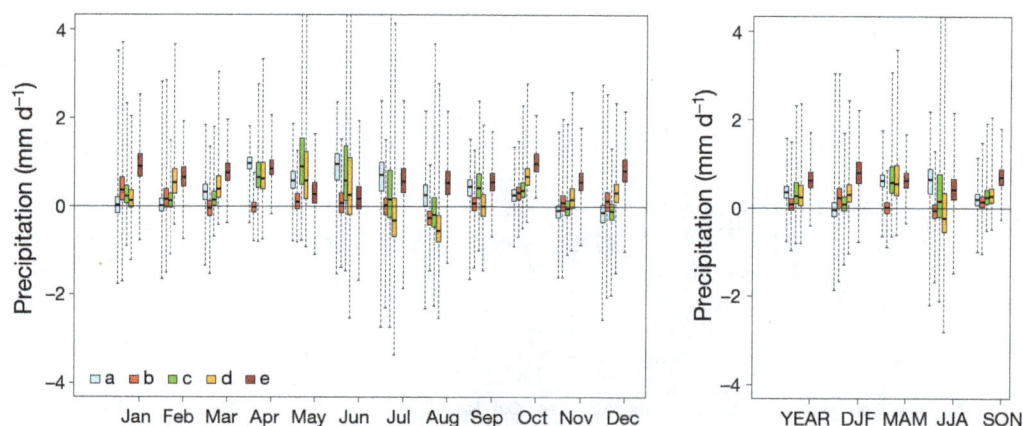

Fig. 3. Precipitation bias, difference of original (uncorrected) model and reality, for 5 experiments, by month (left panel) and by season (right panel): (a) CNRM-CM5_ALADIN (1961–2005), (b) EC-EARTH_RACMO (1961–2005), (c) EC-EARTH_RCA (1970–2005), (d) HadGEM2-ES_RCA (1970–2005), (e) MPI-ESM-LR_CCLM (1961–2005). Boxplots—central line: median; box: interquartile range (IQR); whiskers: outlier limits (1.5 × the IQR)

The number of days with precipitation ≥1 mm are overestimated by RCMs (Fig. 5, left), the same as with precipitation sums. The overestimation ends with number of days with ≥10 mm (with underestimation for JJA and overestimation for MAM). For 20 mm, JJA and DJF are underestimated (resulting in the whole year being underestimated). For 50 mm, all seasons are underestimated (more JJA and DJF). Some examples are given in the Section 3.2.

The initially large differences among the individual experiments (as seen e.g. on Fig. 5) become more consistent after the bias correction. From this, we can conclude that for impact studies in which absolute values play an important role (compared to climatological analysis, which are usually based only on value changes between various periods), bias correction is necessary to obtain meaningful results comparable with the current station (baseline) period.

3.2. Future climate for the Czech Republic

We analyse future climate in this study based on bias-corrected data. To comprehensibly estimate

change in climate for the whole area of the Czech Republic, simple means over all possible grid points were calculated (in the Discussion section, we give information about comparison of various ways of areal averaging). For time series analysis, the values of the individual experiments were smoothed with a 10 yr low-pass Gaussian filter to get rid of incomparable individual yearly values. To better assess possible change based on all the available experiments, an ensemble mean was created from the individual corrected model outputs (see Figs. S2 & S3 in the Supplement), and is further used in this study.

Based on all available experiments, air temperature in the Czech Republic will increase by 2.0°C annually by the end of the 21st century using RCP4.5, or by 4.1°C in the case of the RCP8.5 scenario compared to the reference period (1981–2010). As can be seen in Fig. S2 (in the Supplement), the air temperature will increase similarly to the year 2050 irrespective of the emissions scenarios. The temperature will be about 1°C higher in the period 2021–2040 compared to 1981–2010. We see growing differences between emissions scenarios after the year 2050. Temperatures predicted by RCP8.5 rise steeply, and,

Table 4. Model bias for precipitation (mm d^{-1}) as difference between original (uncorrected) model and reality, areal averages for different altitudes

Altitude (m)	CNRM-CM5_ ALADIN	EC-EARTH_ RACMO	EC-EARTH_ RCA	HadGEM2-ES_ RCA	MPI-ESM-LR_ CCLM
0–300	0.41	0.11	0.18	0.12	0.51
300–600	0.36	0.07	0.33	0.29	0.68
600–900	0.22	−0.01	0.68	0.67	0.78
900–1200	0	−0.25	0.97	0.97	0.66
Above 1200	−0.1	−0.48	0.98	0.98	0.27
Whole Czech Republic	0.34	0.06	0.36	0.32	0.65

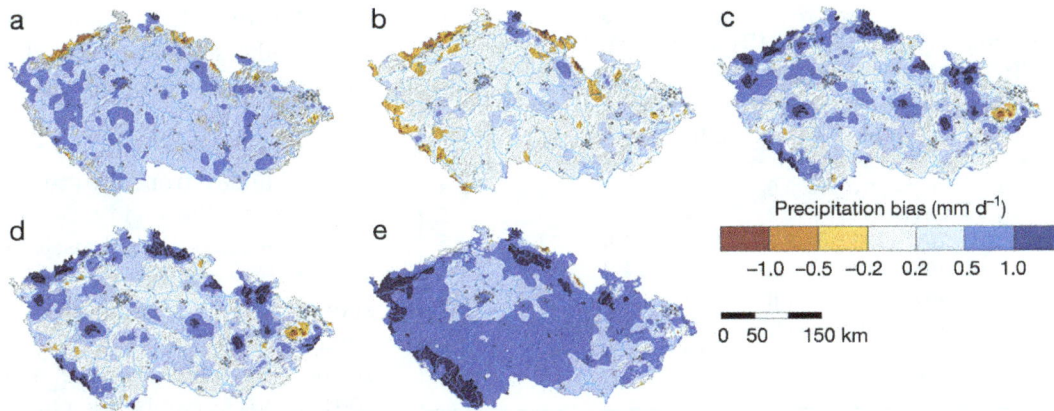

Fig. 4. Precipitation spatial bias for 5 experiments across the Czech Republic: (a) CNRM-CM5_ALADIN (1961–2005), (b) EC-EARTH_RACMO (1961–2005), (c) EC-EARTH_RCA (1970–2005), (d) HadGEM2-ES_RCA (1970–2005), (e) MPI-ESM-LR_CCLM (1961–2005)

for example, the HadGEM2-ES_RCA experiment (having one of the highest trend values) gives, by the end of this century, climate warming of about 5°C compared with the reference period 1981–2010 (Fig. S4 in the Supplement). In contrast, RCP4.5 maintains a practically stable climate from 2061, with a temperature higher by about 2°C compared to the present. According to RCP2.6, the trend will even become negative (and still statistically significant, p = 0.05) by the end of the 21st century. Of the individual seasons, the highest increase in air temperature is modeled for winter. By the end of the 21st century, winter temperature should be higher by about 4.9°C (RCP8.5) (Table 5).

Maximum temperature is to increase mainly in winter, and the least in spring. Absolute maxima reach values of 2.3°C for the year and 3.4°C for winter (RCP4.5) and 4.6°C for the year and 6.0°C for winter (RCP8.5), respectively. Minima are expected to increase even more, again mainly in winter (4.5°C) and then in spring (3.5°C) for RCP4.5 and by 8.3°C (winter) and 8.3°C (spring) for RCP8.5. The minima increase in annual values is similar to those of winter.

Precipitation sums are distinguished by high spatial and temporal variability. This is determined mainly by atmospheric circulation; the amount of precipitation depends on the type of synoptic situation. The complex orography of the Czech Republic has a significant influence as well. Long-term changes in rainfall are not detected. The annual variability is stronger than the trend.

Projection of the precipitation sums based on all 11 experiments shows a slight increase of about 7–13% for RCP4.5 and 6–16% for RCP8.5. Higher amounts of precipitation are observed by the end of the 21st century (Fig. S3 in the Supplement). A statistically significant trend (8.3 mm per 10 yr, p = 0.05) is found for RCP4.5 for the period 2061–2100. The RCP8.5 emissions scenarios give a statistically significant trend of 16 mm per 10 yr in the period 2021–2060 and 13 mm per 10 yr in 2061–2100. RCP2.6 supposes an increase in precipitation only in the first period, 2021–2060 (14.7 mm per 10 yr). The biggest difference is observed for winter precipitation, where the increase could be up by 35% by the end of the 21st century (Table 6). In con-

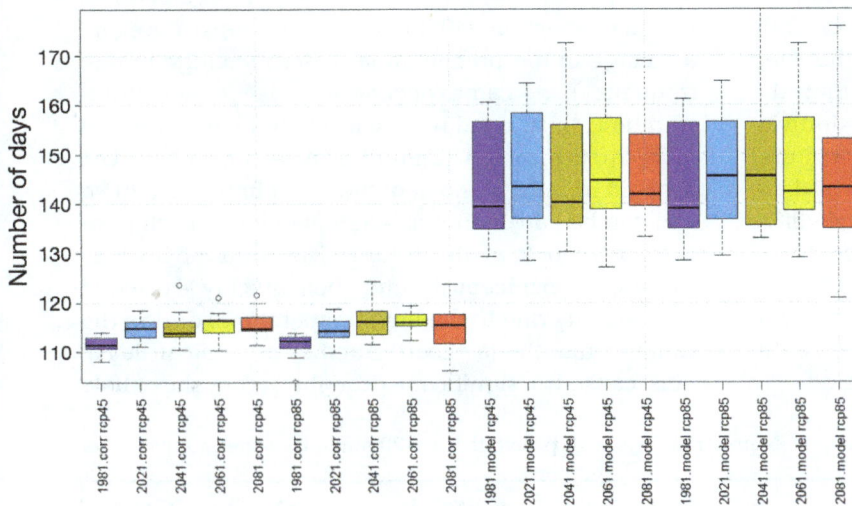

Fig. 5. Boxplots over all 11 experiments for number of days with precipitation ≥1.0 mm (corr: bias-corrected model outputs; model: original model values), for 30 yr (1981–2010) and future 20 yr periods (beginning of periods given on x-axis), and 2 scenarios: RCP4.5 and RCP8.5. Boxplots—central line: median; box: interquartile range (IQR); whiskers: outlier limits (1.5 × the IQR); small circles: outlier values (values beyond the outliers limit). Colours show different decades

Table 5. Difference in air temperature (°C) calculated from 11 experiments for individual periods and seasons (DJF: winter, MAM: spring, JJA: summer, SON: autumn) compared with reference period 1981–2010

Emissions scenario	Period	Year	DJF	MAM	JJA	SON
RCP4.5	2021–2040	0.9	1.1	0.8	0.7	0.8
	2041–2060	1.3	1.4	1.3	1.3	1.1
	2061–2080	1.8	2.2	1.8	1.7	1.5
	2081–2100	2.0	2.4	1.9	1.7	1.7
RCP8.5	2021–2040	1.0	1.1	1.1	0.9	0.9
	2041–2060	1.8	2.1	1.8	1.6	1.8
	2061–2080	2.8	3.3	2.8	2.6	2.6
	2081–2100	4.1	4.9	3.8	3.8	3.9

Table 6. Ratio of precipitation sums calculated from 11 experiments for individual periods and seasons (DJF: winter, MAM: spring, JJA: summer, SON: autumn) compared with reference period 1981–2010 (100%)

Emissions scenario	Period	Year	DJF	MAM	JJA	SON
RCP4.5	2021–2040	106.6	109.3	105.9	105.0	107.4
	2041–2060	107.0	110.5	111.5	100.9	108.7
	2061–2080	110.3	115.9	115.1	104.4	109.5
	2081–2100	112.7	114.0	119.3	107.5	112.4
RCP8.5	2021–2040	106.5	110.6	109.3	103.4	106.2
	2041–2060	112.2	120.4	115.4	105.8	112.3
	2061–2080	113.7	126.1	118.7	104.3	113.8
	2081–2100	116.3	135.1	123.5	102.4	115.9

trast, the smallest change can be expected in summer precipitation.

As can be seen in Fig. S5 (in the Supplement), the changes in the precipitation sums are not spatially consistent. Again using the example of the Had GEM2-ES_RCA experiment, it is shown that the smallest increase should occur in South Moravia, which is among the most important agricultural regions. The differences between the periods and the emissions scenarios are large. Quite a large difference can be observed between the periods 2041–2060 (RCP4.5) and 2061–2080 (RCP8.5). In the first case, similar precipitation sums as for the present are predicted, but with the latter case, significantly higher precipitation sums of >20% are modeled.

During the last decade (in the current climate in the Czech Republic), we have observed a change in precipitation patterns but with no similar change in the average. To capture such behavior, other precipitation characteristics also need to be investigated. We analyzed the number of days with precipitation equal or higher than 1, 10, 20 and 50 mm. No statistically significant trends (for p = 0.05) are observed for the number of days with 1 mm and higher, but for 10, 20 or 50 mm, positive statistically significant (p = 0.05) linear trends were found. The increase of these intense rainfalls is mainly predicted by emissions scenario RCP8.5. For example, the number of days above 10 mm will increase in RCP8.5 by about 0.6 d per 10 yr in the period 2021–2060 and by about 0.5 d per 10 yr in 2061–2100. As can be seen in Table 7, the differences between the individual models are not so large. In the future, there will be about 1 additional day of intense rainfall of ≥20 mm compared to the present.

3.3. Drought in the future

Drought is becoming a very important phenomenon in our region, as it has been more and more frequent in recent years (drought occurred in the Czech Republic in 2012, 2013, 2014 and 2015). The reason for the droughts in the Czech Republic is below-normal amounts of precipitation and/or very high temperatures. The new Euro-CORDEX experiments predict slightly higher sums of precipitation, but, in connection with increased air temperature and a change in the precipitation pattern (change in rain frequency), we can expect an increase in evapotranspiration; thus, conditions will likely favour drought more in the future (Zahradníček et al. 2015). The drought in 2015 was one of the worst in the last 20 yr, and can be considered an example of how such periods could look in the future. The drought in 2015 started inconspicuously and then quickly escalated dramatically due to the high temperatures during the summer months (a record number of tropical days). This caused a significant drought across the whole

Table 7. Number of days with precipitation >20 mm for 3 experiments

Period	RCP	EC-EARTH_ RACMO	HadGEM2-ES_ RCA	MPI-ESM-LR_ CCLM	Czech Republic
1981–2010					4.6
2021–2040	RCP4.5	4.7	5.2	5.5	
	RCP8.5	4.5	5.4	5.0	
2081–2100	RCP4.5	5.3	5.8	5.7	
	RCP8.5	6.2	5.9	5.9	

country. It is for this reason we focus mainly on temperature extremes in this section.

Several consecutive days with high temperatures cause heat waves, which have the potential to deepen the drought. As a threshold for defining a heat wave for our analysis, we used 3 d with temperatures above 30°C. The results are presented in the example of the 3 experiment outputs (Table S1 in the Supplement). In the 1981–2010 period, 3.7 d yr^{-1} with heat waves occurred. No significant difference between RCP4.5 and RCP8.5 is predicted for the near future. The EC-EARTH_RACMO experiment gives an even higher number of days in heat waves for RCP4.5. In any case, the number of days nearly doubles. The large increase in the number of such days and also the increase in the difference between both emissions scenarios occur during the last period of 2081–2100. Compared to the other 2 experiments, HadGEM2-ES_RCA models a significantly higher number of days in heat waves. This experiment predicts 1 mo (33 d) in heat waves for emissions scenario RCP8.5, which is 10 times more than in the baseline period.

A tropical day occurs when the maximum air temperature reaches or exceeds the limit of 30°C. The number of tropical days occurs only a few times a year, but, in the last 2 decades, the number has significantly increased. Such days can be described as uncomfortable for both people and nature. It causes an increase in evapotranspiration and quicker drying of the landscape. If we compare the number of tropical days in the 1960s with the beginning of the 21st century, it is occurring almost twice as often. The number of tropical days has increased mainly in the Moravian lowlands and lowlands around the river Elbe, which are places of importance for agricultural activity (Rožnovský & Zahradníček 2014).

In the first future period (2021–2040), we do not observe a significant increase in the number of tropical days. The values correspond with those in the 2000s and 2010s. Greater variance in the projections between the models and even different emissions scenarios is observed for the end of the century. Emissions scenario RCP4.5 predicts twice the number of tropical days than observed in the period 1981–2010. The EC-EARTH_RACMO experiment predicts about 20% fewer tropical days than the other 2 models (Table S2 in the Supplement). Big differences are modeled by RCP8.5 (Fig. 6). Model HadGEM2-ES_RCA calculated about 50% higher number of tropical days than the other studied experiments. Such a significant increase in these hot days may cause major problems, not only in terms of drought, but also in the population's health, the

energy sector, etc. Interestingly, in 2015, we measured 35–40 tropical days in the Czech Republic, which was higher than the projections of most models for the end of the 21st century.

4. DISCUSSION AND COMPARISON WITH PREVIOUS RCM RESULTS

As mentioned in the Introduction and Section 2.3, the models suffer from biases. Given our knowledge about physical processes in the atmosphere, computational possibilities, etc., results within the same group of models usually have similar problems. Kotlarski et al. (2014) summarizes some of these biases evaluated from the ERA-Interim-driven (Dee et al. 2011) Euro-CORDEX RCMs. They point to a predominant cold and wet bias in most seasons and over most parts of Europe and a warm and dry summer bias over southern and southeastern Europe. The other well-known issue with RCMs is, for instance, systematic underestimation of the dry-day frequency and, on the other hand, overestimation of light (between 0.1 and 1 mm d^{-1}) and heavy precipitation frequency (Themessl et al. 2012). As has been shown in Section 3, we confirm similar bias patterns in our results for the Czech Republic.

In Section 3.2, we presented results based on bias-corrected results and spatial aggregation over all available grid-points. To answer a possible question about the role of location density in the estimation of such an areal average for the Czech Republic, we analyzed averaging based on several versions of input datasets. The datasets are: the simple average of the values of 523 grid points of Euro-CORDEX simulations, and averages over 268 (air temperature) in respect to 787 (precipitation) station locations. We compared several characteristics for a 30 yr period (1981–2010): air temperature, number of tropical days, precipitation sum and number of days with precipitation ≥1 mm. As a reference dataset, we used an areal average based on a 500 m resolution grid layer obtained through geostatistical interpolation, namely, regression kriging applying the dependence of input station data (268 for temperature, 787 for precipitation) on altitude, longitude, slope, exposition and roughness. When comparing the results from these 4 data sources, they are practically the same, with the differences being a maximum of 0.1°C for air temperature and 4% in the case of precipitation or number of days. To conclude, the density of the Euro-CORDEX grid points does not have much effect on the results; it is comparable with other ver-

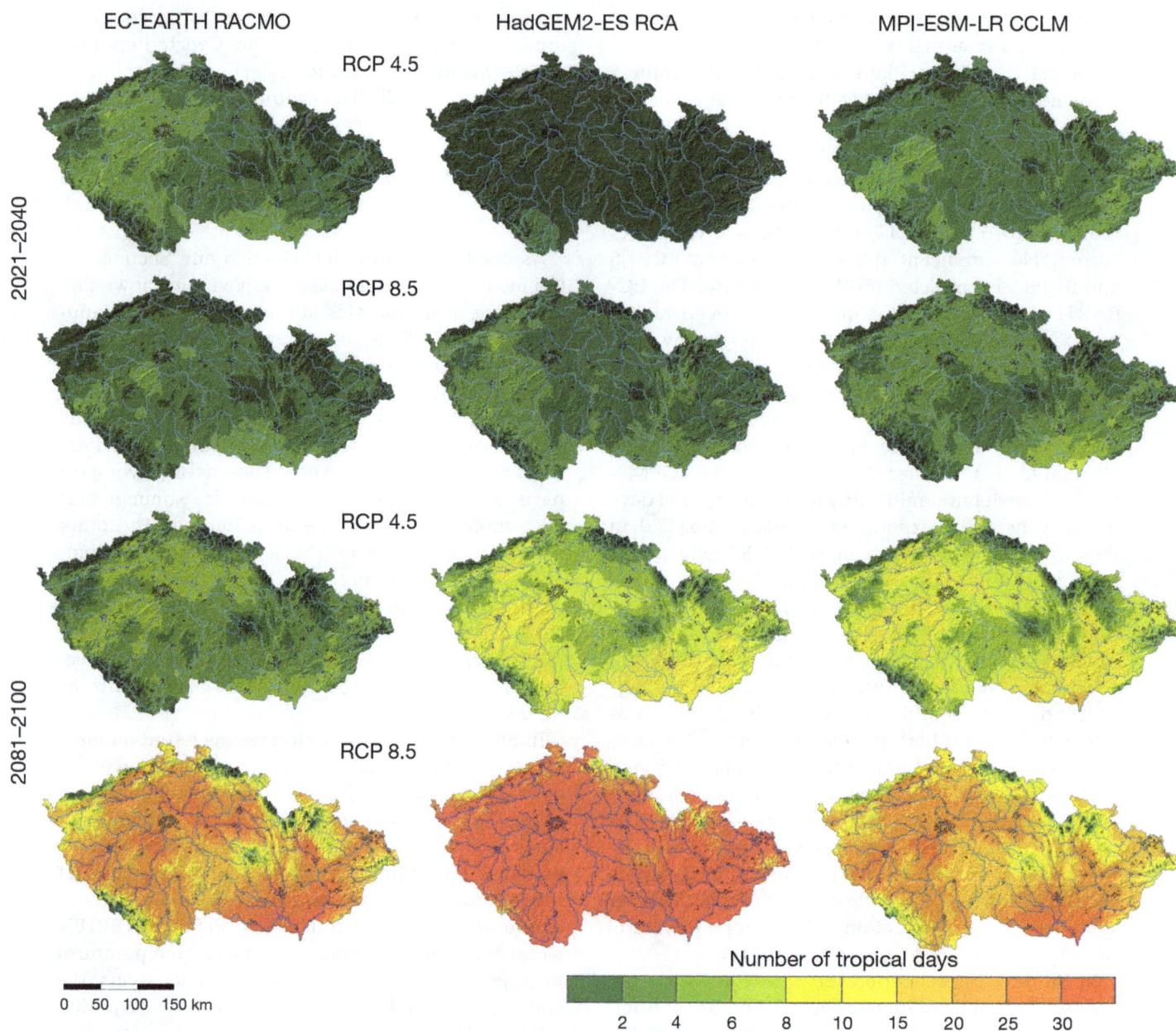

Fig. 6. Spatial differences in number of tropical days between future periods (2021–2040 and 2081–2100) and the present (1981–2010), for 3 experiments (EC-EARTH_RACMO, HadGEM2-ES_RCA, and MPI-ESM-LR_CCLM), and 2 scenarios (RCP4.5 and RCP8.5) across the Czech Republic

sions of national mean average with a different number of points used for the calculation.

The previous studies of future climate in the Czech Republic were mostly based on 3 simulations of 2 RCMs: ALADIN-Climate/CZ in 10 km and 25 km resolutions (ALADIN-10 and ALADIN-25), and RegCM3 in 10 km resolution (RegCM-10). Since these results are still widely used (at least within the Czech Republic), we decided to discuss the differences between the new findings presented in Section 3 of the present paper and the previous studies. Both simulations at 10 km resolution (ALADIN-10 and RegCM-10) were originally prepared within EU FP6

CECILIA (www.cecilia-eu.org/), covered 2 future periods, 2021–2050 and 2071–2100, and followed the path of greenhouse gas emissions according to the IPCC A1B (SRES) emissions scenario. The same emissions scenario was also chosen for the ALADIN-25 km transient simulation, covering the period 1961–2100. The ALADIN-Climate/CZ simulations were directly driven by the stretched version of the ARPEGE GCM. The GCM stretching technique increases the resolution of a GCM over a selected region (to ca. 50 km in the case of ARPEGE) and reduces it on the opposite side of the globe. Thus, GCM stretching reduces the jump in spatial resolution between the driving GCM and the downscaling RCM. In the case of the RegCM-10 experiment, a double nesting from ECHAM GCM via 25 km RegCM3 simulation was used. According to all 3 experiments, air temperature will climb in the future. In the near future, 2021–2050, air temperature in the Czech Republic will rise by about 1.2–1.5°C. This increase is within the range of warming calculated for the RCP4.5 and RCP8.5 emissions scenarios from the Euro-CORDEX models. For the periods 2021–2041 and 2041–2061, the Euro-CORDEX models expect a warming of 0.9–1.3°C for RCP4.5 and 1.0–1.8°C for RCP8.5. For the last 30 yr of the 21st century (2071–2100), older experiments (with the A1B scenario) indicated an air temperature rise of 3.2–3.3°C. This temperature change is close to what simulations based on the RCP8.5 scenario expect (2.8–4.1°C). RCP4.5-based simulations envisage rather milder warming (1.8–2.0°C). The older and newer generations of RCM experiments differ in seasonal temperature change. A slightly higher increase in the temperature is predicted in winter and summer by ALADIN-10 in the period 2021–2050. In contrast, RegCM-10 calculated the change in temperatures during the summer to be significantly smaller than in other seasons. The ALADIN-25 model gives a relatively similar increase for all seasons; the only lower trend is predicted for spring. Both models based on ALADIN-Climate/CZ calculated a higher increase in the temperatures in summer for the distant future (2071–2100). In contrast, RegCM-10 predicts a higher increase in the winter temperature for the period 2071–2100. For both emissions scenarios, the climate models based on Euro-CORDEX calculated a higher growth rate in air temperatures in the winter season for both the near and distant future. The changes in the seasons' air temperatures calculated by Euro-CORDEX are probably closest to the projections of the former RegCM-10 model.

Further, previous experiments calculated a slightly higher annual amount of precipitation on average for the whole Czech Republic for the period 2021–2050. The increase in rainfall was modeled between 4 and 7%. This is relatively consistent with the new results, which predict an increase in annual precipitation of between 7 and 12%. Differences are observed in the projections for seasonal changes. A significant decrease in winter precipitation was indicated by the ALADIN-10 and ALADIN-25 experiments. The decrease in winter precipitation was about 15%. Similar phenomena occurred in the present climate, with the last 5 winters having a lower amount of precipitation and snow. A decrease in winter precipitation is not

modeled by RegCM-10, however. The new results of the Euro-CORDEX experiments do not show a decrease in winter precipitation; on the contrary, they indicate an amount higher by about 9–20%. Previous experiments expected an increase of about 20% in autumn precipitation and about 10% for summer. Conversely, the new results show only a slight increase in summer precipitation, which could be very dangerous in connection with the higher temperature (significantly higher evaporation). Autumn precipitation is predicted by the new models to be higher by only about 7–12%, which is lower than the assumption of previous experiments. The differences between the previous and current versions of the climate models are pronounced in the case of precipitation in the distant future (2071–2100). ALADIN-10 and ALADIN-25 predicted slight decreases in annual precipitation (2–3%). Conversely, new models in accordance with older model outputs RegCM-10 gives annual precipitation increasing by about 10–16%. The ALADIN-Climate/CZ model (ALADIN-10, ALADIN-25), unlike during the first period, did not predict such decline in winter precipitation (4%), while there is a visible modeled decline in summer precipitation (10–12%). The RegCM-10 models an increase in winter, autumn and spring rainfall of up to 20%, but does not predict any significant change in summer precipitation. The outputs of the new experiments predict an ongoing increase in winter precipitation in the distant future, as opposed to the older model outputs by ALADIN-Climate/CZ. Winter precipitation is modeled to be higher by an amount of about 26–35%, especially for emissions scenario RCP8.5. A decline in summer precipitation is not observed in the new model outputs.

5. CONCLUSIONS

In the case of the original (uncorrected) model outputs, the differences between the individual models are large. After bias correction, the absolute values for future climate became more stable and suitable for ongoing analysis within impact studies.

All 11 available Euro-CORDEX experiments were bias-corrected. Most experiments underestimated temperature by up to 2.5°C. Conversely, the increase in air temperature is greater than in reality (for the control run). The found biases are not spatially identical; they vary significantly with altitude. The precipitation sums are overestimated by uncorrected experiments by up to 0.65 mm d^{-1}, but the differences between models are large. The spatial differences of the model biases are not the same for all the experiments, and their analysis by altitude level does not show clear results. Uncorrected RCM outputs would be useless for an impact study of a regional character, since uncorrected RCMs do not capture the conditions for the Czech Republic well.

Air temperature will increase similarly up to 2050 irrespective of the emissions scenario. Temperature will be higher by about 1°C in the period 2021–2040 compared to 1981–2010. After the year 2050, we observe growing differences between the emissions scenarios. For the whole Czech Republic, based on all 11 available experiments, air temperature will increase by 2.0°C (RCP4.5) to 4.1°C (RCP8.5) for the whole year and for the end of the 21st century compared to the reference period (1981–2010). In contrast, RCP2.6 predicts the halting of growth in air temperature by 2060, and even indicates a slow decline. Looking at individual seasons, the highest increase in temperature is modeled in winter. The projections for precipitation gives a slight increase of about 7% (2021–2040) to 13% (the end of the 21st century) for RCP4.5, and 6% to 16% for RCP8.5. The changes are mostly statistically significant (p = 0.05). A stronger trend can be found in the outputs for emissions scenario RCP8.5. The biggest difference is observed in winter precipitation, which could end up with an increase of 35% by the end of the 21st century, whereas the smallest change can be expected in summer precipitation.

Climate change is also reflected in the climatic indices. The experiments predict an increase in the number of tropical days, which is also manifested in the growing number of heat waves. Alarming results by the end of the 21st century are predicated mainly with emissions scenario RCP8.5. These results are not unrealistic, as a similar number of tropical days as predicted by RCP8.5 for the period of 2081–2100 occurred in the year 2015 in the Czech Republic. Higher temperatures can lead to more intense rains from thunderstorms. Significant trends were found in the number of days with precipitation of 10, 20 and 50 mm and higher.

Considering the simulated changes in air temperature and changes in the precipitation regime for the future, we can expect more frequent and severe drought occurrences. In this article, the conclusions are based solely on the climate variables that induce drought, but special drought indices were calculated as well, and we will follow up with articles devoted to their analysis.

Acknowledgements. This work was supported by the Ministry of Education, Youth and Sports of the Czech Republic within the National Sustainability Program I (NPU I), grant no. LO1415. We further acknowledge these projects: P.Š.: project no. LD14043 'Validation and correction of RCMs outputs in the area of the Czech Republic for purposes of impact studies' (Ministry of Education, Youth and Sports, Czech Republic); and project no. 14-12262S (Czech Science Foundation). P.Z. was supported by the project 'Hydrometeorological extremes in Southern Moravia derived from documentary evidence' (Czech Science Foundation, no. 13-19831S), and M.T. was supported by project no. 16-16549S 'Soil moisture and run-off droughts in future climate'.

LITERATURE CITED

Brázdil R, Trnka M (eds) (2015) Sucho v českých zemích: minulost, současnost, budoucnost (Drought in the Czech Lands: past, present, future). Centrum výzkumu globální změny Akademie věd České republiky, v.v.i., Brno

➤ Brázdil R, Trnka M, Mikšovský J, Řezníčková L, Dobrovolný P (2015) Spring-summer droughts in the Czech Land in 1805–2012 and their forcings. Int J Climatol 35: 1405–1421

➤ Brázdil R, Raška P, Trnka M, Zahradníček P and others (2016) The disastrous drought of 1947 in the Czech Lands: its course, trigger-factors, impacts and consequences. Clim Res 70:161–178

➤ Christensen JH, Christensen OB (2007) A summary of the PRUDENCE model projections of changes in European climate by the end of this century. Clim Change 81:7–30

➤ Christensen JH, Boberg F, Christensen OB, Lucas-Picher P (2008) On the need for bias correction of regional climate change projections of temperature and precipitation. Geophys Res Lett 35:L20709

Clarke L, Edmonds J, Jacoby H, Pitcher H, Reilly J, Richels R (2007) Scenarios of greenhouse gas emissions and atmospheric concentrations. Sub-report 2.1A of Synthesis and Assessment Product 2.1 by the US Climate Change Science Program and the Subcommittee on Global Change Research. Office of Biological and Environmental Research, US Department of Energy, Washington, DC

➤ Dee DP, Uppala SM, Simmons AJ, Berrisford P and others (2011) The ERA-Interim reanalysis: configuration and performance of the data assimilation system. QJR Meteorol Soc 137:553–597

➤ Déqué M (2007) Frequency of precipitation and temperature extremes over France in an anthropogenic scenario: model results and statistical correction according to observed values. Global Planet Change 57:16–26

➤ Farda A, Déqué M, Somot S, Horányi A, Spiridonov V, Tóth H (2010) Model ALADIN as regional climate model for Central and Eastern Europe. Stud Geophys Geod 54: 313–332

➤ Frei C, Christensen JH, Déqué M, Jacob D, Jones RG, Vidale PL (2003) Daily precipitation statistics in regional climate models: evaluation and intercomparison for the European Alps. J Geophys Res Atmos 108:4124

➤ Giorgi F, Mearns LO (1991) Approaches to the simulation of regional climate change: a review. Rev Geophys 29: 191–216

➤ Giorgi F, Mearns LO (1999) Introduction to special section: regional climate modelling revisited. J Geophys Res Atmos 104:6335–6352

Giorgi F, Jones C, Asrar G (2009) Addressing climate information needs at the regional level: the CORDEX framework. WMO Bull 58:175–183

Goodess CM, Aganostopoulou C, Bárdossy A, Frei C (2012) An intercomparison of statistical downscaling methods for Europe and European regions — assessing their performance with respect to extreme temperature and precipitation events and the implications for climate change applications. Tech Rep Climatic Research Unit, Norwich

► Gutjahr O, Heinemann G (2013) Comparing precipitation bias correction methods for high-resolution regional climate simulations using COSMO-CLM: effects on extreme values and climate change signal. Theor Appl Climatol 114:511–529

► Hanel M, Vizina A, Maca P, Pavlasek J (2012) A multi-model assessment of climate change impact on hydrological regime in the Czech Republic. J Hydrol Hydromech 60: 152–161

► Hanel M, Mrkvičkova M, Maca P, Vizina A, Pech P (2013) Evaluation of simple statistical downscaling methods for monthly regional climate model simulations with respect to the estimated changes in runoff in the Czech Republic. Water Resour Manage 27:5261–5279

► Hlavinka P, Trnka M, Semerádová D, Dubrovský M, Žalud Z, Možný M (2009) Effect of drought on yield variability of key crops in Czech Republic. Agric For Meteorol 149: 431–442

► Kingston DG, Stagge JH, Tallaksen LM, Hannah DM (2015) European-scale drought: understanding connections between atmospheric circulation and meteorological drought indices. J Clim 28:505–516

► Kotlarski S, Keuler K, Christensen OB, Colette A and others (2014) Regional climate modeling on European scales: a joint standard evaluation of the EURO-CORDEX RCM ensemble. Geosci Model Dev 7:1297–1333

► Maraun D (2013) Bias correction, quantile mapping and downscaling: revisiting the inflation issue. J Clim 26: 2137–2143

► Maraun D, Wetterhall F, Ireson AM, Chandler RE and others (2010) Precipitation downscaling under climate change: recent developments to bridge the gap between dynamical models and the end user. Rev Geophys 48:RG3003

► Mearns L, Arritt R, Biner S, Bukovsky MS and others (2012) The North American Regional Climate Change Assessment Program: overview of Phase I results. Bull Am Meteorol Soc 93:1337–1362

► Meinshausen M, Smith SJ, Calvin K, Daniel JS and others (2011) The RCP greenhouse gas concentrations and their extensions from 1765 to 2300. Clim Change 109: 213–241

Moss R, Babiker M, Brinkman S, Calvo E and others (2008) Towards new scenarios for analysis of emissions, climate change, impacts, and response strategies. IPCC, Geneva

► Pal JS, Giorgi F, Bi X, Elguindi N and others (2007) Regional climate modeling for the developing world: the ICTP RegCM3 and Reg-CNET. Bull Am Meteorol Soc 88: 1395–1409

► Räty O, Räisänen J, Ylhäisi J (2014) Evaluation of delta change and bias correction methods for future daily precipitation: intermodel cross-validation using ENSEMBLES simulations. Clim Dyn 42:2287–2303

► Riahi K, Gruebler A, Nakicenovic N (2007) Scenarios of long-term socio-economic and environmental development under climate stabilization. Technol Forecast Soc Change 74:887–935

Rožnovský J, Zahradníček P (2014) Air temperatures and conditions for recreation. In: Fialová J, Pernicová D (eds) Public recreation and landscape protection — with man hand in hand. Křtiny, 5–6 May 2014, p 27–31

Spinoni J, Naumann G, Vogt J, Barbosa P (2016). Meteorological droughts in Europe: events and impacts—past trends and future projections. Publications Office of the European Union, Luxembourg, EUR 27748 EN, doi:10. 2788/450449

Štěpánek P (2008) AnClim — software for time series analysis. Department of Geography, Faculty of Natural Sciences, Masaryk University, Brno. www.climahom.eu/AnClim.html

Štěpánek P (2010) ProClimDB – software for processing climatological datasets. Czech Hydrometeorological Institute, regional office, Brno. www.climahom.eu/ProcData.html

► Štěpánek P, Zahradníček P, Skalák P (2009) Data quality control and homogenization of air temperature and precipitation series in the area of the Czech Republic in the period 1961–2007. Adv Sci Res 3:23–26

Štěpánek P, Zahradníček P, Huth R (2011) Interpolation techniques used for data quality control and calculation of technical series: an example of Central European daily time series. Id járás 115:87–98

Štěpánek P, Skalák P, Farda A, Zahradníček P (2012) Climate change in the area of the Czech Republic according to various model simulations. In: Kožnarová V, Sulovská S, Hájková L (eds) Proc BIOCLIMATE 2012 — Bioclimatology of Ecosystems, 29–31 August 2012, Ústi nad Labem, Prague

Štěpánek P, Zahradníček P, Farda A (2013) Experiences with data quality control and homogenization of daily records of various meteorological elements in the Czech Republic in the period 1961–2010. Id járás 117:123–141

Stocker TF, Qin D, Plattner GK, Tignor M and others (eds) (2013) Climate change 2013: the physical science basis. Working Group I contribution to the Fifth Assessment Report of the Intergovernmental Panel on Climate Change. Cambridge University Press, Cambridge

► Suklitsch M, Gobiet A, Truhetz H, Awan NK, Göttel H, Jacob D (2010) Error characteristics of high resolution regional climate models over the Alpine area. Clim Dyn 37:377–390

► Taylor KE, Stouffer RJ, Meehl GA (2012) An overview of CMIP5 and the experiment design. Bull Am Meteorol Soc 93:485–498

► Teutschbein C, Seibert J (2013) Is bias correction of regional climate model (RCM) simulations possible for non-stationary conditions? Hydrol Earth Syst Sci 17:5061–5077

► Themessl MJ, Gobiet A, Leuprecht A (2011) Empirical-statistical downscaling and error correction of daily precipitation from regional climate models. Int J Climatol 31: 1531–1544

► Themessl MJ, Gobiet A, Heinrich G (2012) Empirical-statistical downscaling and error correction of regional climate models and its impact on the climate change signal. Clim Change 112:449–468

► Trnka M, Kysely J, Možny M, Dubrovský M (2009) Changes in Central-European soil-moisture availability and circulation patterns in 1881–2005. Int J Climatol 29:655–672

► Trnka M, Brázdil R, Možný M, Štěpánek P and others (2015) Soil moisture trends in the Czech Republic between 1961 and 2012. Int J Climatol 35:3733–3747

UNISDR (United Nations Secretariat of the International Strategy for Disaster Reduction) (2009) Drought risk reduction framework and practices. Contributing to the implementation of the Hyogo Framework for Action. UNISDR, Geneva

van der Linden P, Mitchell JFB (2009) ENSEMBLES: climate change and its impacts: summary of research and results from the ENSEMBLES project. Tech Rep. Met Office Hadley Centre, Exeter

▶ van Vuuren DP, den Elzen MGJ, Lucas PL, Eickhout B and others (2007) Stabilizing greenhouse gas concentrations at low levels: an assessment of reduction strategies and costs. Clim Change 81:119–159

▶ van Vuuren DP, Edmonds J, Kainuma M, Riahi K and others (2011) The representative concentration pathways: an overview. Clim Change 109:5–31

▶ Wang Y, Leung LR, McGregor JL, Lee DK, Wang WC, Ding Y, Kimura F (2004) Regional climate modeling: progress, challenges, and prospects. J Meteorol Soc Jpn 82: 1599–1628

▶ Zahradníček P, Trnka M, Brázdil R, Možný M and others (2015) The extreme drought episode of August 2011/ May 2012 in the Czech Republic. Int J Climatol 35: 3335–3352

14

Assessing the combined hazards of drought, soil erosion and local flooding on agricultural land

Miroslav Trnka[1,2,*], Daniela Semerádová[1,2], Ivan Novotný[3], Miroslav Dumbrovský[4], Karel Drbal[5], František Pavlík[6], Jan Vopravil[3], Pavla Štěpánková[5], Adam Vizina[5], Jan Balek[1,2], Petr Hlavinka[1,2,], Lenka Bartošová[1], Zdeněk Žalud[1,2]

[1]Global Change Research Centre, Czech Academy of Sciences, Belidla 986/4a, Brno 60300, Czech Republic

[2]Department of Agrosystems and Bioclimatology, Mendel University in Brno, Zemedelska 1, Brno 61300, Czech Republic

[3]Research Institute for Soil and Water Conservation, Žabovřeská 250, 156 27 Praha 5 – Zbraslav, Czech Republic

[4]Faculty of Civil Engineering, Department of Landscape Water Management, Brno University of Technology, 60200 Brno, Czech Republic

[5]T. G. Masaryk Water Research Institute, Mojmirovo nam. 16, Brno 61200, Czech Republic

[6]State Land Office, Husinecká 1024/11a, 130 00 Praha 3, Czech Republic

ABSTRACT: Present-day agriculture faces multiple challenges, including ongoing climate change that is at many locations combined with soil degradation. The deterioration of soil properties through unsustainable agricultural practices and changing climate could lead to a fall in productivity beyond the point of no return with devastating effects on ecosystem services in large areas. Identifying areas with the highest hazard levels should therefore be a top priority. The key hazards for agricultural land in the Czech Republic considered in this study include the occurrence of water stress in the topsoil layer during both the first and second half of the growing season, the proportion of fast-drying soils, the risk of sheet and ephemeral gully erosion and the risk of local floods originating primarily from agricultural land. The results clearly marked regions where primary attention should be given to reduce the level of the hazards and/or to increase cropping capacity. These regions were found to be concentrated in the southeastern and northwestern lowland areas. Typical areas with the highest hazard levels were identified: regions with low precipitation and a high proportion of soils with a degraded or naturally occurring low water-holding capacity, and those with steeper than average slopes and terrain configurations in relatively large catchment areas that have urbanized countryside landscapes located at their lower elevations. Despite some limitations, the methods presented in this paper can be applied generally as the first step in developing strategies for efficient reduction of hazard levels.

KEY WORDS: Soil moisture · Sheet erosion · Ephemeral gully erosion · Critical point · Fast-drying soil · Vulnerability · Climate change

1. INTRODUCTION

Food security and its relationship with ongoing climate change featured prominently in the 5th Assessment Report of IPCC Working Group II. The chapter on food security (Porter et al. 2014) has been dis-cussed in great detail, and its conclusions provide a set of key messages for world political leaders (IPCC 2014). The urgency of prioritizing agronomy research that allows humankind to produce sufficient amounts of high-quality food in a sustainable way is driven by multiple pressures. First, there is the sheer

*Corresponding author: mirek_trnka@yahoo.com

challenge of increasing food production by 60–70 %
by 2050 to feed a global population estimated to increase to over 9 billion. Second, that challenge must
be achieved using less water and energy than is used
today, as agriculture is under pressure to decrease its
water, carbon and energy footprints to become sustainable in the long term. Third, the challenge of
ongoing climate change showing warming trends
across the globe is already leading to changes in the
distribution of climate variability and extremes
(Rahmstorf & Coumo 2011, Gourdji et al. 2013, Liu &
Allan 2013), while a high uncertainty remains in the
relationship between global warming and climate
variability (Huntingford et al. 2013). There is also
concern about deteriorating food quality, which was
highlighted by Myers et al. (2014), that projects a
lower protein content in key C3 crops than exists at
present, or risks posed by the future occurrences of
pests and diseases in some key agricultural production regions (e.g. Svobodová et al. 2014). At the same
time, there is ever-growing concern with regard to
soil fertility. According to the EEA (1998), damage to
Europe's soils from modern human activities is increasing and leading to irreversible soil loss due to
erosion, local and diffuse contamination and the sealing of soil surfaces. In western and northern Europe,
the sealing of soil surfaces due to increased urbanization and new infrastructure is the main cause of soil
degradation. Land area in the Czech Republic is
about 10.6 % urban, with a yearly increase of about
0.4 %. In the Mediterranean, soil erosion is the main
cause of soil loss. The areas with the most severe soil
loss from both wind and water erosion include the
Balkan Peninsula, the Black Sea region, and also
some central European countries such as the Czech
Republic and Slovakia (EEA 1999). The European
Union (EU) Mediterranean countries also have
severe soil erosion problems, which in some cases
are at a critical stage and could lead to desertification. Alarming is the possibility of areas currently not
at risk (e.g. parts of the Mediterranean or the Alps)
reaching advanced degradation status that cannot be
reversed within 2 or 3 generations, with some areas
having reached this status over 2 decades ago (e.g.
Van Lynden 1994). Central Europe is also showing
signs of serious problems in this regard.

Previous studies assessing the agricultural impacts
of climate change have demonstrated that the effects
depend on the crops being grown, cropping season and region (e.g. Olesen et al. 2007), and very
few have considered cropping system responses to
changes in the frequency and severity of climatic extremes (e.g. Ruiz-Ramos et al. 2011). However, it is

well known that the impacts of such extreme events
can be substantial (e.g. Reyer et al. 2013). Studies by
Trnka et al. (2014, 2015a) showed that, despite the
large uncertainty in climate projections within the
CMIP5 ensemble, the overall frequency of adverse
events is much more likely to increase than to decrease across the European domain, including Central Europe. It is also obvious that soil degradation is
intertwined with the increased drought risk as
degraded soils tend to have smaller infiltration rates
and much smaller water-holding capacity.

Existing research confirms that climate conditions
in April–June have the most profound impact on the
yield of field crops (e.g. Hlavinka et al. 2009, Kolář et
al. 2014). The interannual yield variance explained
by the interannual changes in a single climatic variable could reach 50 % on a regional scale (Brázdil et
al. 2009, Trnka et al. 2016, this issue). A recent analysis indicated that climatic variability influences yield
variability in Central Europe more now than it did in
the late 19th and early 20th centuries (e.g. Trnka et
al. 2012). Most of the explanatory power of the model
was derived from a negative sensitivity to temperature and drought. Interestingly, the biggest changes
in drought sensitivity from the late 19th to the late
20th century were found in the regions where processes (wind and water-driven) of soil erosion and of
overall soil degradation are considered severe. An
awareness of this problem has led to increased attention being paid to the issues of drought vulnerability
and soil degradation in recent years (e.g. Trnka et al.
2015b, Zahradníček et al. 2015, etc.). Many farmers
partially mitigate drought impacts through crop selection, irrigation, and modified tillage practices, but
in many cases, they struggle economically, as the
economic returns, especially in dry years, are extremely poor (Vopravil et al. 2012). Until recently, at
the state level, the emphasis of disaster management
has been largely on the response to and recovery
from droughts, with little or no attention paid to
drought mitigation, preparedness, prediction and
monitoring. The situation is better in relation to soil
erosion and local flood risks due to existing EU rules
(Panagos et al. 2015), but these measures are usually
not intertwined with measures aimed at decreasing
the impact of droughts. Increased losses from droughts
(as observed in 2000, 2003, 2012 and 2015) suggest a
growing societal vulnerability to this hazard. At the
same time, erosion risks and the rate of soil degradation are still unsustainable in some areas despite a
number of existing measures being taken (Vopravil
et al. 2012). It has also been felt by both the Agrarian
Chamber (representing the great majority of farmers

in the Czech Republic) and some governmental organization (e.g. the State Land Office) that major revisions are needed of existing policies related to the management of soil degradation frequently associated with intensive rains and increased drought risks due to climate change. Therefore, a multidisciplinary task force was formed and supported, called the Master Plan of Landscape Water Management of the Czech Republic (F. Pavlik pers. obs.). It has been recognized that there must be an inevitable shift in policies to change drought management from a reactive, crisis-management approach to a proactive, risk-management approach. At the same time, these measures must complement and support existing policies aimed at suppressing processes of soil degradation. While the former requires detailed monitoring, early warning and planning between events (e.g. Wilhite 2000), the latter can be achieved only through a comprehensive and regionally tailored and coordinated set of long-term policies. The final goal is to lower the risks in real terms arising from the impacts of drought and soil degradation for the whole of the Czech Republic through the implementation of an efficient strategy. Among other factors, this strategy includes selecting areas and 'actors' (i.e. farmers) facing the highest risk and providing them with the proper support as early as possible to prevent a situation from deteriorating beyond the 'point of no return'.

The development and implementation of the methodological framework into an operational procedure considers the overall risk for the Czech Republic's rural landscape as being a function of hazard, exposure and vulnerability (Giupponi et al. 2015). The ultimate goal of the whole procedure is the quantification of these risks, putting into place policies and measures leading to risk reduction, and ensuring their adoption in practice through the use of demonstration areas, as well as through technical and financial assistance. This may be achieved only when hazards, vulnerability and exposure are known, allowing for the calculation of the expected damages related to the risks associated with different hazardous scenarios. Clearly, the first step required is a proper analysis of the hazards themselves.

While according to Kappes (2012), multi-hazard assessment may be understood as an assessment of 'the totality of relevant hazards in a defined area', the present study uses the concept of 'more than one hazard'. The research presented here focuses on drought and soil degradation hazard assessments for the agricultural landscape of the Czech Republic. While the potential indicators for describing the

hazards to agricultural land are numerous, their applicability depends both on their policy relevance and the availability of data. We attempt to provide concept hazard and vulnerability to key decision-makers (agricultural producers, natural resource managers, and others), as well as provide them with spatially explicit information. The goal of this study was to develop a method for assessing hazards posed by drought, erosion and floods using geographic processing techniques. The objectives were: (1) identify key indicators that define agricultural drought, soil erosion and local flood hazards for agricultural land across the Czech Republic; (2) evaluate the weight of the indicators that contribute to the combined hazard; and (3) classify and map the combined hazard.

2. MATERIALS AND METHODS

2.1. Setting

The first step towards the implementation of the proposed framework is the identification of the context in which it will be applied. This involves identifying regions with the highest hazards so that the next steps of risk assessment (vulnerability and exposure assessments) and policy application can be targeted to these regions (Fig. 1). Fig. 1a–c shows the Czech Republic's main orographic features and soil characteristics that, together with climate conditions, help to illustrate why land degradation and drought need to be assessed together. The country's mean altitude is 430 m (CSO 2005), and it lacks large mountain chains. However, the combination of the country's morphology, which is dominated by uplands and highlands, the occurrence of areas with considerable slopes even at low altitudes, and the comparatively large field blocks explain why over half of agricultural and especially arable land is exposed to strong water and wind erosion (Janeček et al. 2012). The fact that the most intensively farmed regions are also dotted with high numbers of settlements increases the number of areas in which floodwater from relatively small catchments over the agricultural land poses a frequent hazard for these settlements. The hazard is also higher than in comparable regions of Austria, due to much larger field blocks that contain a single crop. Most Czech agriculture is rain-fed, with irrigation being available for <4 % of the area (Batysta et al. 2015) and only ~1.5 % being actively used (Batysta et al. 2015). While the Czech Republic's mean annual precipitation of

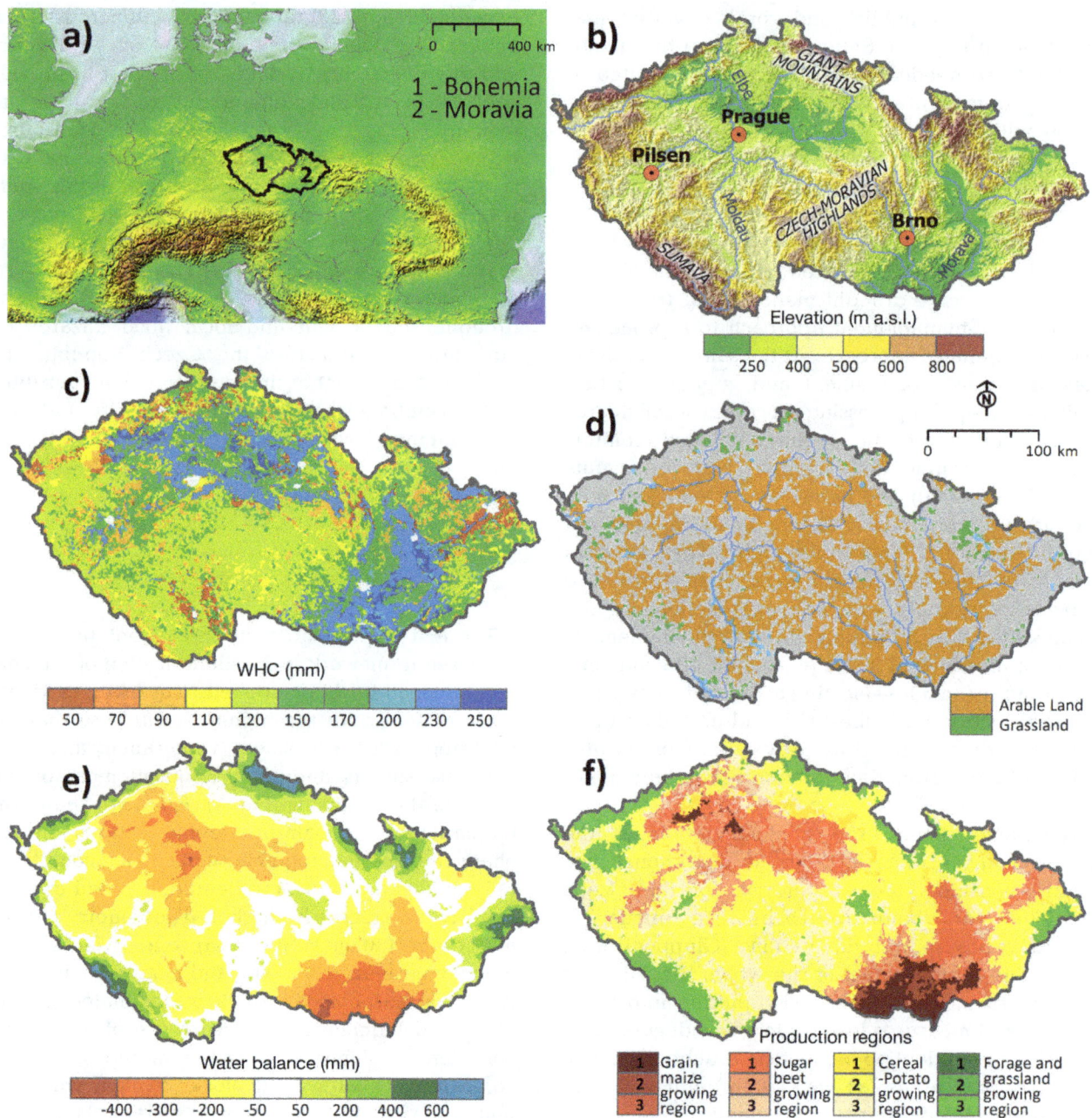

Fig. 1. (a) Overall location of the Czech Republic case study region in Central Europe. (b) Main orographic features of the country. (c) Water-holding capacity (WHC) of the soil in the first 100 cm of the soil profile. (d) Main areas of arable land and grassland for which the assessment was primarily carried out, with other land-uses colored grey. (e) Difference between sum of annual precipitation and annual potential evapotranspiration. (f) Agrometeorological zoning of the country into 4 main production zones. In each production zone 3 sub-classes are distinguished: (1) prime agricultural soil; (2) could be used for agricultural production with limitations; (3): not suitable as agricultural land

700 mm would generally be sufficient, some key producing regions in the northwest and southeast show much lower precipitation totals (450 and 500 mm, respectively). Given the high interannual variability and shifts in the distribution of precipitation, in many seasons production is severely limited by the availability of water. In fact, rain-fed agriculture is only

sustainable thanks to a generally favorable distribution of precipitation and to soil that holds enough water to allow crop survival through episodes of drought that are sometimes prolonged (Hlavinka et al. 2009). According to Trnka et al. (2009), the June–August sum of precipitation for the Czech Republic is on average higher than 1/3 of the annual rainfall total (ranging from 27 to 43%). The driest season that accounts for less than 1/5 of the annual precipitation is winter (December–February). Winter precipitation (if in the form of snow) allows for a recharging of the soil profile just before the growing season, reducing the effects of severe droughts in the early part of growing season, which is critical for crop yield formation (e.g. Hlavinka et al. 2009). As Fig. 1e shows, the majority of the lowlands show higher potential evapotranspiration than precipitation, in particular in the areas where precipitation is the lowest. This deficit is the highest in the southeast, which in addition to low levels of precipitation, also has the highest temperatures and the most hours of sunshine compared to the rest of the country. Climate and soil conditions together with the terrain predetermines the agricultural use of each area. The Czech Republic is traditionally divided into 4 production regions, with those growing sugar beets being the most productive (Fig. 1f), followed by maize-producing regions.

2.2. Indicators of hazards

The potential indicators that could be used for the assessment of hazards are numerous. Because the focus of this study was on assessing combined hazards for agricultural land, we focused on the indicators that in our view can best be used to quantify these hazards. An analysis of the literature, suggestions from specialists, and data availability formed the fundamental assumptions underlying the methodology we used.

In the first step towards the assessment of the combined hazard (Fig. 2) for the agricultural lands analyzed, we identified the following hazards as being the most critical ones:

(a) Drought occurring during the growing season;
(b) Pre-existing poor soil conditions decreasing the ability of the soil to hold water (fast-drying soils);
(c) Increased susceptibility to water erosion, including the occurrence of concentrated runoff pathways;
(d) Pre-existing infrastructure and/or settlements in the path of the concentrated runoff pathways.

It was also clear that the occurrence of one hazard at a greater intensity might contribute to the effects of other hazards becoming more severe. For example, prolonged droughts lead to the disintegration of the soil structure and to a decrease in the vitality of the crop cover, which makes the area more vulnerable to soil erosion (Fig. 2).

2.2.1. Agricultural drought during the growing season

We selected the number of days during which soil moisture was <30% of the relative soil water content (i.e. the percentage to which water fills the soil pores between the so-called wilting point and field capacity) in the topsoil, which was defined as the layer between 0.0 and 0.4 m. The calculation procedure has been explained in detail by Hlavinka et al. (2011) and was further tested by Trnka et al. (2015a,b) and carried out for agricultural land (Fig. 1d). It relies on the SoilClim model (Hlavinka et al. 2011) based on the model of Allen et al. (1998), and accounts for the following factors and processes:

• Water accumulation in snow cover and subsequent melting
• Water-holding capacity of the soil
• Influence of the slope and aspect on the energy balance
• Influence of the type of the vegetation on daily evapotranspiration, interception and runoff
• Dynamically changing properties of the plant cover based on the phenology phase estimated through thermal time (including sowing, harvest, leaf area index, rooting depth, and crop height)
• Influence of underground water and shallow water tables

As the indicator of drought hazard, we selected the median number of days per season (based on 1991–2014 data) with a saturation of the surface soil layer below 30%, which is a good proxy for drought damage to agricultural crops according to our observations (e.g. Zahradníček et al. 2015). In general, this value could be considered as the level below which the physiological processes of the plant begin to be significantly limited by a lack of water (e.g. Larcher 2003). While decreases in the relative saturation of the soil below 50% slows a plant's intake of water, at values below 30% it is no longer able to produce sufficient turgor, and growth stagnates. The calculations were performed in 500 m grids covering the whole of the Czech Republic (Trnka et al. 2015b). Based on the drought-yield relationship, we divided the growing season into 2 parts, April–June and July–September. In the former, mostly spring- and winter-sown cereals (usually harvested in July) are known to be affected the most (e.g. Hlavinka et al. 2009), while the latter season represents the time period in which latter-maturing crops (e.g. maize, potatoes or sugar beets) can be negatively affected.

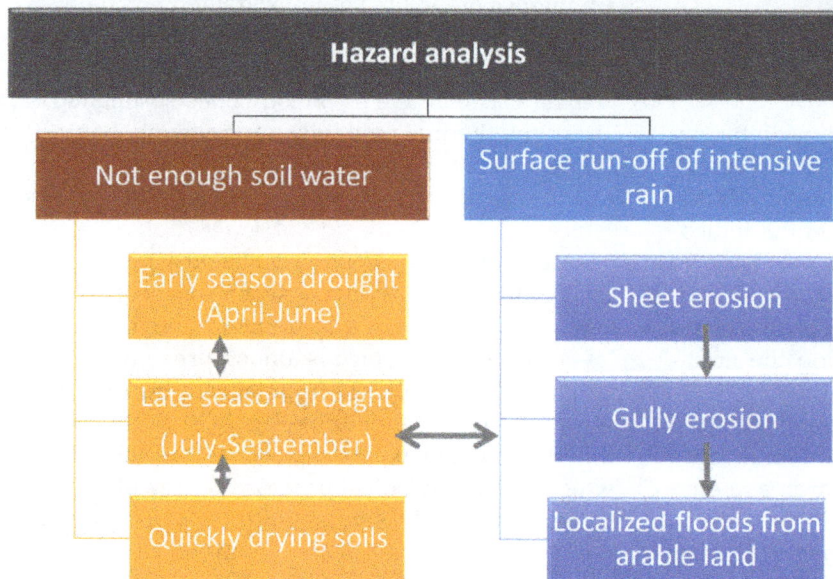

Fig. 2. Overview of individual hazard indicators (arrows show their interactions), and photographs illustrating the observed impacts: (a) poor growth and leaf folding of maize as a consequence of drought in August 2015, (b) soil degradation enhanced by wind erosion due to drought in southeast Czech Republic, (c) example of a fast-drying soil profile (see Section 2.2.2), (d) impacts of sheet erosion, (e) ephemeral gully erosion in maize crops, (f) impact of a flash flood primarily produced by an agricultural area

2.2.2. Fast-drying soils

While the occurrence of drought is primarily driven by climate, and the definitions used account for the influence of terrain, vegetation cover and soil water-holding capacity, it was felt that specific local soil conditions needed to be taken into account. Therefore, the proportion of fast-drying soils that tend to be particularly negatively affected by drought was considered. Fast-drying soils are the result of soil processes that are driven by a long-term lack of water in the soil and the intensive mineralization of organic

matter, which leads to a decrease in soil fertility and its water-holding capacity. In the Czech Republic, this issue is of concern in the northwestern and southeastern parts of the country. Including fast-drying soils as a hazard indicator is justified by the fact that the extent of this area has been expanding over the past decades, and the occurrence of fast-drying soils in a region indicates a heightened hazard. The expansion of fast-drying soils is driven by erosion, and many areas with very fertile soils <100 yr ago (e.g. chernozems) are presently fast-drying soils consisting of an underlying loess (giving at least a hope of restoring soil fertility) or sand from the original bottom of the sea that was once present in these areas. The process is accelerated by ongoing climate change connected with the increasingly frequent occurrence of long periods of drought and also by unsuitable tillage practices with a low re-supply rate of organic matter to the soil. Determining the occurrence of fast-drying soils was performed through an evaluation of a high-resolution (5 × 5 m) map of the soil conditions, based on information obtained from the soil database that is maintained and permanently updated by the Research Institute for Soil and Water Conservation (RISWC), which includes extensive data on soil and associated components of the environment. Only agricultural land (Fig. 1d) was considered.

2.2.3. Sheet, interrill and rill soil erosion

The first indicator of an erosion hazard for agricultural land (Fig. 1d) focuses predominantly on so-called sheet erosion (i.e. the transport of loosened soil particles by overland flow). We used an approach based on the universal soil loss equation (USLE) (Wischmeier & Smith 1978), which accounts for rainfall erosivity factor (R), soil erodibility factor (K), topographic factors (slope and length) and cropping management factors (C and P). The topography factors were estimated according to the modified equation of Desmet & Govers (1996) using a 5 × 5 digital elevation model. The efficiency factor of erosive rainfall was set at R = 40 MJ ha^{-1} cm h^{-1} (Janeček et al. 2012), and the C factor was based on the actual crop proportions at the same resolution as the slope and length estimates. After estimating annual soil loss, those 5 × 5 m grids showing an annual potential loss higher than 4 t ha^{-1} (i.e. the nationally enforced limit) were marked as those with a significantly higher than permissible erosion rate.

2.2.4. Rill and ephemeral gully erosion

In addition to the classic erosion furrows on the surface slopes of arable land, there are also so-called rills and ephemeral gullies present, which differ from the classic erosion furrows because of their cross-sectional area (>1 square foot or >0.09 m^2) (Morgan 2005). These features tend to appear in places where the basin shape leads to a concentration of the outflowing surface water. They can either follow the flow path of excess water, or they can follow linear landscape elements such as land boundaries, furrows created by agricultural practices or unpaved country roads. The term ephemeral expresses the temporariness of these elements, which are rehabilitated by tillage of the growing season, but they tend to reappear in the same place in the next growing season under the 'right' farming and weather conditions.

For the analysis of ephemeral gully erosion hazards, the method of plotting potential paths of runoff concentration at a resolution of 5 m was used. This method is based on the modeling of flow accumulation from drainage areas, the interpretation of the nature of the terrain and the visual interpretation of aerial photos of the affected land blocks. Contributing areas were used to automatically generate the direction and accumulation of runoff over a digital terrain model with manual correction using raster topographic maps and aerial orthophotos (Dumbrovský 2011).

2.2.5. Localized floods originating from agricultural land

Catastrophic floods with tragic consequences in the Odra river basin in 2009 and similar well-documented events in the following years vividly demonstrate that settlements can be significantly affected in places where there is no (and has not been) any known permanent stream (Drbal 2009). Drbal & Dumbrovský (2009) reported that even a contributing area of 5 ha is sufficient to generate a flow that can cause severe damage to property. The causal factors critical for the formation of a concentrated runoff were determined based on the number of recent flood events from torrential rainfall, and parameters were set to estimate so-called 'critical points'. A critical point (CP) was defined as the point where the trajectory of the concentrated runoff penetrates into the municipality. CPs were thus determined based on the intersection of a municipality (urban) boundary with concentric lines of a track drainage area contributing to a region ≥0.3 km^2. As the area affected by torrential rainfall tends to be limited, the contributing area was also limited to 10 km^2. Torrential rainfalls, while very localized, occur fairly frequently between April and September and in particular over the summer months. However, the spatiotemporal localization of torrential rainfalls or even the mapping of return probabilities is not possible with the present dataset. Therefore, in this analysis we assumed that torrential rain could occur at any location in the Czech Republic.

Table 1. Indicators for the individual hazards

Hazard	Indicator as used at the cadaster unit level	Grid resolution (m)	Unit	Reference period
Early season drought	Median number of days with soil moisture below 30 % of maximum available soil water-holding capacity at the surface layer (0–0.4 m) in April–June	500	Days per season	1991–2014
Late season drought	Median number of days with soil moisture below 30 % of maximum available soil water-holding capacity at the surface layer (0–0.4 m) in July–September	500	Days per season	1991–2014
Fast-drying soils	Proportion of fast-drying soils per unit of arable land in the cadaster	5	%	Continuously updated
Sheet erosion	Proportion of the arable land in the cadaster unit with significantly higher than permissible erosion rate (>4 t ha^{-1} yr^{-1})	5	%	Continuously updated
Ephemeral gully erosion	Contributing area to ephemeral gully erosion pathways from arable land in the cadaster unit	5	Hectare	Continuously updated
Localized flood from arable land	Proportion of the cadaster unit area belonging to the contributing area for critical points	5	%	Continuously updated

2.3. Multiple hazard analysis

After selecting the key hazards, the indicators that best represented them were formulated. Table 1 lists the indicators for the individual hazards, while Fig. 2 illustrates the impacts that the indicators represent. The original quantification of the indicators was based on different resolutions, with data on drought occurrence being available as a 500 × 500 m grid and the remaining indicators being calculated at a resolution of 5 m due to the importance of the local terrain conditions. As the study aimed at identifying the areas with the highest hazard level for policy-making purposes, the indicators were aggregated at the level of the cadaster unit, which is the smallest administrative unit in the Czech system. Territory in the Czech Republic is composed of 13 091 cadaster units with a mean area of 6 km^2. For each cadaster unit, the value of each indicator was calculated. All indicators were normalized using a z-score approach. It is one of the most commonly used normalization procedures in which all indicators are converted into a common scale with an average of zero and a standard deviation of one. The scale described in Table 2 was applied to communicate the results to the stakeholders. As 6 indicators were used, weighting was considered to express the relative importance of individual indicators to calculate a composite hazard index. Weights are essentially value judgments, and thus, they are essentially subjective and can make the objectives underlying the construction of a composite index explicit (e.g. Giupponi et al. 2015). In this case, we applied equal weights to all indicators. In the final

determination of the regions with the highest hazards, we combined an averaging of the z-scores due to their transparency with an identification of regions affected by at least 2 of the 6 indicators with a z-score below −2. The latter was performed to limit the shortcomings of the averaging approach, as a bad score in one criterion can be offset by a good score in another one, even if there is no interaction among the criteria. Apart from the cadaster units, the so-called 4th-order catchments were considered as an alternative spatial unit for aggregating the hazard analysis results. There are almost 9000 of these 4th-order catchments in the Czech Republic, which have a fairly variable area, and a mean area of 9.6 km^2.

2.4. Mapping

The final result of the combination of factors was a numeric value, which was calculated through the 'union' mathematical function in ERDAS Imagine GIS by a simple averaging of the z-scores of all 6 indicators. As a second approach, the frequency with

Table 2. z-score table used to interpret the standardized values of the indicators, as used in Figs. 4, 5 & 7

Indicator interpretation	z-score range
Above average	0–0.5
Markedly above average	>0.5 and <1.0
Highly above average	1.0–1.5
Very highly above average	>1.5 and <2.0
Extremely above average	2.0 and higher

which a given cadaster unit had a z-score <-2.0 for a particular indicator was also mapped. For the analysis, heavily urbanized cadasters were excluded, as were the cadaster units with a large proportion of surface mines (especially in the northwestern part of the country) and water bodies. In general, a low value for a z-score is an indicator that in the given cadaster unit, the hazard value is proportionally higher than in the rest of the country. Therefore, a very low combined z-score signals the occurrence of multiple hazards. The occurrence of a z-score below -2.0 means that for a given indicator, the cadaster unit belongs to several dozen cadaster units with extremely high levels for that particular hazard. These two approaches were then combined in the final map. Finally, we compared the newly developed classification of the cadaster units according the level of multiple hazards with the extent of the less-favored areas (LFA) as defined in accordance with European Commission (EC) regulation 1305/2013, which is used to support areas with existing natural constraints.

2.5. Climate change scenarios

To evaluate the impact of climate change on the values of the selected indicators, we tested the change in the number of days with drought stress in the topsoil layer during the period April–June. For each 500 m grid, the weather data were modified based on the expected climate change conditions for the region. To be able to assess the development of conditions during 2021–2040, we modified 1981–2010 daily weather data using a delta approach and 5 climate models. These models were selected as representations of mean values (IPSL: model of the Institute Pierre Simon Laplace, France) and to best capture the variability of expected changes in precipitation and temperature (BNU: Beijing Normal University, China; MRI: Meteorological Research Institute, Japan; CNRM: National Centre for Meteorological Research, France; and HadGEM: Hadley Centre Global Environment Model, UK). These models were picked from 40 climate models available in the CMIP5 database (Taylor et al. 2012). These projections used the Representative Concentration Pathway (RCP) 4.5 greenhouse gas concentration trajectory and a climatic sensitivity of 3.0 K. Before using all meteorological data as input for subsequent steps, the SnowMAUS model (Trnka et al. 2010) was applied to estimate the appearance of snow cover. In this way, daily precipitation totals were modified to better match the real timing and amount of water infiltration into the soil considering probable snow accumulation, melting and sublimation.

3. RESULTS AND DISCUSSION

3.1. Agricultural drought and the proportion of fast-drying soils

As shown in Fig. 3a,b, the median number of days under drought conditions in the most affected areas is >30 if we consider the whole growing season. For most of the arable land, the value is >10 d yr^{-1}, with the maxima being achieved not only in the southeast of the Czech Republic but also in parts of the northeast, in the northwest area of so-called central Bohemian region and in the southwest. The areas with a high incidence of dry days also include the southern edge of the Czech and Moravian Highlands in the center of the country (Fig. 1a). If we consider the z-score results (Fig. 4a,b), it is clear that the highest drought hazard is indicated over a fairly large and continuous area in the lowlands of the southeast of the Czech Republic and through a number of smaller regions in the northwest, north-central and, to a lesser degree, in the southwest and northeast. The drought hazard depends on which part of the growing season it is. In the first half of the growing season, the west of the Czech Republic tends to be affected more (Fig. 4a), while the southeast follows an opposite pattern, with drought hazards in the east of the Czech Republic increasing significantly from July to September (Fig. 4b).

Hazards posed by a high proportion of fast-drying soils are limited to 2 principal areas at the west and southeast of the country. While partly overlapping with areas that have an increased agricultural drought risk, these areas are not identical. Fast-drying soils are not present at a number of areas potentially influenced by agricultural drought, notably those in the agricultural lowlands of the southwest, north-central and northeast of the country. Obviously, combining a high proportion of fast-drying soils with a high probability of a greater number of drought days increases the hazard level in the respective cadaster units. This criterion also reflects a gradual decline in the fertility area due to a prolonged lack of water, unsustainable rates of erosion and organic matter depletion caused by human activity, and it reflects decreases in soil fertility that are not fully captured by the indicators of agricultural drought, as these calculations did not consider the processes of land degradation leading to the estab-

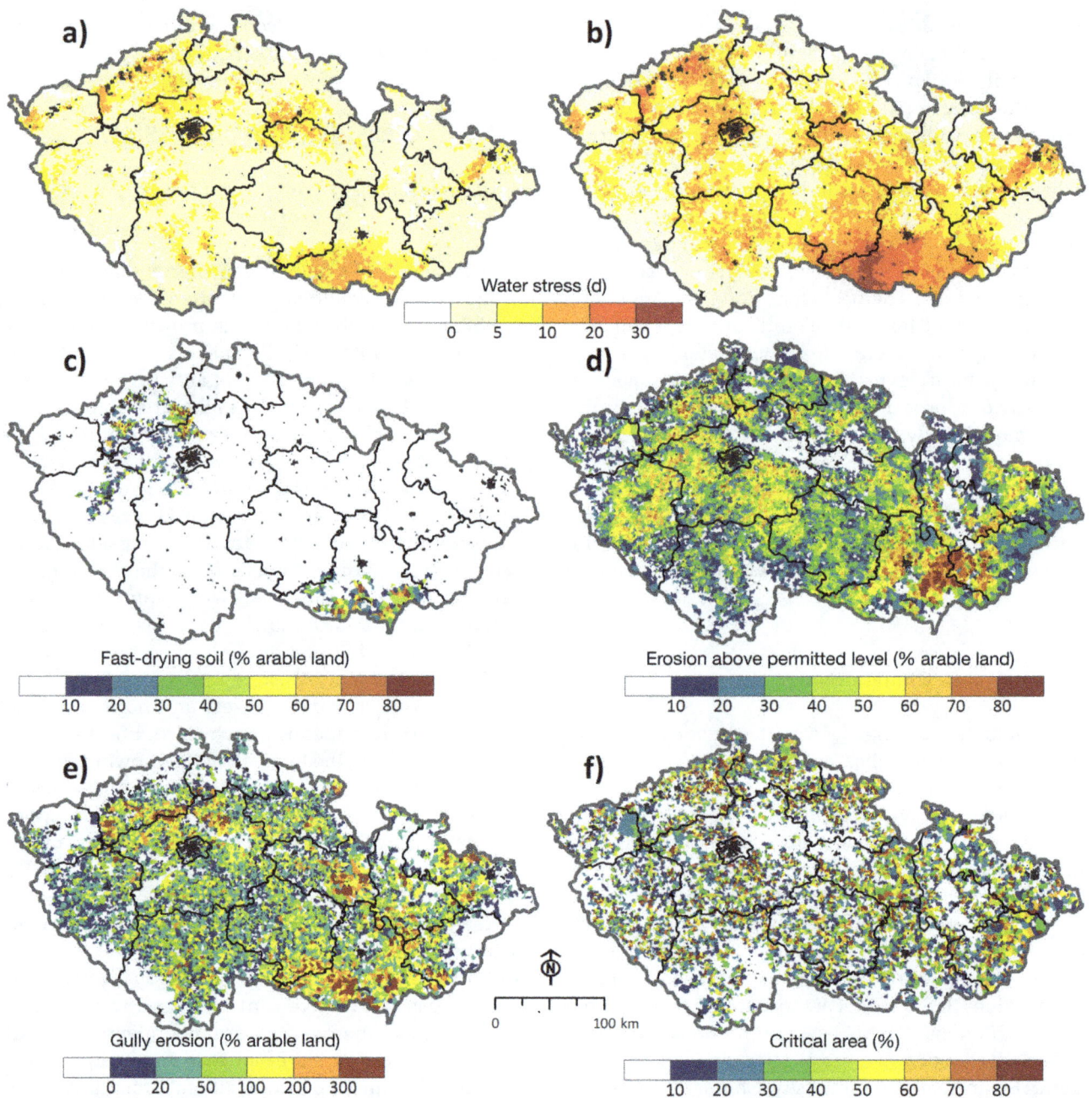

Fig. 3. Map of indicators used for hazard assessment (see Table 1) at the cadaster unit level: (a,b) number of days with water stress in the topsoil (0–40 cm) in (a) April–June and (b) July–September, (c) proportion of arable land in the fast-drying soils category, (d) percentage of arable land within the cadaster unit with erosion rate above permissible levels, (e) proportion of area contributing to ephemeral gully erosion pathways from arable land, (f) proportion of cadaster unit area belonging to contributing area for critical flood points

lishment of fast-drying soils over the last 3 decades. Figs. 3c & 4c clearly show that in terms of fast-drying soils, the most affected areas are concentrated in the southeastern Czech Republic and in an even larger area to the west of Prague and around Pilsen. These areas are likely still expanding, and their growth is proportional to the intensity of the erosion processes.

3.2. Sheet and ephemeral gully erosion

The hazard posed by sheet erosion affects the majority of the key agricultural areas, with the exception of flat areas around large rivers (Elbe and Morava). It is clear that a high percentage of the arable land is at risk of a higher than permissible rate of soil loss (>4 t ha^{-1} yr^{-1}), which is a major cause for

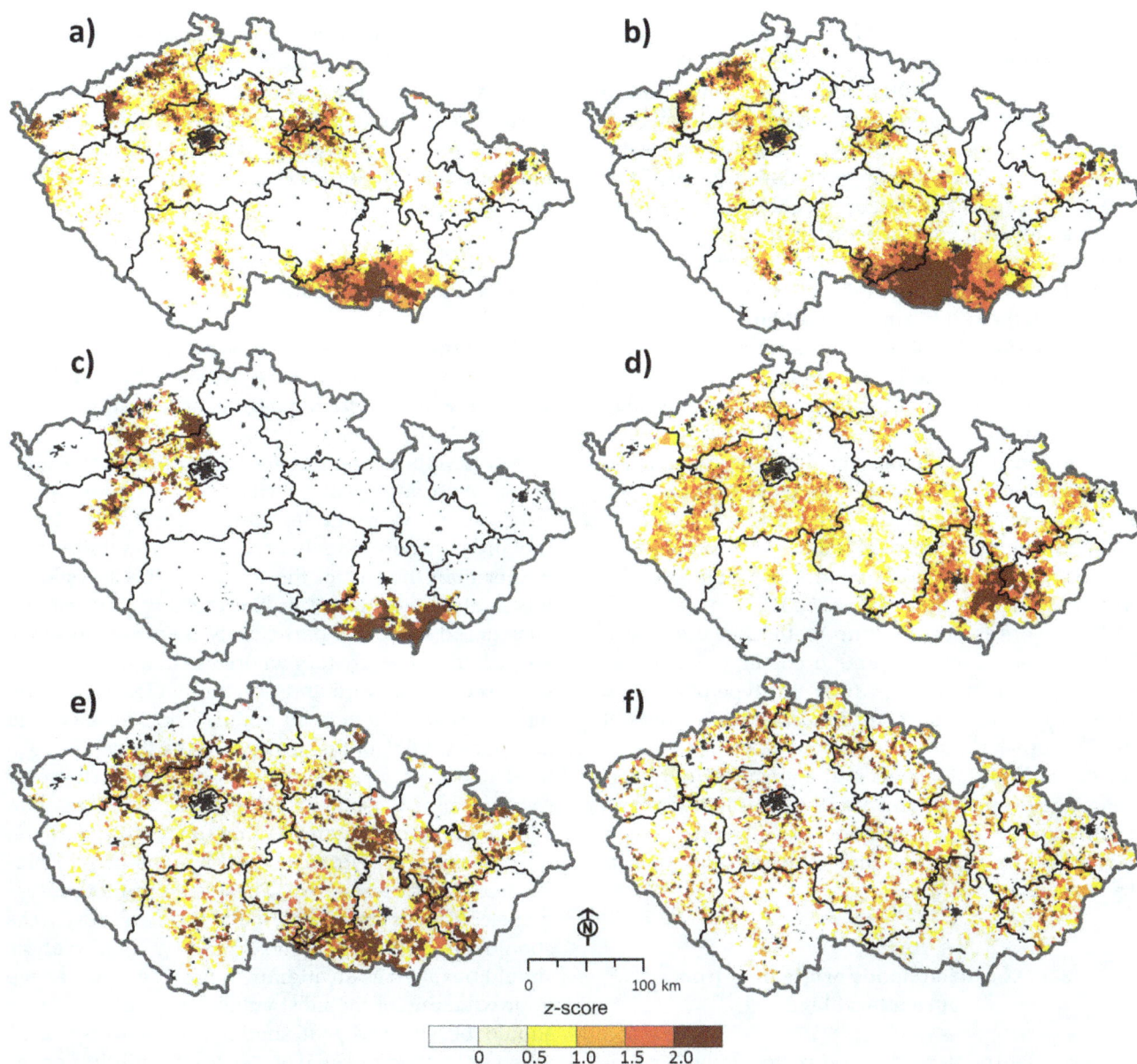

Fig. 4. Hazard indicators expressed as z-score at the cadaster unit level: (a) days with water stress in the topsoil (0–40 cm) in (a) April–June and (b) July–September, (c) arable land in the fast-drying soils category, (d) proportion of arable land within each cadaster unit with erosion rate above permissible levels, (e) proportion of drainage area within each cadaster unit contributing to ephemeral gully erosion pathways from arable land, (f) proportion of cadaster unit area belonging to the contributing area for critical points

concern (Fig. 3d). The problem is particularly pronounced in the southeast of the country (albeit in different regions than those affected by drought and fast-drying soils), as Fig. 4d indicates. In each region, one can find 'hot-spots' that are at a markedly greater risk of sheet erosion than the surrounding areas but without a notable 'center', as in the case of the southeast region. On the other hand, the lowest hazard level (apart from areas already listed) is found in the southwest (Fig. 4d).

Compared to the hazard of sheet erosion, the hazard posed by ephemeral gully erosion is more evenly spread across the country. The identification and plotting of potential paths of concentration runoff were based on the modeling of flow accumulation from their collection (contributing) areas, interpretation of the nature of the terrain and a visual interpretation of aerial photos for the affected fields of arable land. We identified >29 000 paths of potential ephemeral gully erosion (pathways for concentrated flow) with a total length of nearly 11 963 km, and for each ephemeral gully erosion path, the contributing area of the arable land was calculated. It is obvious that in some cadasters, the contributing areas are very high (over 200 or 300 ha), indicating potentially large water flows in cases of extreme precipitation (Fig. 3e). Compared to the areas affected by sheet erosion, the areas where ephemeral gully erosion hazards are the greatest tend to be more concentrated, creating an 'arc' in the east of the country and then a 'belt' in the northwest. Using the USLE, land users often underestimate erosion on agricultural fields because it does not account for the loss of soil from ephemeral gullies. It is estimated that ephemeral gully erosion is responsible for up to 20–40 % of the total volume of the sediment from erosion. This is approximately the same magnitude as that of sheet erosion, and has thus far not been accounted for in any hazard analysis. Bennett et al. (2000) presented estimates for the percentage of total soil loss from agricultural watersheds due to ephemeral gullies ranging from 20 to 100 %.

While there is currently no method in the Czech Republic for identifying and predicting the occurrence of sediment delivery from this type of erosion, it needs to be seriously considered, due to its share of the total erosion rate. It is obvious that high erosion rates caused both by sheet and ephemeral gully erosion are, in combination with frequent droughts and unsuitable agronomy practices, the leading causes behind the gradual spreading of the presence of fast-drying soils.

3.3. Localized floods originating from agricultural land

Analysis of the hazard posed by local floods originating from arable land included 6248 municipalities across the Czech Republic, with >35 437 intersections of potential water flow paths with built-up area boundaries, and 9261 were identified as critical points and thus potentially dangerous (Drbal 2009). In total, the contributing areas of these critical points constitute >23 % of the total area of the Czech Republic. For the analysis, the proportion of agricultural land in each cadaster unit that belongs to the contributing area of a so-called 'critical point' (hazard profiles of flash flood risks on boundaries of built-up areas) was considered as the indicator. Unlike for sheet or ephemeral gully erosion, other agricultural land use apart from just arable land was considered in this analysis. Out of all the indicators, the results from this analysis show that the hazards due to flood-

ing are the most evenly spread across the country (Fig. 3f), and it is not possible to pinpoint any particularly vulnerable regions (Fig. 4f). It is critically important that this hazard be included, as it leads to direct threats to property (Fig. 2) and human lives, and could be at least partly addressed by proper agronomic practices.

3.4. Combining the individual hazard indicators

The percentage of territory where the hazard level is highly above average or worse is 8 % (Fig. 5a). Within the multi-criteria analysis, we simultaneously examined how a large part of the territory of the Czech Republic meets at least one of the criteria for an extreme degree of risk (Fig. 5b). This combined approach provides, in our view, a good overview of the areas where the hazard level is significantly higher than the rest of the territory. The last step of this analysis was to define the territory that may be considered to be at a particularly high risk. As such, we considered a territory where the average value of the z-scores was >1.5 and/or where at least 2 criteria had z-scores >2.0 to be at a high risk. These criteria are met by 4.5 % of the territory in the Czech Republic. The percentage is slightly higher when 4th-order river basins are used instead of cadaster units due to a higher mean basin area within the regions with the highest risks. As Fig. 5c shows, 2 areas can be pinpointed as the most at risk. These most vulnerable regions constitute areas where attention and resources should be given the highest priority. This analysis should be taken as an attempt to provide an objective approximation of the most vulnerable regions, and it should be stressed that similar results were found when only 4 indicators were applied (omitting ephemeral gully erosion and combing the 2 drought-day indicators into a single indicator) or when basins were used instead of cadaster units. While there was a great deal of general agreement between the individual approaches, the local differences between the versions suggest that local knowledge should also be applied and that the delimitation of the endangered areas should be inclusive rather than rigid.

The multi-criteria analysis presented here tries to define particularly endangered areas in the Czech Republic in terms of drought hazard and exposure to the risks of erosion (Fig. 6). This analysis indicates relatively 'typical' hazard areas, which include the regions of northwest Bohemia and south Moravia. Northwest Bohemia in particular is at risk from hydrological drought, and these areas have the smallest

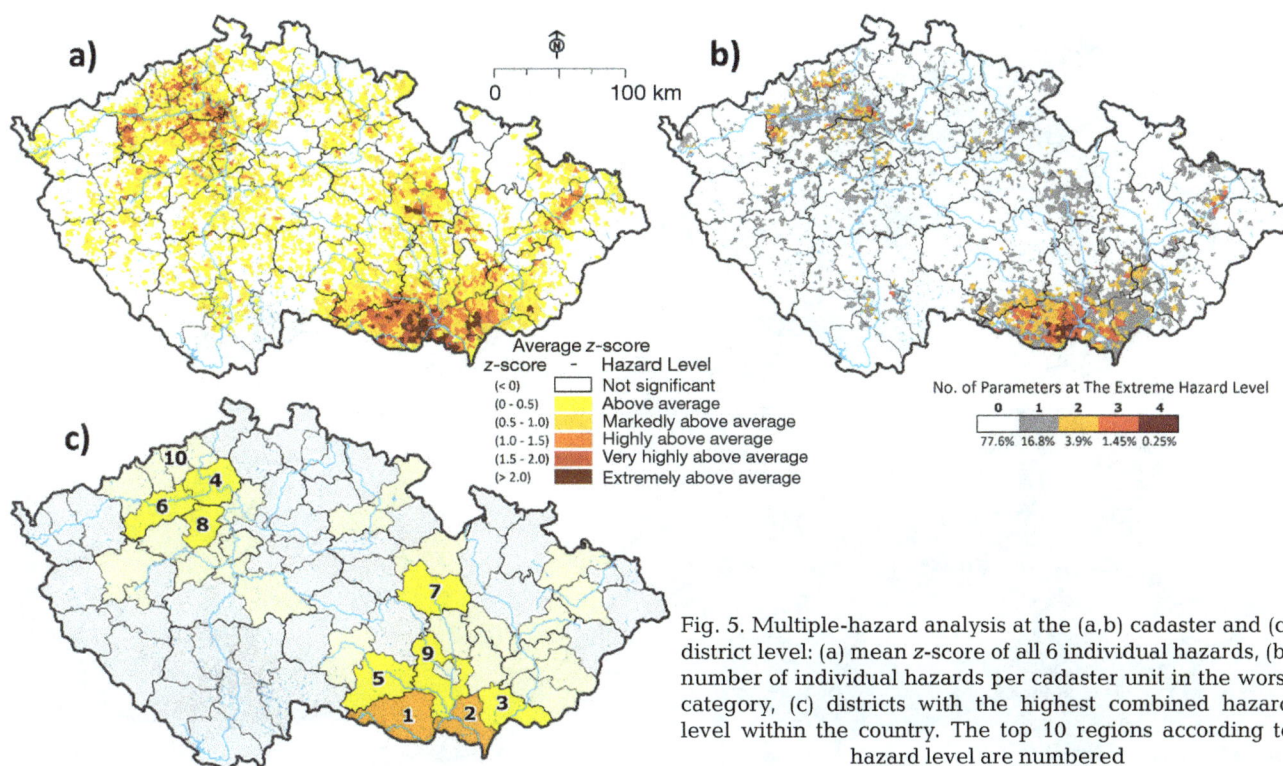

Fig. 5. Multiple-hazard analysis at the (a,b) cadaster and (c) district level: (a) mean z-score of all 6 individual hazards, (b) number of individual hazards per cadaster unit in the worst category, (c) districts with the highest combined hazard level within the country. The top 10 regions according to hazard level are numbered

availability of water resources. These territories also have agroclimatic conditions typical of the corn and sugar beet-growing regions (Fig. 6b) that include the most fertile regions of the country. At the same time, the cadaster units identified as the most vulnerable using the multiple hazard approach show only a 10 % overlap with the presently defined LFAs, as Fig. 6a indicates. LFA is a term used within the EU (and defined according a set of EU-based criteria) to describe an area with natural handicaps (lack of water, climate, short crop season and tendencies of depopulation), or that is mountainous or hilly, as defined by its altitude and slope. As the LFAs are defined through the use of more or less common European criteria, it leads to a paradox whereby the regions with the highest combined hazards from drought and soil degradation receive significantly less support than the LFA areas. As LFA regions have seen significant improvement in their agroclimatic conditions and overall productivity thanks to climate change over past 2 decades, it has led to an imbalance that the LFA introduction has attempted to rectify.

3.5. Change in climate conditions

The estimated risks posed by the hazards discussed here are not likely to remain stable in the near future. We demonstrate this in the case of the number of drought days in April–June for the period 2021–2040, assuming an RCP 4.5 emission scenario that predicts a fairly modest increase in CO_2 concentrations. All 5 global circulation models show a marked increase in z-score levels compared to baseline (Fig. 7). The rate of occurrence of the most extreme level (z-score >2.0) is, under the baseline climate, 5.7 %, and the average for 2021–2040 is estimated at 14.2 % (with the range of the 5 GCMs considered being 10.9–20.1 %), which is an almost 3-fold increase in the frequency of this most extreme category. Such changes would mean profound increases in the overall drought hazard. In the southeast, the expansion of the highest hazard area occurs in a northward direction, while in the west, the expansion covers the Elbe River lowland. Both areas are presently considered to be the most fertile regions in the country. An additional factor of concern is the occurrence of drought 'spots' across the entire country, with the only exception being the northeast region. The increased incidence of drought in these sites is driven primarily by a lower soil water-holding capacity. These results indicate that hazard levels are not static and are likely to change in the future. In addition, this dynamic (i.e. hazard levels in relation to climate change) must be considered when areas most at risk are defined. What was surprising, how-

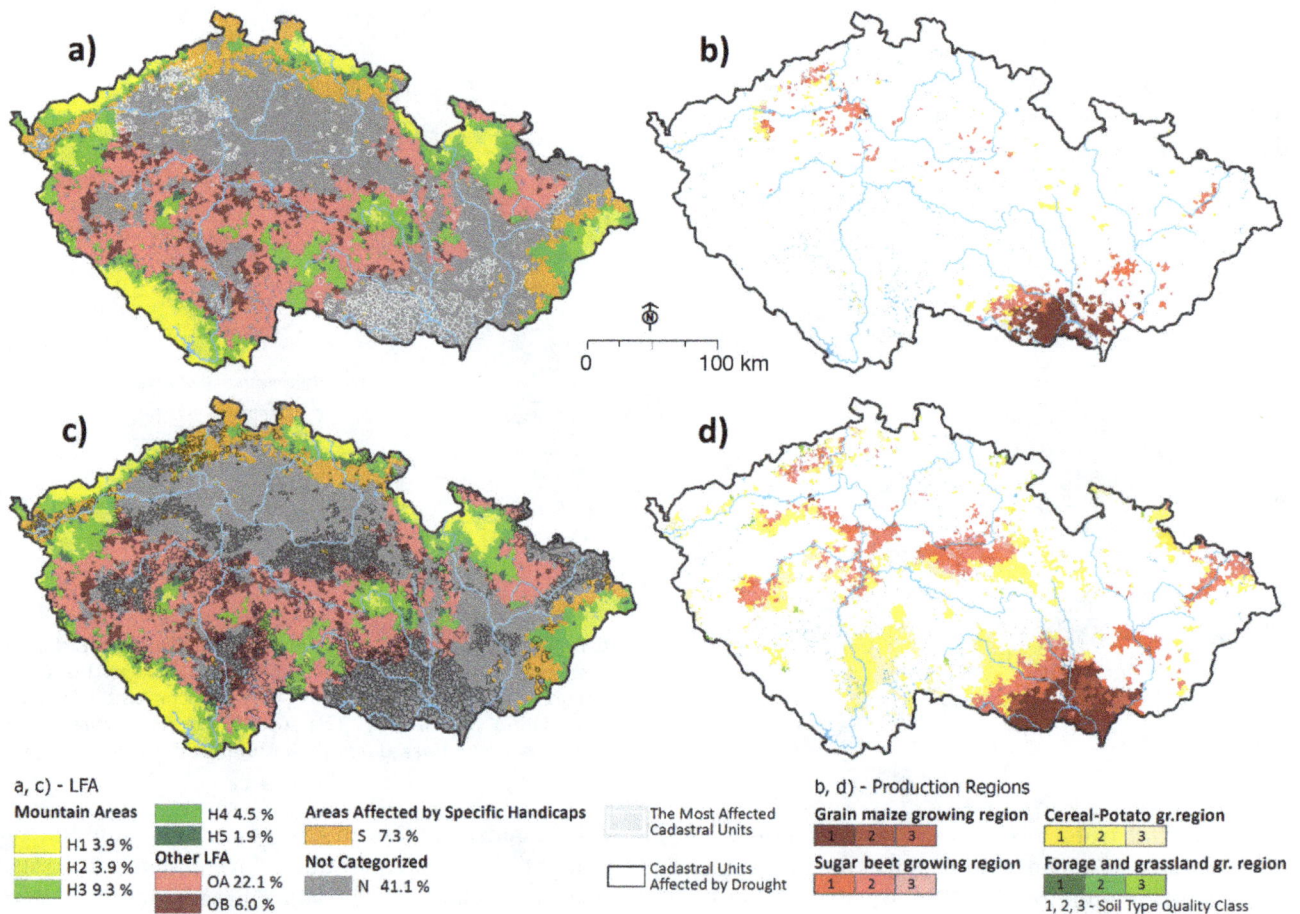

Fig. 6. (a,b) Drought risk and (c,d) multiple hazards. Comparison of cadaster units most at risk from (a) drought and (c) multiple hazards with less-favored areas (LFAs). Agroclimatological production regions belonging to cadaster units with (b) high drought and (d) multiple hazards. Areas marked H1-5 are limited by the altitude and slopes; areas OA and OB by slopes and soil conditions; areas marked S are primarily limited by need to protect water resources

ever, was the magnitude of the predicted changes that could occur over such a short time-frame in the near future. The probability of extreme drought increases considerably under predicted future climate conditions, and these changes may occur much more quickly than is generally anticipated. This leaves relatively little time for a response.

3.6. Uncertainty in hazard classifications

Apart from the presented approach using the 6 hazard indicators and cadaster units, 2 other approaches were tested. One relied on only 4 indicators (drought stress was considered for the entire April–September period, and gully erosion was omitted), and we also considered using 4th-order catchments instead of cadaster units. The use of 4 indicators would increase the weight of each one, while making the approach simpler. The overall area affected by the highest risk levels was slightly smaller but was consistent with the finding presented above (this paragraph). However, it was felt that drought effects in particular were underestimated when using only 4 indicators. As Fig. 4a,b shows, there are considerable differences between the drought hazards in the periods April–June and July–September. When these 2 indicators were considered as a single April–September indicator, some of the areas with known and persistent drought hazards in the northwest of the Czech Republic were left out of the evaluation. Similarly, not including ephemeral gully erosion as a factor led to the omission of some areas where major damage has occurred, as documented in Fig. 2.

The analysis presented here shows the urgent need for explicitly accounting for climate change. This is fairly easy in the case of drought-day-based indicators, as the methodology is flexible and de-

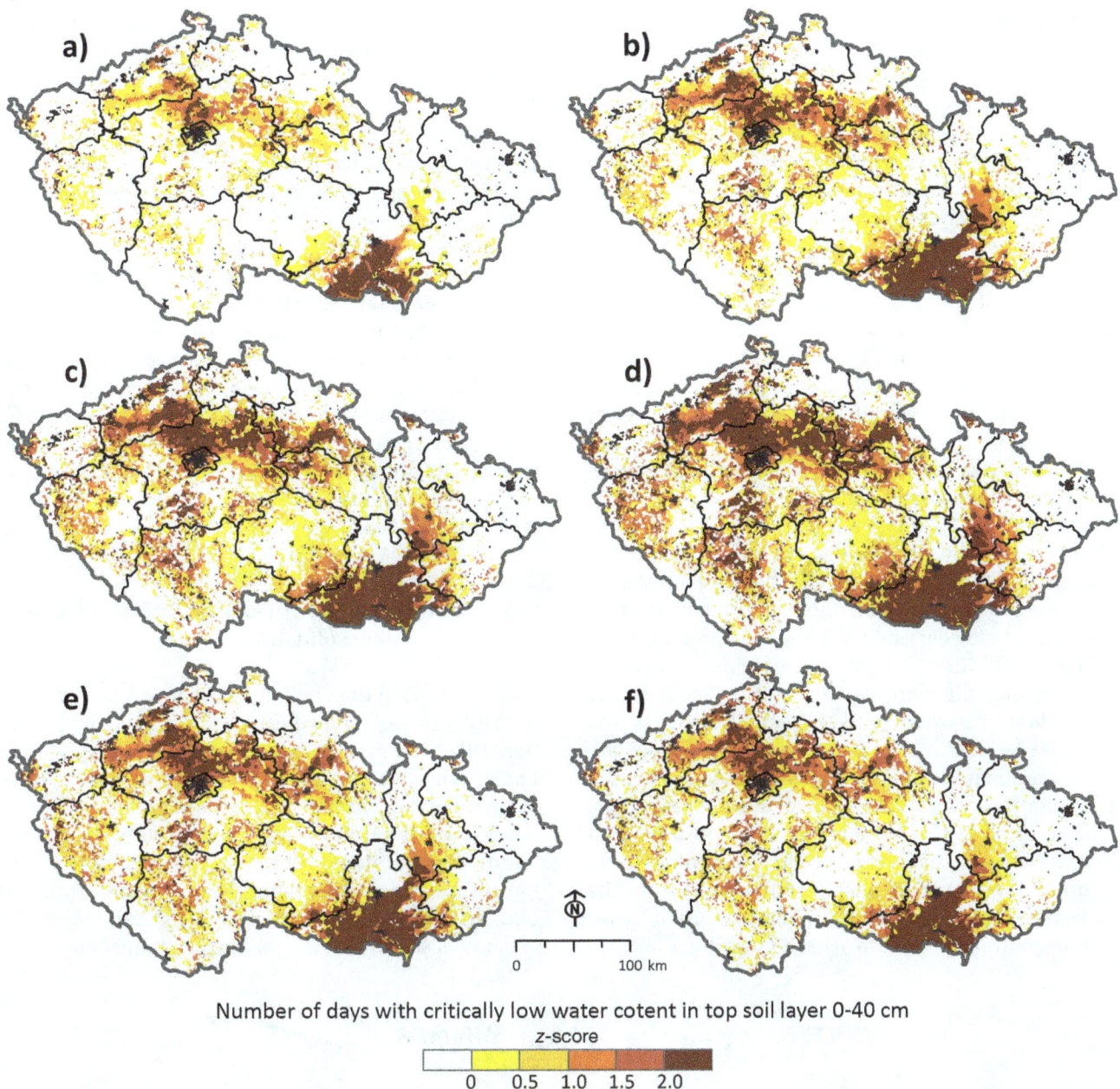

Number of days with critically low water cotent in top soil layer 0-40 cm

z-score

| 0 | 0.5 | 1.0 | 1.5 | 2.0 |

Fig. 7. Hazard of water stress in April–June in the topsoil (0–40 cm) expressed as *z*-scores at the cadaster unit level. (a) The period 1991–2014 was used as the baseline, and *z*-scores for (b–f) the period 2021–2040 were estimated from this baseline. The following models are represented: (b) IPSL, (c) HadGEM, (d) MRI, (e) CNRM, (f) BNU (see Section 2.5 for abbreviations)

signed to handle the effects of climate change (e.g. Trnka et al. 2015a). However, it will be more of a challenge for the remaining 4 indicators. In the case of fast-drying soils, the dynamics of persistently drying soil profiles speeding up soil degradation should be accounted for. Similarly, in the case of sheet and ephemeral gully erosion, changes in the probability of major precipitation events need to be considered, as do changes in the phenological calendar. While

there is a general view that the occurrence of higher-intensity events is more likely under future climate conditions (increasing the potential erosion from each event), there is also clear data from previous studies showing the protective effect of vegetation cover. All of these factors need to be analyzed and researched prior to conducting multi-hazard assessments for determining the effects of future climate conditions.

3.7. Analyzing water resources in areas with the highest hazard

The Czech Republic is situated in a region with annual precipitation that ranges from 450 mm in dry regions to 1300 mm in mountainous regions; however, as the country is located on the continental divide, its water resources are driven primarily by rainfall. Previous research (e.g. Hlavinka et al. 2009, Trnka et al. 2012) has shown that areas with considerable lack of water in the top layer of the soil might hamper agriculture, as yields are closely related to the water balance (e.g. Hlavinka et al. 2009, Trnka et al. 2012). Occasional water shortages do not usually result from the overall unavailability of water resources, but rather from the spatiotemporal variability of water supply/demand and the high degree of water resource exploitation.

However, as Hanel et al. (2012) and Trnka et al. (2015b) have indicated, water availability is likely to change due to the projected changes in temperature and precipitation (i.e. an increase in temperature over the whole year and no change in annual precipitation, but with a decrease in precipitation in the summer and an increase in winter). Fig. 8 shows interactions between shifts in rainfall amounts and evapotranspiration. Higher precipitation totals are more than matched by the increase in actual evapotranspiration, as estimated with the BILAN model (Vizina et al. 2015). In summer, although precipitation decreases, the increase in actual evapotranspiration is not as large as would be expected from the increase in temperature (and hence potential evapotranspiration), because it is limited by available water. The observed changes in the difference between precipitation and potential evapotranspiration are shown in Fig. 8 and are becoming more negative in spring and summer.

Trnka et al. (2015b) and Hanel et al. (2012) used different approaches, but agreed on changes in the amount of water fixed in snow. This influences both runoff and the speed of the snow melt, and underground water recharge. An important factor for the changes in runoff is a shift in the snow melt from early spring to winter.

The combination of reduced precipitation and increased temperature leads to measures that attempt to protect water resources. Practical experience indicates that the most robust and effective measures are those that increase the water supply (in our case specifically, the reconstruction of old—or the design of new—reservoirs or water transfer systems) in high-hazard areas.

3.7. Using multi-hazard analysis results in land consolidation process

The approach developed in this paper can be used in the process of land consolidation. It is a multifunctional tool for sustainable development of the land. Land consolidation spatially and functionally arranges the land in the public's interest and consolidates or splits parcels while ensuring its accessibility. Land consolidation provides the conditions for improving the environment, land resource protection, and water management and for improving the ecological stability of the landscape. Land consolida-

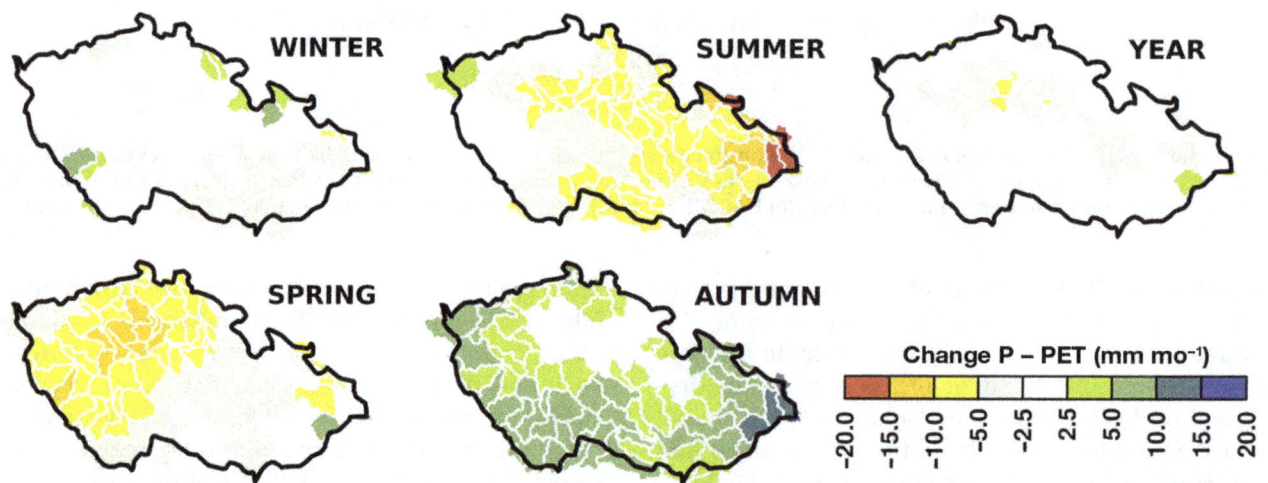

Fig. 8. Changes in observed difference in precipitation (P) and potential evapotranspiration (PET) between the periods 1961–1980 and 1981–2005 for 3rd-order river basins

tion is performed pursuant to Act No. 139/2002 Coll.

This consolidation is the only process in the Czech Republic to solve complex issues in rural areas, including the realization of measures that are in the public's interest. It also addresses considerable issues relating to the property of citizens and legal entities. Therefore, it is obvious that this complex set of issues cannot be managed without objective sources of information. The identification of the most vulnerable areas in the Czech Republic through a multi-hazard analysis is an important source of information in guiding the prioritization of the land consolidation process and its spatial targeting for the State Land Office. In this way, the State Land Office will receive unique material that can be used to improve their ability to mitigate the impacts of climate change. In addition, it will be able to effectively participate in the establishment of a legislative and economic framework that could possibly realize adaptation measures acceptable to agricultural entities.

4. CONCLUSIONS

The mapping of multiple hazards for agricultural land is intended as a first but crucial step in the assessment of the vulnerability of the agricultural sector to the occurrence of drought and extreme precipitation events under the present conditions and under the predicted future climate conditions in the Czech Republic. The map presented here synthesizes a variety of data and serves as an indicator of areas deserving more detailed attention. The key hazards for agricultural land in the Czech Republic include the occurrence of water stress in the topsoil layer during both the first and second half of the growing season, the proportion of fast-drying soils, the risk of sheet and ephemeral gully erosion and the risk of local floods originating primarily from agricultural land. The generation of z-scores was used as a standardization method, and a combination of an equal-weighting scheme and z-scores below -2.0 signaling the most extreme values were used in drafting the final output map. The final output map also shows results aggregated at the district level to clearly mark regions where primary attention should be given to reduce the level of the hazards and/or to increase cropping capacity. These regions are concentrated in the southeastern and northwestern lowland areas. As for typical areas with the highest hazard levels, we can identify regions with below-average precipitation and a high proportion of soils with a degraded or naturally occurring low water-holding capacity, and

those with steeper than average slopes and terrain configurations in relatively large catchment areas that have urbanized landscapes located at their lower elevations. This study also allows for the definition of cases in which data quality limits the usefulness of such hazard mapping. While state of the art digital elevation models were used, the information on the actual soil status had to rely on data from complex soil surveys carried out in the 1970s. While these data have been constantly updated, the last comprehensive campaign completely assessing soil status was carried out approximately 40 yr ago. As the next step in this research, farms in the areas with the highest hazard levels were selected as sites to conduct detailed and thorough assessments of the hazards present and to perform a complete vulnerability analysis. Based on this pilot study of farms and ground-level validation of the concept, a national vulnerability map will be prepared that will also include social aspects of vulnerability.

Despite some limitations, the methods presented in this paper serve as a step forward in developing techniques for reducing hazard levels at the individual cadaster unit, especially in the process of land adjustment, which aims at improving the organization, productivity and sustainability of agricultural production and the optimization of ecosystem services. Our results also point to the fact that the present definition of the LFA does not match the areas threatened by increased drought and erosion hazards, and that other mechanisms should be introduced to support sustainable and viable farming in these regions. This is especially important given that many of the areas at considerable risk are in the regions that are nominally the most productive land in the country and that have the highest taxes levied per hectare of land and the lowest level of support. As we have also shown, ongoing climate change will considerably change the hazard levels compared to those present today. Drought and soil loss are likely to become even more dominant factors affecting production in the future, and therefore, the adaptation of the agrarian sector to these coming conditions is critical. In addition, understanding the present hazard levels can and should lead to adjustments in agricultural practices and to the selection of more appropriate cropping patterns to obtain maximum financial yields during years with normal precipitation, and to reduce declines in crop yields and income loss during drought years, while at the same time conserving the soil. The multiple-hazards map presented here can help decision-makers visualize hazards and communicate them to farmers, natural resource managers and others. The education of the Czech Republic's decision-makers about these multiple hazards has already begun, and includes both grassroots as well as responsible decision-makers, both in the executive and legislative branches of the government.

Acknowledgements. The support of the State Land Office is greatly appreciated by all authors. M.T., P.H., J.B., D.S. and Z.Ž. acknowledge funding from the CzechAdapt Norwegian funds project funded through the Czech Ministry of Finances (no. EHP-CZ02-OV- 30 1-014-2014) and funding from the Ministry of Education, Youth and Sports of the Czech Republic within the National Sustainability Program I (NPU I), grant no. LO1415. We thank the National Agency for Agricultural Research of the Czech Republic, project no. QJ1610072 'System for monitoring and forecast of impacts of agricultural drought', for supporting the purchase of some datasets. We thank Mrs. Svatava Maradová, MBA (head of the State Land Office) and Mr. Václav Hlaváček (vice-president of the Czech Agrarian Chamber) for their comments and improvement of the study.

LITERATURE CITED

Allen RG, Pereira LS, Raes D, Smith M (1998) Crop evapotranspiration: guidelines for computing crop water requirements. FAO Irrigation and Drainage Paper No. 56. FAO, Rome

Batysta M, Hruška M, Jirásková I, Leibl M and others (2015) Situační a výhledová zpráva Půda. Ministry of Agriculture of the Czech Republic, Prague (in Czech)

Bennett SJ, Casali J, Robinson KM, Kadavy KC (2000) Characteristics of actively eroding ephemeral gullies in an experimental channel. Trans ASAE 43:641–649

Brázdil R, Trnka M, Dobrovolný P, Chromá K, Hlavinka P, Žalud Z (2009) Variability of droughts in the Czech Republic, 1881–2006. Theor Appl Climatol 97:297–315

CSO (Czech Statistical Office) (2005) Statistical yearbook of the Czech Republic. Scientia, Prague

Desmet PJJ, Govers G (1996) A GIS procedure for automatically calculating the USLE LS factor on topographically complex landscape units. J Soil Water Conserv 51: 427–433

Drbal K (ed) (2009) Vyhodnocení povodní v červnu a červenci 2009 na území České republiky (Methodology for flood risk evaluation). Water Research Institute of TGM, Ministry of Environment of the Czech Republic, Prague. www.vuv.cz/files/pdf/problematika_povodni/povoden-2009_cinnost_povodnove_sluzby_a_slozek_izs.pdf

Drbal K, Dumbrovský M (2009) Metodický návod pro identifikaci KB. Ministry of Environment of the Czech Republic, Brno, Prague. www.povis.cz/html/download_smernice.htm

Dumbrovský M (2011) Vymezení přispívajících ploch nad závěrovými profily erozně ohrožených drah odtoku na orné půdě pro potřeby Rámcové směrnice pro vodní politiku 2000/60/ES

EEA (European Environment Agency) (1998) Europe's environment: the second assessment. EEA, Aarhus

EEA (European Environment Agency) (1999) Environment in the European Union at the turn of the century. EEA, Copenhagen. www.eea.europa.eu/publications/92-9157-202-0/SOER_1_1999.pdf/view

Giupponi C, Mojtahed V, Gain AK, Biscaro C, Balbi S (2015) Integrated risk assessment of water related disasters. In: Paron P, Di Baldassare G (eds) Hydro-meteorological hazards and disasters. Elsevier, Amsterdam, p 163–200

Gourdji SM, Sibley AM, Lobell DB (2013) Global crop exposure to critical high temperatures in the reproductive period: historical trends and future projections. Environ Res Lett 8:024041

Hanel M, Vizina A, Máca P, Pavlásek J (2012) A multi-model assessment of climate change impact on hydrological regime in the Czech Republic. J Hydrol Hydromech 60: 152–161

Hlavinka P, Trnka M, Semerádová D, Dubrovský M, Zalud Z, Mozný M (2009) Effect of drought on yield variability of key crops in Czech Republic. Agric For Meteorol 149: 431–442

Hlavinka P, Trnka M, Balek J, Semerádová D and others (2011) Development and evaluation of the SoilClim model for water balance and soil climate estimates. Agric Water Manage 98:1249–1261

Huntingford C, Jones PD, Livina VN, Lenton TM, Cox PM (2013) No increase in global temperature variability despite changing regional patterns. Nature 500:327–330

IPCC (2014) Summary for policymakers. In: Field CB, Barros VR, Dokken DJ, Mach KJ and others (eds) Climate change 2014: impacts, adaptation, and vulnerability. A. Global and sectoral aspects. Contribution of Working Group II to the Fifth Assessment Report of the Intergovernmental Panel on Climate Change. Cambridge University Press, Cambridge, p 1–32

Janeček M, Dostál T, Kozlovsky-Dufková J, Dumbrovský M and others (2012) Erosion control in the Czech Republic-handbook. Czech University of Life Sciences, Prague (in Czech)

Kappes MS, Keiler M, von Elverfeldt K, Glade T (2012) Challenges of analyzing multi-hazard risk: a review. Nat Hazards 64:1925–1958

Kolář P, Trnka M, Brázdil R, Hlavinka P (2014) Influence of climatic factors on the low yields of spring barley and winter wheat in Southern Moravia (Czech Republic) during the 1961–2007 period. Theor Appl Climatol 117:707–721

Larcher W (2003) Physiological plant ecology, 4th edn. Springer-Verlag, Berlin

Liu C, Allan RP (2013) Observed and simulated precipitation responses in wet and dry regions 1850–2100. Environ Res Lett 8:034002

Morgan RPC (2005) Soil erosion and conservation, 3rd edn. Blackwell Publishing, Oxford

Myers SS, Zanobetti A, Kloog I, Huybers P and others (2014) Increasing CO_2 threatens human nutrition. Nature 510: 139–142

Olesen JE, Carter TR, Díaz-Ambrona CH, Fronzek S and others (2007) Uncertainties in projected impacts of climate change on European agriculture and terrestrial ecosystems based on scenarios from regional climate models. Clim Change 81:123–143

Panagos P, Borrelli P, Poesen J, Ballabio C and others (2015) The new assessment of soil loss by water erosion in Europe. Environ Sci Policy 54:438–447

Porter JR, Xie L, Challinor AJ, Cochrane K and others (2014) Food security and food production systems. In: Field CB, Barros VR, Dokken DJ, Mach KJ and others (eds) Climate change 2014: impacts, adaptation, and vulnerability. Contribution of Working Group II to the Fifth Assessment Report of the Intergovernmental Panel on Climate

Change. Cambridge University Press, Cambridge, p 485–533

► Rahmstorf S, Coumo A (2011) Increase of extreme events in a warming world. Proc Natl Acad Sci USA 108:17905–17909

► Reyer CPO, Leuzinger S, Rammig A, Wolf A and others (2013) A plant's perspective of extremes: terrestrial plant responses to changing climatic variability. Glob Change Biol 19:75–89

► Ruiz-Ramos M, Sánchez E, Gallardo C, Mínguez MI (2011) Impacts of projected maximum temperature extremes for C21 by an ensemble of regional climate models on cereal cropping systems in the Iberian Peninsula. Nat Hazards Earth Syst Sci 11:3275–3291

► Svobodová E, Trnka M, Dubrovsky M, Semeradova D, Eitzinger J, Stepanek P, Zalud Z (2014) Determination of areas with the most significant shift in persistence of pests in Europe under climate change. Pest Manag Sci 70:708–715

► Taylor KE, Stouffer RJ, Meehl GA (2012) An overview of CMIP5 and the experiment design. Bull Am Meteorol Soc 93:485–498

► Trnka M, Dubrovský M, Svoboda M, Semerádová D, Hayes M, Žalud Z, Wilhite D (2009) Developing a regional drought climatology for the Czech Republic. Int J Climatol 29:863–883

► Trnka M, Kocmánková E, Balek J, Eitzinger J and others (2010) Simple snow cover model for agrometeorological applications. Agric For Meteorol 150:1115–1127

► Trnka M, Brázdil R, Olesen JE, Zahradníček P and others (2012) Could the changes in regional crop yields be a pointer of climatic change? Agric For Meteorol 166–167: 62–71

► Trnka M, Rötter RP, Ruiz-Ramos M, Kersebaum KC, Olesen JE, Žalud Z, Semenov MA (2014) Adverse weather conditions for European wheat production will become more frequent with climate change. Nat Clim Change 4:637–643

► Trnka M, Hlavinka P, Semenov MA (2015a) Adaptation options for wheat in Europe will be limited by increased adverse weather events under climate change. J R Soc Interface 12:20150721

► Trnka M, Brázdil R, Možný M, Štěpánek P and others (2015b) Soil moisture trends in the Czech Republic between 1961 and 2012. Int J Climatol 35:3733–3747

► Trnka M, Olesen JE, Kersebaum KC, Rötter RP and others (2016) Changing regional weather—crop yield relationships across Europe between 1901 and 2012. Clim Res 70:195–214

Van Lynden GWJ (1994) The European soil resource: current status of soil degradation causes, impacts and need for action. Council of Europe, Strasbourg

Vizina A, Horáček S, Kašpárek L, Hanel M (2015) Nové možnosti modelu BILAN. VTEI 57:4–5

Vopravil J, Rožnovský J, Hladík J, Khel T and others (2012) Možnosti řešení degradace půdy a její ovlivnění změnou klimatu na příkladu aridních oblastí. Ministry of Agriculture of the Czech Republic, Prague

Wilhite DA (2000) Drought as a natural hazard: concepts and definitions, Chapter 1. In: Wilhite DA (ed) Drought: a global assessment. Natural Hazards and Disasters Series. Routledge Publishers, London, p 3–18

Wischmeier WH, Smith DD (1978) Predicting rainfall erosion losses: guide to conservation planning. US Department of Agriculture, Agriculture Handbook 537. US Government Printing Office, Washington, DC

► Zahradníček P, Trnka M, Brázdil R, Možný M and others (2015) The extreme drought episode of August 2011–May 2012 in the Czech Republic. Int J Climatol 35: 3335–3352

15

Global warming-induced changes in climate zones based on CMIP5 projections

Michal Belda*, Eva Holtanová, Jaroslava Kalvová, Tomáš Halenka

Charles University in Prague, Dept. of Meteorology and Environment Protection, 18200, Prague, Czech Republic

ABSTRACT: Climate classifications can provide an effective tool for integrated assessment of climate model results. We present an analysis of future global climate projections performed in the framework of the Coupled Model Intercomparison Project Phase 5 (CMIP5) project by means of Köppen-Trewartha classification. Maps of future climate type distributions were created along with the analysis of the ensemble spread. The simulations under scenarios with representative concentration pathway (RCP) 4.5 and RCP8.5 showed a substantial decline in ice cap, tundra, and boreal climate in the warming world, accompanied by an expansion of temperate climates, dry climates, and savanna, nearly unanimous within the CMIP5 ensemble. Results for the subtropical climate types were generally not conclusive. Changes in climate zones were also analyzed in comparison with the individual model performance for the historical period 1961–1990. The magnitude of change was higher than model errors only for tundra, boreal, and temperate continental climate types. For other types, the response was mostly smaller than model error, or there was considerable disagreement among the ensemble members. Altogether, around 14% of the continental area is expected to change climate types by the end of the 21st century under the projected RCP4.5 forcing and 20% under the RCP8.5 scenario.

KEY WORDS: Köppen-Trewartha climate classification · Coupled Model Intercomparison Project Phase 5 · CMIP5 · Global climate model · Climate type change · Representative concentration pathways

1. INTRODUCTION

The outputs of state-of-the-art global climate models are currently available within the Coupled Model Intercomparison Project Phase 5 (CMIP5, Taylor et al. 2012b), which served as the basis for the IPCC's Fifth Assessment Report, published in September 2013 (available online at www.ipcc.ch). Besides the global climate models (GCMs) themselves, which were improved, e.g. toward higher resolution and in some cases by including new processes and interactions such as so-called Earth System Models (ESMs), the methodology regarding the construction of projection scenarios also changed in comparison to previous GCM experiments (CMIP3 GCMs, Meehl et al. 2007). For the core CMIP5 GCM experiments, 4 representative concentration pathways (RCPs) with radiative forcing ranging from 2.6 to 8.5 W m^{-2} in the year 2100 were chosen, designated as RCP2.6, RCP4.5, RCP6.0, and RCP8.5 (Moss et al. 2010).

Each generation of climate models must inevitably be subject to tests of how realistic the models are in simulating the observed climate characteristics in the recent past. The climate classifications can serve, inter alia, as effective tools for analysis of model performance. The Köppen classification (Köppen 1923, 1931, 1936, Geiger 1954) or Köppen-Trewartha classification (KTC, Trewartha 1968, Trewartha & Horn 1980) have most often been used for this purpose. The climate types are based on long-term climatological means of near-surface air temperature and precipitation that are easily obtained from the outputs of GCMs. The KTC provides a slightly more detailed description of climate type distributions than the original

*Corresponding author: Michal.Belda@mff.cuni.cz

Köppen scheme (de Castro et al. 2007). Belda et al. (2014) reviewed the KTC and its differences from the original Köppen scheme, and analyzed observed patterns in climate types and their changes during the 20th century.

Climate types derived from GCM projections of future climate are useful for a variety of sectors and scientific fields. They provide an idea of what changes can be expected in the areas of individual climate types. Due to their strong relationship with the distribution of natural vegetation zones (e.g. Trewartha & Horn 1980, Bailey 2009), it is possible to assess the development of different ecoregions, even though further information on e.g. edaphic and topografic properties (Baker et al. 2010, Hargrove & Hoffman 2004) is needed for such assessments.

Projected changes in climate types have previously been analyzed in various studies using different climate models and emission scenarios. Lohmann et al. (1993) assessed the outputs of the atmospheric general circulation model ECHAM3 using the Köppen classification, and derived shifts in climate zones in greenhouse gas warming simulations over 100 model years. They projected a retreat of the permafrost climate and an extension of both the tropical rainy climate and dry climate.

Kalvová et al. (2003) applied the Köppen classification to simulations of 4 GCMs, namely HadCM2, ECHAM4, CSIRO-Mk2b, and CGCM1, for the present and future periods. They confirmed the results described by Lohmann et al. (1993) regarding tropical and dry climates and described a decline in the area of boreal and cold climates.

More recently, Rubel & Kottek (2010) created a series of digital world maps of Köppen climate types for the period 1901–2100 based on observed data (CRU TS2.1, GPCC Version 4) and 20 simulations of 5 GCMs (each GCM with 4 Special Report on Emission Scenarios [SRES] emission scenarios). In the case of the emission scenario with the highest rate of emission increase (A1FI), the results showed an increase in the areas covered by tropical, dry, and temperate climates, and a decrease in the coverage of cold and boreal climates. Projected changes for the milder emissions scenario (B1) were significantly smaller.

Baker et al. (2010) compared KTC types over China for historical (1961–1990) and projected future climates (2041–2070) simulated by HadCM3 for the SRES A1FI scenario. They showed that the spatial patterns of climate change resulted in a northern migration of warmer climatic types as well as a slight expansion in the high-latitude desert and arid shrubland regions in northwestern China.

Mahlstein et al. (2013) used simulations of 13 CMIP5 GCMs for determination of Köppen-Geiger climate types and analyzed their changes during 1900–2098. They found that under the RCP8.5 forcing, for which the mean warming reaches about 4.5°C by the end of the 21st century (Rogelj et al. 2012), approximately 20% of the global land area would undergo a shift in the original climate zones. Frost climates are projected to largely decline, some arid climatic zones are expected to expand, and large parts of the global land area with cool summers will experience a change to climates with hot summers. However, Mahlstein et al. (2013) also emphasized large model uncertainties and reported that the pace of the climate type shifts increases with increasing global mean temperature.

Feng et al. (2012) analyzed observed and projected climate changes and their impact on vegetation for the area north of 50° N over the period of 1900–2099 using the KTC scheme. To estimate the future changes, they used the simulations of 16 CMIP3 GCMs for 3 SRES emission scenarios (B1, A1B, and A2). Their results showed a decrease in areas classified as tundra, ice cap, and subarctic continental climates, and an expansion of the temperate and boreal oceanic climates. Moreover Feng et al. (2012) projected that arid, warm temperate, and snow and polar climates will successively shift to the north in the northern hemisphere.

Feng et al. (2014) focused on shifts in KTC climate types in 1900–2100. In contrast to Feng et al. (2012), the analysis was done for the whole global land area and model simulations of 20 CMIP5 GCMs for RCP4.5 and RCP8.5 pathways. Feng et al. (2014) found that during the 21st century, the KTC types would shift toward warmer and drier types, with the largest changes in the northern hemisphere north of 30° N. They also concluded that temperature changes are the dominant factor causing the projected shifts in climate types during the 21st century.

Here we used the KTC to assess changes in climate type areas simulated by a suite of 30 CMIP5 GCMs for the period of 2006–2100 and 2 RCPs (RCP4.5 and RCP8.5). Our study follows previous papers, i.e. Belda et al. (2014) mentioned above and Belda et al. (2015), wherein we assessed the performance of 43 CMIP5 GCMs in simulating the KTC climate types in the reference period 1961–1990. One of the main conclusions of Belda et al.'s (2015) analysis was that models generally had problems capturing the rainforest climate type Ar (see Table 2 for climate types), mainly in Amazonia. The desert climate type BW was underestimated by half of the models. Boreal climate type *E* was overestimated by many models, mostly spreading over to the areas of observed tundra type *Ft*. Further, Belda et al. (2015) indicated that CMIP5 GCMs did not show any clear tendency to improve the representation of climate types with increasing spatial resolution.

In addition to previous analyses of CMIP5 models in terms of Köppen classification by Mahlstein et al. (2013) and Feng et al. (2014), here we use the largest possible set of GCMs, describe the temporal evolution of KTC types for individual GCMs, and present simulated changes in the context of model performance for the present climate. We also add an analysis of future climate uncertainty in terms of ensemble spread throughout the scenario simulations.

Various supplementary graphical products, including figures describing the model performance of CMIP5 GCMs used by Belda et al. (2015) are available at http://kfa.mff.cuni.cz/projects/trewartha/.

2. DATA AND METHODS

2.1. Data

A suite of CMIP5 GCM simulations is employed here, selected based on the availability of data for both RCP4.5 and RCP8.5 scenarios. Basic information on all model simulations incorporated here is presented in Table 1. The data are available at http://cmip-pcmdi.llnl.gov/cmip5/; we used monthly mean surface air temperature and precipitation to classify the KTC types. The outputs from the experiment denoted as 'historical' were used for the reference period 1961–1990. For the future time period 2006–2100, we considered 2 alternative simulations, RCP4.5 and RCP8.5. RCP4.5 assumes radiative forcing of 4.5 W m^{-2} at stabilization after 2100, whereas RCP8.5 represents a 'rising pathway' with radiative forcing higher than 8.5 W m^{-2} after 2100. For more details on RCPs, see Moss et al. (2010). Where more ensemble members were available, we chose the ensemble member r1i1p1 (considered a baseline simulation of the subensemble for the puposes of this analysis) (Taylor et al. 2012a).

As one of the indicators of uncertainty in the climate signal, errors in the historical experiment during the reference period were considered in terms of KTC types based on monthly mean surface air temperature and precipitation provided by the Climatic Research Unit (CRU) TS 3.22 dataset (Harris et al. 2014, hereafter TS3) available in spatial resolution of 0.5° × 0.5° over global land areas excluding Ant-

arctica. As a part of the uncertainty analysis, a comparison of the classification based on 2 versions of CRU (TS 3.22 and TS 3.1.10) and the University of Delaware dataset version 4.01 (UDEL; Willmott & Matsuura 2001) was performed with the conclusion that the differences between these datasets are considerably smaller than the spread of the model simulations, and thus the impact of the choice of the observational dataset on GCM performance evaluation is negligible.

2.2. Methods

The KTC system (Trewartha & Horn 1980, Belda et al. 2014) has 6 main climate groups. Five of them (*A*, *C*, *D*, *E*, and *F*) are basic thermal zones. The sixth group, *B*, is the dry climatic zone that cuts across the other climate types, except for the polar climate *F*. Similarly to original Köppen classification scheme, the main climate types are determined according to long-term annual and monthly means of surface air temperature and precipitation amounts. The dryness threshold distinguishing group *B* is based on the definition by Patton (1962). A brief summary of climate types and subtypes is provided in Table 2.

The KTC climate types were calculated in the original model grids for the reference period 1961–1990 and for running 30 yr periods during the 21st century, beginning with 2006–2035 until 2071–2100 or 2070–2099 (as data for some of the model runs are only available until the year 2099). Land areas falling into each climate type/subtype were expressed in terms of relative areas, i.e. as a percentage of the whole global land area (excluding Antarctica). Simulated changes of KTC types for both RCP4.5 and RCP8.5 were assessed in several different ways. An overall picture of the multi-model ensemble evolution in time is provided as medians and 10th and 90th percentiles of changes of relative areas with respect to the values simulated for the reference period 1961–1990.

Further, we pay special attention to 3 selected time periods denoted as near future (2006–2035), mid-century (2020–2050), and far future (2071–2100 or 2070–2099 based on the simulation period). We demonstrate changes in selected climate type areas for each of these periods simulated by individual GCMs together with model errors in the reference period indicating the reliability of the climate change signal. All changes are expressed in percentage of area simulated by the respective GCMs in the reference period. The model errors are defined as differences in

Table 1. CMIP5 global climate models analyzed in this study with model versions explained (where applicable)

No.	CMIP5 model	Resolution	Modeling center/model versions
1	ACCESS1.3	1.88° × 1.24°	Commonwealth Scientific and Industrial Research Organisation (CSIRO) and Bureau of Meteorology, Australia
2	CanESM2	2.8° × 2.8°	Canadian Centre for Climate Modelling and Analysis
3	CCSM4	1.25° × 0.94°	National Center for Atmospheric Research
4	CESM1-BGC	1.25° × 0.94°	Community Earth System Model Contributors BGC: BioGeoChemistry CAM5: Community Atmospheric Model v5 FV2: Finite volume 2degree
5	CESM1-CAM5	1.25° × 0.94°	
6	CESM1-CAM5.1-FV2	2.50° × 1.88°	
7	CNRM-CM5	1.4° × 1.4°	Centre National de Recherches Météorologiques; Centre Européen de Recherche et Formation Avancées en Calcul Scientifique
8	CSIRO-Mk3.6.0	1.9° × 1.9°	CSIRO; Queensland Climate Change Centre of Excellence
9	FGOALS-g2	2.81° × 3.00°	LASG, Institute of Atmospheric Physics, Chinese Academy of Sciences and CESS,Tsinghua University
10	GFDL-CM3	2.5° × 2°	Geophysical Fluid Dynamics Laboratory
11	GFDL-ESM2G	2.5° × 2°	
12	GFDL-ESM2M	2.5° × 2°	
13	GISS-E2-H	2.5° × 2°	NASA Goddard Institute for Space Studies H: Hycom Ocean Model R: Russell Ocean Model CC: interactive terrestrial carbon cycle, ocean biogeochemistry
14	GISS-E2-H-CC	2.5° × 2°	
15	GISS-E2-R	2.5° × 2°	
16	GISS-E2-R-CC	2.5° × 2°	
17	HadGEM2-AO		Met Office Hadley Centre AO: aerosols, ocean & sea-ice CC: AO+terrestrial carbon cycle, ocean biogeochemistry ES: CC+chemistry
18	HadGEM2-CC	1.875° × 1.25°	
19	HadGEM2-ES	1.875° × 1.25°	
20	INM-CM4	2° × 1.5°	Institute for Numerical Mathematics
21	IPSL-CM5A-MR	2.5° × 1.3°	Institut Pierre-Simon Laplace MR: Medium resolution LR: Low resolution
22	IPSL-CM5B-LR	3.75° × 1.9°	
23	MIROC5	1.4° × 1.4°	Atmosphere and Ocean Research Institute (The University of Tokyo), National Institute for Environmental Studies, and Japan Agency for Marine-Earth Science and Technology
24	MIROC-ESM	2.8° × 2.8°	Japan Agency for Marine-Earth Science and Technology, Atmosphere and Ocean Research Institute (The University of Tokyo), and National Institute for Environmental Studies CHEM: added atmospheric chemistry
25	MIROC-ESM-CHEM	2.8° × 2.8°	
26	MPI-ESM-LR	1.9° × 1.9°	Max Planck Institute for Meteorology MR: Medium resolution LR: Low resolution
27	MPI-ESM-MR	1.9° × 1.9°	
28	MRI-CGCM3	1.125° × 1.125°	Meteorological Research Institute
29	NorESM1-M	2.5° × 1.9°	Norwegian Climate Centre M: intermediate resolution ME: M+carbon cycle
30	NorESM1-ME	2.5° × 1.9°	

Table 2. Definition of Köppen-Trewartha classification (KTC) climate types according to Trewartha & Horn (1980), with dryness threshold defined by Patton (1962). *Tmo*: long-term monthly mean air temperature; *Tcold* (*Twarm*): monthly mean air temperature of the coldest (warmest) month; *Pmean*: mean annual precipitation (cm); *Pdry*: mean precipitation of the driest summer month; *R*: Patton's precipitation threshold, defined as $R = 2.3T - 0.64Pw + 41$, where T is mean annual temperature (°C) and Pw is the percentage of annual precipitation occurring in winter

Type Subtype	Criteria Precipitation/temperature regime
A	$Tcold > 18°C$; $Pmean > R$
Ar	10 to 12 mo wet; 0 to 2 mo dry
Aw	Winter (low-sun period) dry; >2 mo dry
As	Summer (high-sun period) dry; rare in type A climates
B	$Pmean < R$
BS	$R/2 < Pmean < R$
BW	$Pmean < R/2$
C	$Tcold < 18°C$; 8 to 12 mo with $Tmo > 10°C$
Cs	Summer dry; at least 3 times as much precipitation in winter half-year as in summer half-year; $Pdry < 3$ cm; total annual precipitation < 89 cm
Cw	Winter dry; at least 10 times as much precipitation in summer half-year as in winter half-year
Cf	No dry season; difference between driest and wettest month less than required for *Cs* and *Cw*; $Pdry > 3$ cm
D	4 to 7 mo with $Tmo > 10°C$
Do	$Tcold > 0°C$
Dc	$Tcold < 0°C$
E	1 to 3 mo with $Tmo > 10°C$
F	All months with $Tmo < 10°C$
Ft	$Twarm > 0°C$
Fi	$Twarm < 0°C$

simulated and observed CRU TS3.22 areas expressed as a percentage of observed values.

For illustration of the geographical distribution of changes simulated for the far-future period, maps of projected distributions of KTC types are shown for both RCPs. Climate zones in the future period were calculated based on temperature and precipitation scenarios constructed using the delta method (Deque 2007). Ensemble mean values in the periods 2070–2099 and 1961–1990 were used to calculate deltas that were then added (multiplied) to the present climate state represented by the temperature (precipitation) from the CRU TS3.22 database. KTC was then applied, which provided spatial distributions of climate zones in the scenarios.

3. RESULTS

The geographical distribution of observed KTC types and its simulated changes are illustrated in Figs. 1–3. The climate change signal patterns are similar for both scenarios, with stronger manifestation under stronger forcing of RCP8.5. In the northern hemisphere, the most remarkable feature is the northward shift of the border between *Dc* and *E* types, with an increase in the area of *Dc* and shrinking of the *E* type. The shift of the southern border of *Dc* is not as evident; only in Europe, an eastward shift of the *Dc–Do* border is projected, inducing expansion of the *Do* area over western and central Europe. Further, a global feature is shrinking of the *Ft* area, not only in the high latitudes, but also in high-elevation regions of the Himalayas and the Andes. In South America, the *Ft* type is projected to disappear by the end of the century under both RCPs. Another distinct pattern of change in South America is the expansion of the dry types *BW* and *BS*. In Africa and Australia, the GCMs project an increase in the *BW* area and shrinking of the *C* types. In southeastern Asia, our results suggest an expansion of the *Aw* type, which might be connected to increased strength of the Indian summer monsoon as documented e.g. by Menon et al. (2013).

The values of multi-model medians of simulated changes, 10th and 90th percentiles, and the range between them for the period 2071–2100 under both RCP4.5 and RCP8.5 forcings are summarized in Table 3.

The KTC climate types can be divided into 3 groups (decreasing area, increasing area and no conclusive change) based on the temporal behavior of simulated continental areas belonging to respective KTC types during the 21st century under the RCP4.5 and RCP8.5 forcing. The first group comprises boreal climate *E*, tundra *Ft*, and ice cap climate *Fi* that, according to the GCMs analyzed in our study, are expected to retreat. These 3 types occur at high latitudes or altitudes.

All GCMs simulate a decrease in the continental coverage (Antarctica not included in the analysis) of ice cap climate *Fi* (Fig. 4), which is clearly seen for the multi-model median (M-MED). Under the RCP4.5 forcing, the relative area of *Fi* decreases to 73% (Table 3) of the value simulated for the reference period 1961–1990. In the case of RCP8.5, the decrease is even stronger, as the *Fi* area decreases to 52% of its reference value. The decrease to less than 90% is already expected in the period 2006–2035 for both scenarios. The multi-model spread of simulated

CRU TS3.22 KTC 1961–1990

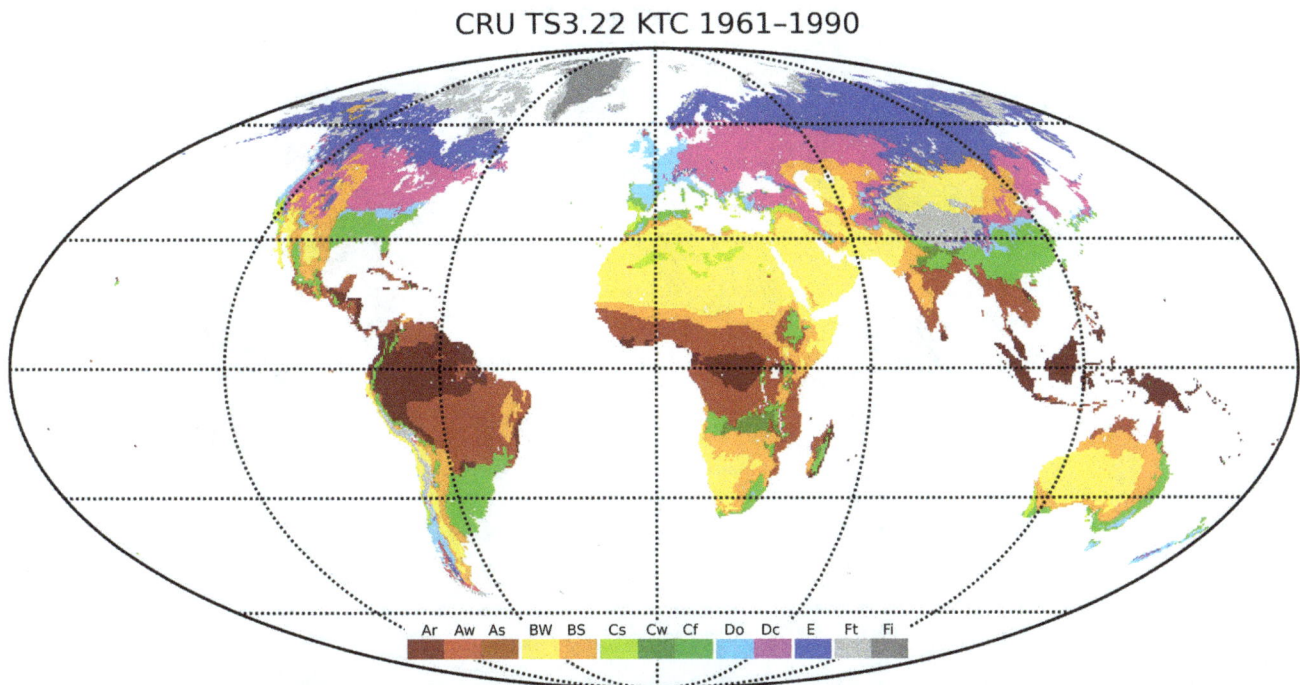

Fig. 1. Köppen-Trewartha climate types derived from observations (CRU TS3.22) for the period 1961–1990

CRU TS3.22+CMIP5 KTC 2070–2099

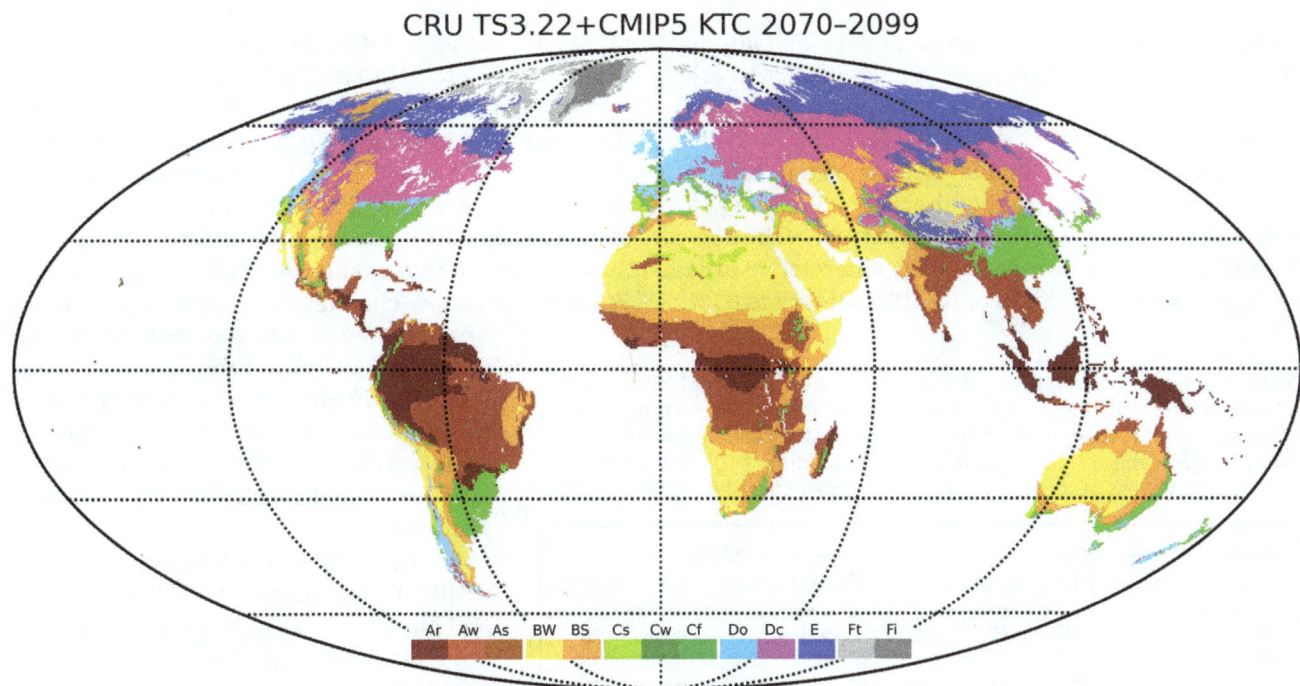

Fig. 2. Köppen-Trewartha climate types for the period 2070–2099, derived from the CRU TS3.22 observational dataset and the CMIP5 ensemble RCP4.5 scenario using the delta method

changes is quite large and is the same for both RCPs. In the case of RCP4.5, the decrease is often fastest during the first half of this century; in the second half it is rather slow, whereas the stronger forcing of RCP8.5 leads to more pronounced decline during the whole century. The decrease in relative area occupied by type *Fi* is solely due to transformation to tundra, *Ft*. Regarding the comparison of simulated changes to model errors in the reference period, all GCMs (except for CSIRO-Mk3-6-0 and MIROC-

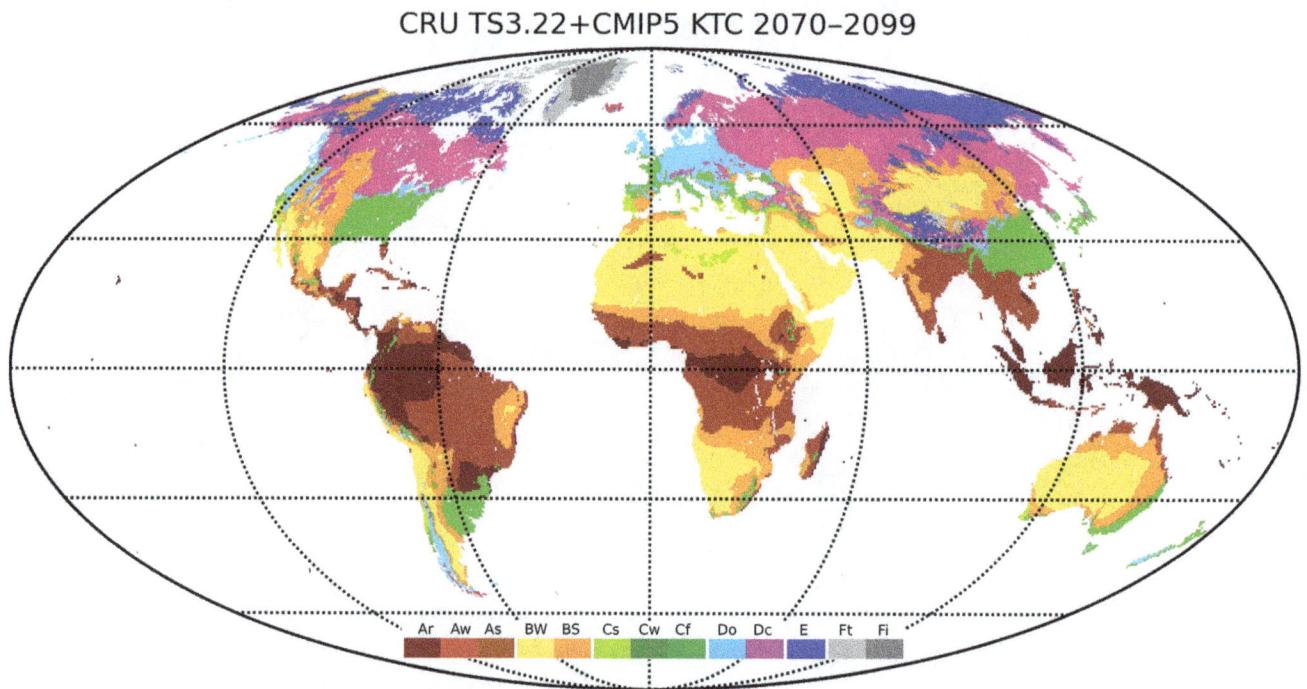

CRU TS3.22+CMIP5 KTC 2070–2099

Fig. 3. Köppen-Trewartha climate types for the period 2070–2099, derived from the CRU TS3.22 observational dataset and the CMIP5 ensemble RCP8.5 scenario using the delta method

ESM) overestimate the observed area of *Fi*. Only in 37 out of 180 cases (3 periods, 2 RCPs, 30 GCMs) are the projected changes larger (in absolute value) than model errors (Fig. 5).

Similarly, all GCMs simulate a decrease in tundra climate type *Ft*, and under the RCP4.5 forcing, a faster rate of change occurs in the first half of the century (Fig. 4). According to M-MED, the relative area of *Ft* decreases by the end of the 21st century to 63 %

(42 %) for RCP4.5 (RCP8.5), and the multi-model range is larger for RCP8.5 (Table 3). Models MIROC-ESM, MIROC-ESM-CHEM, and INMCM4 simulate the slowest decline in *Ft* (Fig. 5), even though the first 2 of these GCMs are the most sensitive to *Fi* changes. The largest change in *Ft* for the far future period is simulated by GFDL-CM3, which shows a decrease to 30 % of the reference under RCP4.5 and 19 % under RCP8.5. Projected changes are larger than model errors in 60 % of all cases for RCP4.5, and in 90 % for RCP8.5 at the end of the century. The *Ft* climate type is expected to transform into boreal climate *E*, although under RCP8.5, transitions of smaller areas to *Dc* and *Do* climate types are also simulated.

According to the outputs of all analyzed CMIP5 GCMs (except for GFDL-ESM2G and NorESM1-ME), the continental area occupied by boreal climate *E* is also expected to decrease (Fig. 5, Table 3). Time evolution of *E* type area in the running 30 yr periods according to individual GCMs shows a gradual monotonic decrease or only small fluctuations (Fig. 4). The exception is model CESM1-CAM5-1-FV2, which for both RCPs shows a negative peak around the year 2061 preceded by a steep decrease after 2045 and fol-

Table 3. Multi-model statistics of the percentage changes of Köppen-Trewartha classification (KTC; see Table 2 for definitions) climate type areas in the future with respect to the reference period (1961–1990) for the RCP4.5 and RCP8.5 scenarios. M-MED: multi-model median, p10: 10th percentile, p90: 90th percentile, range: range between p10 and p90

KTC type	RCP4.5				RCP8.5			
	M-MED	p10	p90	Range	M-MED	p10	p90	Range
Fi	73	57	85	28	52	40	68	28
Ft	63	45	77	32	42	28	66	38
E	83	67	97	30	64	36	86	50
Dc	115	106	127	21	130	116	145	29
Do	112	96	122	26	115	100	134	34
BW	108	103	117	14	113	108	123	15
BS	108	100	120	20	113	101	131	30
Aw	117	100	125	25	120	101	137	36
Ar	103	96	109	13	103	89	116	27
Cf	95	85	100	15	98	80	106	26
Cw	41	7	79	72	20	7	59	52
Cs	122	77	156	79	121	52	214	162

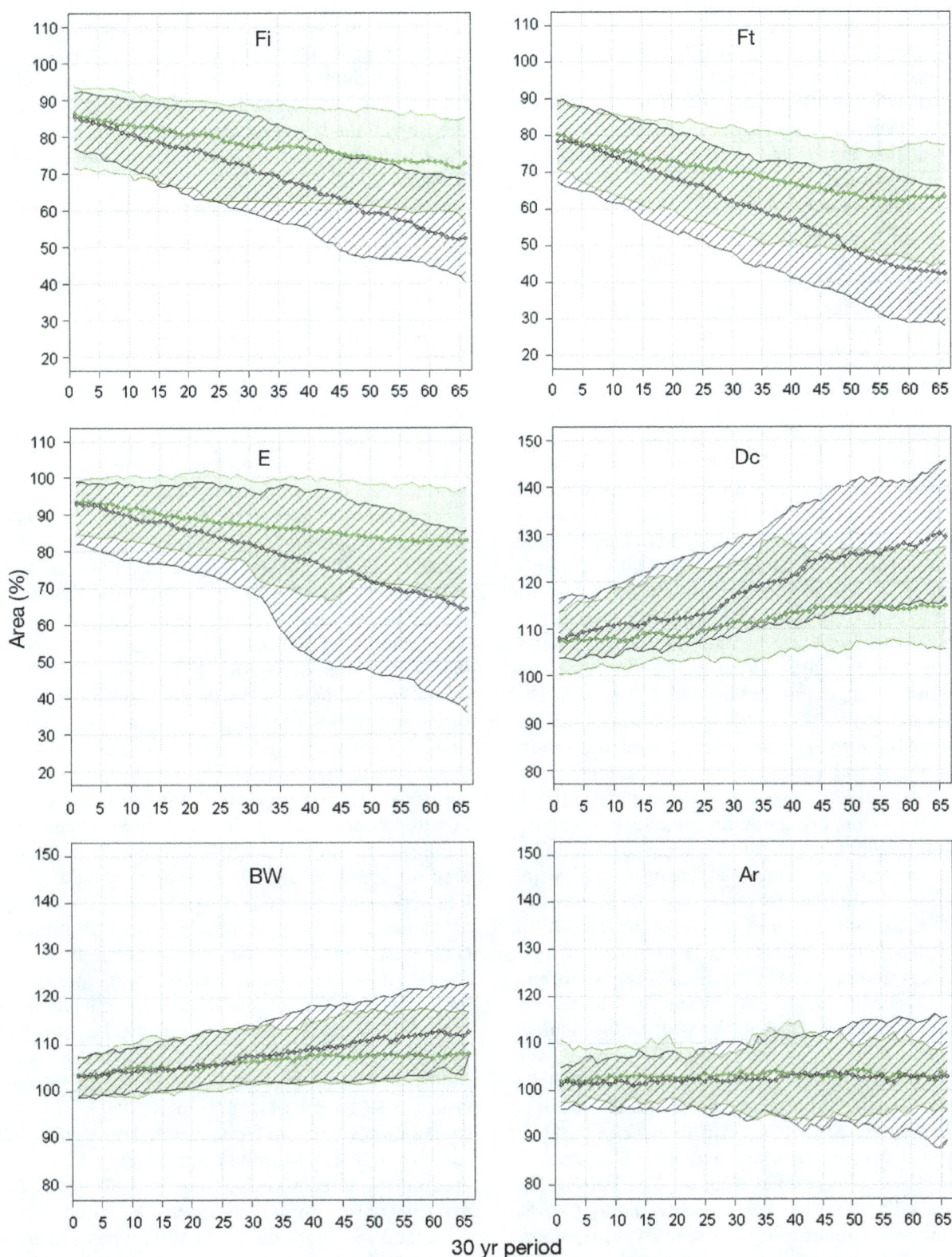

Fig. 4. Temporal evolution of continental area belonging to selected climate types (*Fi, Ft, E, Dc, BW,* and *Ar*; see Table 2 for definitions) for moving 30 yr periods throughout the 21st century relative to the reference period 1961–1990 (100% means no change); x-axis: 30 yr periods (period 1 is 2006–2035, period 66 is 2071–2100); squares: multi-model medians calculated from the ensemble of 30 selected CMIP5 GCMs (green for RCP4.5, black for RCP4.5); green area (diagonal hatching): values between the 10th and 90th percentiles of the multi-model ensemble for RCP4.5 (RCP8.5)

lowed by a steep rise until 2073 and a moderate decrease afterwards (not shown). This pattern both affects the spread of the results and is reflected in some other types (*Do, Dc, Ar*). Even though all GCMs simulated the observed area of type *E* with the smallest errors, their reactions to radiative forcing are quite diverse. The spread of the multi-model ensemble is larger for the stronger forcing of RCP8.5 than for RCP4.5 (Table 3). The decrease in continental area for type *E* by the end of the century seems to be the most convincing (in comparison to model errors, Fig. 5) of all climate types that are expected to decrease. Boreal climate transforms mainly to temperate continental climate *Dc*. For RCP8.5 the losses, generally from the southern extent of type *E* in the northern hemisphere, are >4 times larger than the gains of the area from tundra climate *Ft*.

The second group of KTC types consists of *Dc, Do, BW, BS,* and *Aw* that are all expected to increase their relative continental areas, according to most of the GCMs and both RCPs. All GCMs considered in our study (except for CESM1-CAM5-1-FV2) give a gradual expansion of continental temperate climate *Dc* during 2006–2100. Based on M-MED, the relative area occupied by *Dc* for RCP4.5 (RCP8.5) increases to approx. 115% (130%) of the area in the reference period by the end of the century (Fig. 5, Table 3). The stronger forcing of RCP8.5 leads to a higher increase in *Dc* area but also to a somewhat larger multi-model spread (Fig. 4, Table 3). Regarding the model errors, the GCMs tend to overestimate the observed area of *Dc*, but the errors are generally smaller in comparison to other KTC types. In the far-future period under RCP8.5, most of the simulated changes are larger than corresponding model errors (Fig. 5). The expansion of *Dc* is given mainly by the transition from *E*; for RCP8.5, a small portion also comes from *Ft*.

The expected increase in dry climate types *BW* and *BS* is not as convincing and well-marked as the increase in type *Dc*. According to M-MED, the relative continental area of desert climate *BW* grows by the end of the century to ~108% (113%) for RCP4.5 (RCP8.5) (Fig. 4, Table 3). Some of the GCMs, e.g. FGOALS-g2 and MRI-CGCM3, give a similar relative continental area for *BW* at the end of the 21st century as in the reference period (Fig. 5). The patterns in temporal behavior differ considerably among GCMs. Some models simulate a steady rise in *BW* area, others project a slight decrease during first decades followed by an increase or an increase followed by a short decline and a final rise. However, the multi-model spread of changes simulated for the end of the 21st century is one of the lowest of all KTC types. Both the

simulated increase and the multi-model spread are larger for the RCP8.5 scenario. The simulated changes are larger than model errors for 50% (30%) of the GCMs for RCP8.5 (RCP4.5) in 2071–2100. Regarding the transitions between climate types, the *BW* gains the area mainly from *BS*. However, a small part of the *BW* area transforms into *BS*.

Our findings for steppe climate type *BS* are similar to *BW*. Most of the GCMs simulate a larger or similar relative continental area for *BS* at the end of the century with respect to the reference period (see Fig. S15 in the Supplement at www.int-res.com/articles/suppl/c071p017_supp.pdf). The multi-model median of changes represents an increase to 108% (113%) for RCP4.5 (RCP8.5) (Fig. S3, Table 3). For about half of the GCMs, under RCP8.5 in the far future, the expected change is greater than the model error. The expected climate changes lead to transition of *Cf* and *Aw* into *BS* and from *BS* into *BW*.

For savanna climate type *Aw*, most GCMs project a moderate expansion with M-MED of 117% (120%) for RCP4.5 (RCP8.5) (Fig. S2, Table 3). An exception is the model CanESM2, which projects a slight decrease in *Aw* area. Model errors are smaller than simulated changes for 2071–2100 according to 30% (50%) of simulations under RCP4.5 (RCP8.5). Similar to the case of boreal climate *E*, even though model performance in simulating *Aw* in the reference period is relatively good, the reactions to radiative forcing differ considerably among models. Part of the continental area occupied by *Aw* undergoes a transition to *BS* and a part of *Cf* area transforms into *Aw*.

Expected temporal evolution of relative continental area occupied by oceanic temperate climate *Do* differs between individual GCMs. Some of them project an increase in the area, others project an initial decrease and then a slow rise to approximately the same *Do* extent as simulated for the reference period. The time development of the 10th percentile (Fig. S8) shows that some GCMs even project a decrease in *Do* area in the far future, especially for RCP4.5. M-MED shows an overall change to 112% of the reference area for RCP4.5 and 115% for RCP8.5. Simulated changes in *Do* are smaller than model errors (Fig. S20), except for IPSL-CM5A-MR and HadGEM2-AO. Regarding the transitions between climate types, *Do* is expected to transform mainly into *Cf*.

Until now we have dealt with KTC types that are expected to decline or increase their area according to most CMIP5 GCMs, even though the sensitivity of the models was different and multi-model spread was quite large in some cases. Results for the remaining KTC types are less conclusive. Regarding the tropical

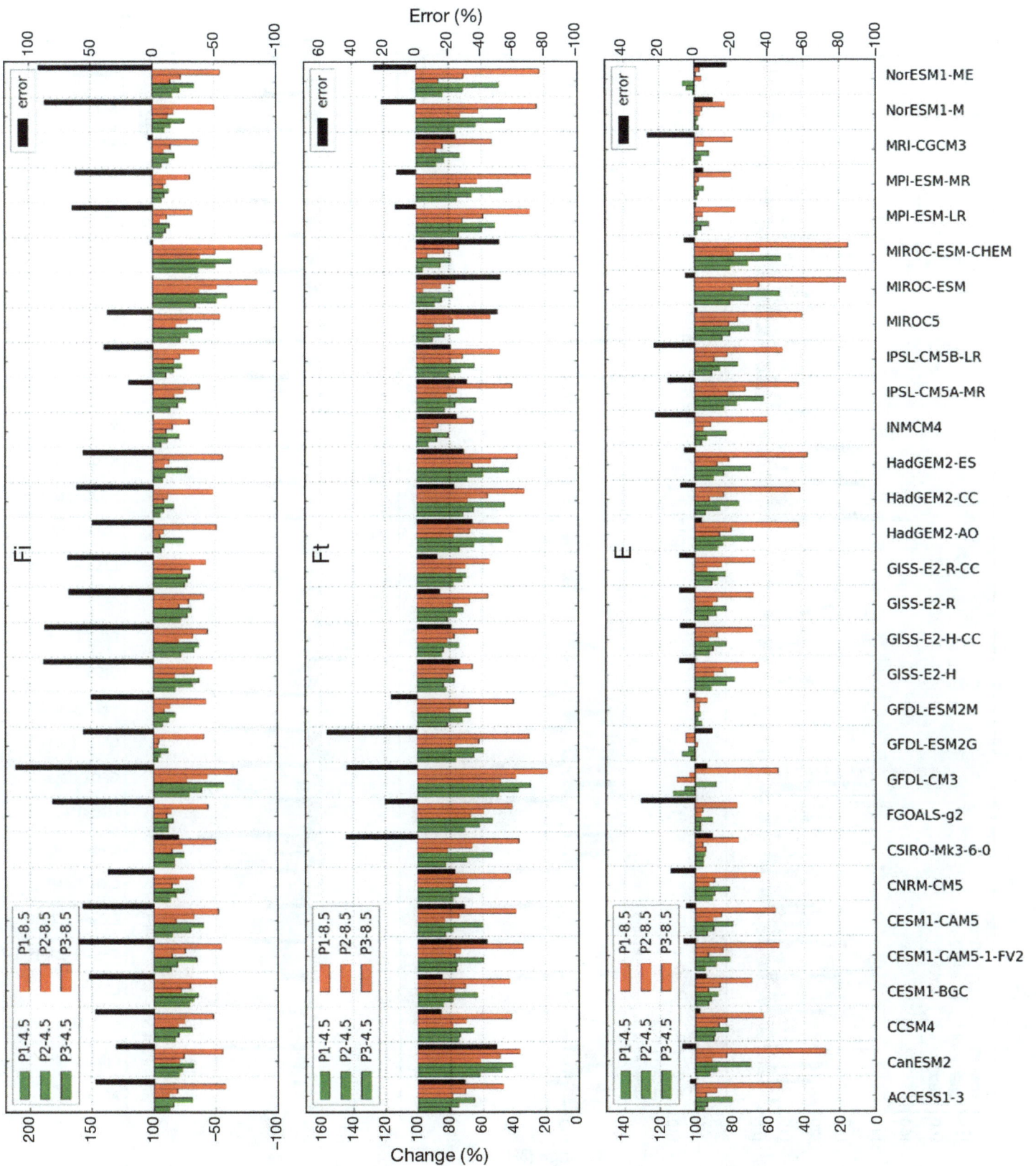

Fig. 5. Changes in relative continental areas of selected Köppen-Trewartha classification (KTC) climate types (*Fi, Ft, E, Dc, BW, Ar*; see Table 2 for definitions) projected for the periods 2006–2035 (P1), 2021–2050 (P2), and 2071–2100 (P3) relative to the reference period 1961–1990 (100% means no change) based on the ensemble of 30 selected CMIP5 GCMs for RCP4.5 (green) and RCP8.5 (red); error: model error in the reference period expressed as the difference between simulated and observed (based on CRU TS3.22) relative area in the percentage of the observed value

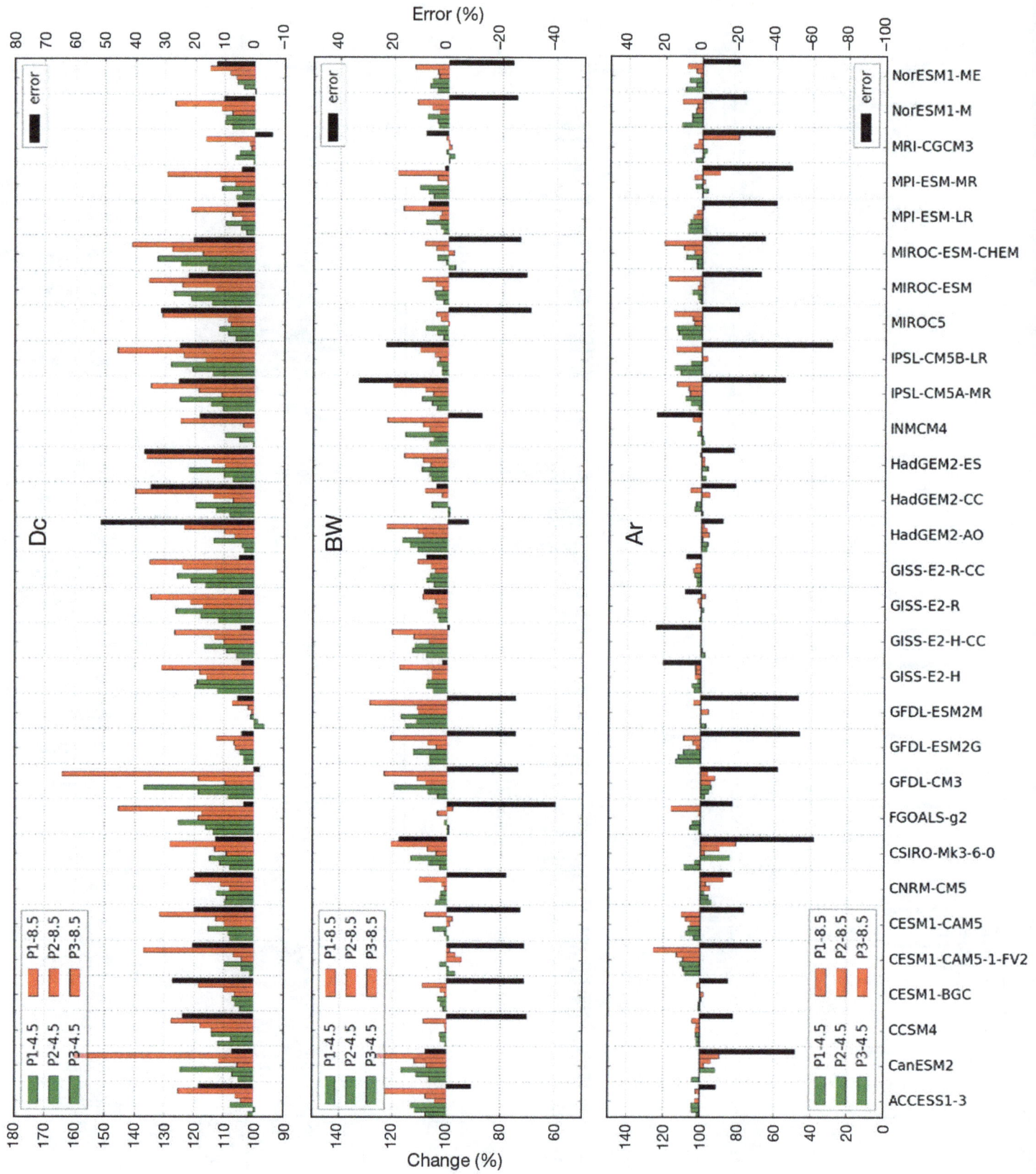

Fig. 5. (continued)

rainforest climate *Ar*, the ensemble does not show any significant signal, with ambiguous signs of change for individual GCM simulations (Fig. 5, Table 3). For both RCPs, the spread is rather small, similar to *BW* and *Cf*. The model errors are larger than simulated changes for all GCMs. The changes of *Ar* are given by transitions from *Aw* and *Cf* and into *Aw*.

Most of the GCMs simulate a decline in the area occupied by the subtropical humid climate *Cf* in 2006–2035, to ca. 95% of reference value according to M-MED. Thereafter, M-MED does not vary considerably, even though the multi-model spread grows throughout the century (Fig. S5). Simulated changes are mostly smaller than model errors, except for GISS-E2-R, GISS-E2-R-CC, and CanESM2 (Fig. S17). The subtropical humid climate *Cf* transforms mainly to *Aw* and *BS*. The area of type *Cf* increases due to gains from *Do*, *Dc*, and *BS*.

We do not discuss the results for *Cw* and *Cs*, as they both occupy a small fraction of global land area, and the spread of the model results is quite large. Therefore, it is difficult to draw any conclusions about their projected changes.

Overall, the GISS models and MRI-CGCM3, AC-CESS1-3, GFDL-ESM2M, and NorESM1 have the least pronounced response to radiative forcing. For RCP8.5, these models simulate changes of ca. 16% of continental area (not including Antarctica). On the other hand, MIROC-ESM, MIROC-ESM-CHEM, GFDL-CM3, and CanESM2 show the largest KTC type changes. According to these GCMs, more than 30% of the considered land area will undergo a change of KTC type by the end of the 21st century. However, for individual KTC types, the models simulating the largest or smallest changes differ. For example, MIROC-ESM and MIROC-ESM-CHEM show the largest reduction in *Fi* but the slowest decline of *Ft*. It is noteworthy that GCMs developed in the same modeling center do not necessarily yield similar results. For example, GFDL-CM3 shows the most sensitive response of *Dc* area to radiative forcing, whereas GFDL-ESM2M gives a change of only 1% (7%) for RCP4.5 (RCP8.5) at the end of the century.

4. DISCUSSION AND CONCLUSIONS

We assessed changes in the global distributions of Köppen-Trewartha climate types throughout the 21st century as simulated by a suite of 30 CMIP5 global climate models for 2 representative concentration pathways, RCP4.5 and RCP8.5. Ice cap climate *Fi*, tundra *Ft*, and boreal climate *E* are expected to

decline (Fig. 4). On the other hand, the relative continental area occupied by temperate climates *Dc* and *Do*, dry climates *BW* and *BS*, and savanna climate *Aw* will increase (with a few exceptions). The results for 2 remaining climate types, *Ar* and *Cf*, are less convincing; the changes are rather small, and the models do not even agree on the sign of the changes. Nevertheless, most of the GCMs simulate a slight decrease or increase at the beginning of the 21st century and very small changes thereafter. The types *Cs* and *Cw* cover only a small portion of the total continental area, and simulated changes have a large spread; therefore we will not discuss these types in detail.

Our conclusions about a decrease in *Fi* and *Ft* area and an increase in *Dc* and *Do* extent are consistent with the expected rise in near-surface air temperature and are in agreement with results described by other recent studies based on CMIP5 GCMs, e.g. by Feng et al. (2014), and also by studies for the previous generation of GCMs, e.g. Rubel & Kottek (2010).

Regarding the temporal evolution of relative continental areas covered by specific KTC types during the 21st century based on M-MED of simulated changes, a distinct difference in comparison to the reference period is already apparent for the first 30 yr time window of 2006–2035, and in most cases (except for *Ar* and *Cf*), the magnitude of simulated changes increases throughout the century. This pattern is more pronounced for RCP8.5. The course of simulated changes is not always smooth; for example, under RCP4.5 forcing, the decrease in area covered by *Ft* is faster during the first half of the century, while for RCP8.5 the decline is more stable. Similarly, under RCP4.5, the rate of increase/decrease of *BW*, *Dc*, and *Ar* is slower in the last third of the century. This might be partly due to differences in the RCPs; RCP4.5 represents a stabilization scenario with radiative forcing reaching its maximum in the second half of the 21st century; in contrast, under RCP8.5, radiative forcing increases throughout the whole 21st century (IPCC 2013). However, the influence of RCPs cannot be simply generalized. For example, the expansion of *Do* shows almost the same rate under both RCPs.

Considering the projections given by individual GCMs, for some KTC types the GCMs agree on the sign and general pattern of changes; however, the sensitivity of models to the radiative forcing differs for different KTC types (Fig. 5). For example, the models MIROC-ESM and MIROC-ESM-CHEM give the smallest change in *Ft* and the largest change in *Fi* (Fig. S24). The course of simulated changes is quite smooth for some of the GCMs, while for others it exhibits wave-like behavior, breaks, and jumps even when using 30 yr running means.

The magnitude of changes for 3 selected time periods (near future, mid-century, far future) were compared to model errors in the reference period 1961–1990 (Fig. 5). The errors were evaluated from comparison of relative land areas of KTC types derived from GCM simulations and from CRU TS3.22 observations; in this way, they could be interpreted as biases as well. Regarding the end of the 21st century (far future), only for 3 out of 12 KTC types, viz. *Ft*, *E*, and *Dc*, the changes are higher than the model errors (according to most of GCMs under RCP8.5; under RCP4.5, half of the GCMs show that changes are higher than errors for *Dc*, and ca. 75% of GCMs indicate that this is the case for *E* and *Ft*). Thus, considering the model errors, the simulated decrease in relative continental area is clearly pronounced in the case of boreal climate *E* and tundra climate *Ft*, and the increase is pronounced in the case of continental temperate climate *Dc*. Regarding the expected decrease in *Fi* area, the simulated changes are larger than model errors according to only one-third of the GCMs for both RCP 8.5 and RCP 4.5. Further, in case of savanna climate *Aw*, dry climates *BW*, and steppe climate *BS*, the simulated changes in the far future are larger than model errors according to about half of the GCMs for both RCPs, except for dry climate *BW* under RCP4.5 (one-third of the models) and steppe climate *BS* under the same scenario (only 13%).

For *Cf* and *Do*, the simulated changes are larger than model errors according to only 6 and 4 out of 30 GCMs, respectively. The type *Ar* is the only KTC type for which the simulated changes are smaller than model errors for both RCPs and all 3 time periods. We found no straightforward relationship between the model performance and the strength of the climate signal in projected changes.

Besides a comparison of simulated changes to model errors, we assessed the uncertainty stemming from necessary choices in GCM structure. We used the range between the 10th and 90th percentile of the multi-model ensemble to assess this uncertainty. The smallest multi-model spread of simulated changes is seen for the *BW* type (Table 3), the largest for *E* and *Ft*. The simulated changes of *Ar*, *Cf*, and *Do* types are ambiguous in the sense that the multi-model ranges include a 'zero change'.

Considering the changes in relative areas for the KTC types all together, the lowest sensitivity to radiative forcing under RCP4.5 is seen for MRI-CGCM3, with 8% of total continental area undergoing a KTC

type change until the end of the 21st century. Then follows a group of models with simulated changes of <12% (3 GISS GCMs, Nor-ESM1-M, Nor-ESM1-ME, GFDL-ESM2M, ACCESS1-3). For RCP8.5, the lowest sensitivity was found for all GISS models and MRI-CGCM3, with changes of 16–17%. The largest sensitivity was found for MIROC-ESM and MIROC-ESM-CHEM, with nearly a quarter of the global continental area (without Antarctica) showing changed KTC types under RCP4.5 and about 35% under RCP8.5. According to M-MED for RCP4.5 (RCP8.5), 14% (22%) of the continental area is expected to change its climate type by the end of the century. Our results are in agreement with Mahlstein et al. (2013), who projected that approximately 20% of global land will experience a change in climate type until 2100 under RCP8.5 forcing, although their study was based on a smaller number of models than ours. According to Feng et al. (2014), a larger portion of the continental area is expected to undergo a change in climate type (31% for RCP 4.5 and 46% for RCP8.5).

Regarding the shifts in the 6 major climate types, the changes projected for the far-future period under RCP8.5 based on our results and the study of Feng et al. (2014) are summarized in Table 4. The values are shown as a percentage of the respective KTC type area in the reference period 1961–1990. The simulated changes are more distinct for types *D*, *E*, and *F* than for other KTC types. For these 3 types, 25–50% of the reference area is expected to shift to another KTC type. This likely points to a more important influence of air temperature changes than precipitation changes on the KTC type shifts, which was shown by Feng et al. (2014) and Mahlstein et al. (2013).

Table 4. Comparison of multi-model statistics aggregated for the main Köppen-Trewartha classification (KTC) climate types (*A–F*, see Table 2 for definitions) in this study (rows 1–5) and in Feng et al. (2014, rows 6–9). Values are given as a percentage of the respective KTC type area in the reference period 1961–1990. M-MED: multi-model median, M-mean: multi-model mean, SD: standard deviation, F: Feng et al. (2014) values

		A	B	C	D	E	F
1	M-MED	17.4	13.2	−11.2	24.7	−37.4	−50.0
2	M-mean	14.9	15.6	−13.5	26.3	−39.4	−61.0
3	SD	10.1	6.3	12.7	9.3	22.3	27.4
4	M-mean − SD	4.9	9.3	−26.2	17.0	−61.7	−88.5
5	M-mean + SD	25.0	21.9	−0.8	35.5	−17.1	−33.6
6	Fmean	11.6	15.9	−13.4	40.0	−50.4	−59.2
7	FSD	4.0	5.3	7.9	13.7	16.9	10.8
8	Fmean − FSD	7.6	10.6	−21.3	26.3	−67.3	−70.0
9	Fmean + FSD	15.6	21.2	−5.5	53.7	−33.5	−48.4

For the purpose of comparing our results to the study of Feng et al. (2014), the multi-model mean (M-mean) and standard deviation (SD) were calculated for the 6 main climate types (Table 4). M-mean differs from M-MED most prominently for the F climate type, for which the SD has also the highest value. The values of M-mean according to our results and Feng et al. (2014) are fairly similar except for D and E, where Feng et al. (2014) found larger changes than presented in our analysis. The SD values according to Feng et al. (2014) are all smaller than our SD values, except for type D. In our study, we prefer the median and range between the 10th and 90th percentile to characterize the distribution of the multi-model ensemble, as the distribution of simulated KTC type changes is generally not symmetrical.

There are several possible reasons for the differences between our results and the results of Feng et al. (2014). The analyses are based on different groups of GCMs which may significantly influence the results, as individual models show different changes in KTC types in reaction to a given forcing. Further, we used model outputs in the original model grids, but Feng et al. (2014) applied a downscaling procedure to the GCM outputs. The fact that we investigated all continental areas excluding Antarctica, whereas Feng et al. (2014) only considered global continents north of 60° S, and the different observational datasets used could also play a role, although, as discussed previously, the differences are very small compared to the ensemble spread.

A change in relative continental areas of climate types is not the only expected impact of climate change. Potential geographical shifts are also very important. Our results indicate a poleward shift of Ft, E, Do, Dc, and Cf types (not shown). On the other hand, Ar and Aw types, which are found near the equator, did not experience any latitudinal shift. Regarding the dry climates, the GCMs do not agree entirely, especially in the case of BW. A detailed analysis of these shifts, however, is beyond the scope of this study and will be the subject of future investigations.

Acknowledgement. We acknowledge the World Climate Research Programme's Working Group on Coupled Modelling, which is responsible for CMIP, and we thank the climate modeling groups for producing and making available their model outputs. For CMIP the US Department of Energy's Program for Climate Model Diagnosis and Intercomparison provides coordinating support and led development of software infrastructure in partnership with the Global Organization for Earth System Science Portals. The CRU TS 3.22 dataset was provided by the Climatic Research Unit, University of East Anglia. This study was supported by project UNCE 204020/2012, funded by Charles University in Prague, and by research plan no. MSM0021620860 funded by the Ministry of Education, Youth and Sports of the Czech Republic. In addition, the work is part of the activity under the Program of Charles University PRVOUK No. 02 'Environmental Research'.

LITERATURE CITED

Bailey RG (2009) Ecosystem geography: from ecoregions to sites, 2nd edn. Springer, New York, NY

ä Baker B, Diaz H, Hargrove W, Hoffman F (2010) Use of the Köppen-Trewartha climate classification to evaluate climatic refugia in statistically derived ecoregions for the People's Republic of China. Clim Change 98:113–131

▶ Belda M, Holtanová E, Halenka T, Kalvová J (2014) Climate classification revisited: from Köppen to Trewartha. Clim Res 59:1–13

▶ Belda M, Holtanová E, Halenka T, Kalvová J, Hálvka Z (2015) Evaluation of CMIP5 present climate simulations using the Köppen-Trewartha climate classification. Clim Res 64:201–212

de Castro M, Gallardo C, Jylha K, Tuomenvirta H (2007) The use of a climate-type classification for assessing climate change effects in Europe from an ensemble of nine regional climate models. Clim Change 81:329–341

Deque M (2007) Frequency of precipitation and temperature extremes over France in an anthropogenic scenario: model results and statistical correction according to observed values. Global Planet Change 57:16–26

▶ Feng S, Ho CH, Hu Q, Oglesby RJ, Jeong SJ, Kim BM (2012) Evaluating observed and projected future climate changes for the Artic using the Köppen–Trewartha climate classification. Clim Dyn 38:1359–1373

▶ Feng S, Hu Q, Huang W, Ho CH, Li R, Tang Z (2014) Projected climate regime shift under future global warming from multi-model, multi-scenario CMIP5 simulation. Global Planet Change 112:41–52

Geiger R (1954) Klassifikationen der Klimate nach W. Köppen. In: Bartels J, ten Bruggencate P (eds) Landolt-Börnstein: Zahlenwerte und Funktionen aus Physik, Chemie, Astronomie, Geophysik und Technik (alte Serie), Vol 3. Springer, Berlin, p 603–607

▶ Hargrove WW, Hoffman FM (2004) Potential of multivariate quantitative methods for delineation and visualization of ecoregions. Environ Manag 34:S39–S60

▶ Harris I, Jones PD, Osborn TJ, Lister DH (2014) Updated high-resolution grids of monthly climatic observations — the CRU TS3.10 dataset. Int J Climatol 34:623–642

IPCC (Intergovernmental Panel on Climate Change) (2013) Climate Change 2013: the physical science basis. Cambridge University Press, Cambridge

▶ Kalvová J, Halenka T, Bezpalcová K, Nemešová I (2003) Köppen climate types in observed and simulated climates. Stud Geophys Geod 47:185–202

Köppen W (1923) Die Klimate der Erde. Grundriss der Klimakunde. Walter de Gruyter & Co., Berlin

Köppen W (1931) Grundriss der Klimakunde. Walter de Gruyter & Co., Berlin

Köppen W (1936) Das geographische System der Klimate. In: Köppen W, Geiger R (eds) Handbuch der Klimatologie. Gebrüder Borntraeger, Berlin, p C1–C44

Lohmann U, Sausen R, Bengtsson L, Cubasch U, Perlwitz J,

Roeckner E (1993) The Köppen climate classification as a diagnostic tool for general circulation models. Clim Res 3:177–193

▶ Mahlstein I, Daniel JS, Solomon S (2013) Pace of shifts in climate regions increases with global temperature. Nat Clim Change 3:739–743

▶ Meehl GA, Covey C, Delworth T, Latif M and others (2007) The WCRP CMIP3 multi-model dataset: a new era in climate change research. Bull Am Meteorol Soc 88: 1383–1394

▶ Menon A, Levermann A, Schewe J, Lehmann J, Frieler K (2013) Consistent increase in Indian monsoon rainfall and its variability across CMIP-5 models. Earth Syst Dyn 4:287–300

▶ Moss RH, Edmonds JA, Hibbard KA, Manning MR and others (2010) The next generation of scenarios for climate change research and assessment. Nature 463: 747–756

Patton CP (1962) A note on the classification of dry climate in the Köppen system. Calif Geogr 3:105–112

▶ Rogelj J, Meinshausen M, Knutti R (2012) Global warming under old and new scenarios using IPCC climate sensitivity range estimates. Nat Clim Change 2:248–253

▶ Rubel F, Kottek M (2010) Observed and projected climate shifts 1901–2100 depicted by world maps of the Köppen-Geiger climate classification. Meteorol Z 19:135–141

Taylor K, Balaji V, Hankin S, Juckes M, Lawrence B, Pascoe S (2012a) CMIP5 Data Reference Syntax (DRS) and controlled vocabularies. http://pcmdi-cmip.llnl.gov/xmip5/docs/cmip5_data_refernce_syntax.pdf (accessed 7 Nov 2016)

▶ Taylor K, Stouffer RJ, Meehl GA (2012b) An overview of CMIP5 and the experiment design. Bull Am Meteorol Soc 93:485–498

Trewartha GT (1968) An introduction to climate. McGraw-Hill, New York, NY

Trewartha GT, Horn LH (1980) Introduction to climate, 5th edn. McGraw Hill, New York, NY

Willmott CJ, Matsuura K (2001) Terrestrial air temperature and precipitation: monthly and annual time series (1950–1999). http://climate.geog.udel.edu/~climate/html_pages/README.ghcn_ts2.html

Permissions

All chapters in this book were first published in CR, by Inter-Research; hereby published with permission under the Creative Commons Attribution License or equivalent. Every chapter published in this book has been scrutinized by our experts. Their significance has been extensively debated. The topics covered herein carry significant findings which will fuel the growth of the discipline. They may even be implemented as practical applications or may be referred to as a beginning point for another development.

The contributors of this book come from diverse backgrounds, making this book a truly international effort. This book will bring forth new frontiers with its revolutionizing research information and detailed analysis of the nascent developments around the world.

We would like to thank all the contributing authors for lending their expertise to make the book truly unique. They have played a crucial role in the development of this book. Without their invaluable contributions this book wouldn't have been possible. They have made vital efforts to compile up to date information on the varied aspects of this subject to make this book a valuable addition to the collection of many professionals and students.

This book was conceptualized with the vision of imparting up-to-date information and advanced data in this field. To ensure the same, a matchless editorial board was set up. Every individual on the board went through rigorous rounds of assessment to prove their worth. After which they invested a large part of their time researching and compiling the most relevant data for our readers.

The editorial board has been involved in producing this book since its inception. They have spent rigorous hours researching and exploring the diverse topics which have resulted in the successful publishing of this book. They have passed on their knowledge of decades through this book. To expedite this challenging task, the publisher supported the team at every step. A small team of assistant editors was also appointed to further simplify the editing procedure and attain best results for the readers.

Apart from the editorial board, the designing team has also invested a significant amount of their time in understanding the subject and creating the most relevant covers. They scrutinized every image to scout for the most suitable representation of the subject and create an appropriate cover for the book.

The publishing team has been an ardent support to the editorial, designing and production team. Their endless efforts to recruit the best for this project, has resulted in the accomplishment of this book. They are a veteran in the field of academics and their pool of knowledge is as vast as their experience in printing. Their expertise and guidance has proved useful at every step. Their uncompromising quality standards have made this book an exceptional effort. Their encouragement from time to time has been an inspiration for everyone.

The publisher and the editorial board hope that this book will prove to be a valuable piece of knowledge for researchers, students, practitioners and scholars across the globe

List of Contributors

Ying Xu, Ying Shi and Zhou Botao
National Climate Center, China Meteorological Administration, 100081 Beijing, China

Xuejie Gao
Climate Change Research Center, Institute of Atmospheric Physics, Chinese Academy of Sciences, 100029 Beijing, China

Lei Li, Deli Wang and Mingyan Tan
Shenzhen National Climate Observatory, Meteorological Bureau of Shenzhen Municipality, Shenzhen Key Laboratory of Severe Weather in South China, Shenzhen 518040, PR China

Pak Wai Chan
Hong Kong Observatory, 998000 Kowloon, Hong Kong SAR

Markku Rummukainen
Centre for Environmental and Climate Research, Lund University, Sölvegatan 37, 223 62 Lund, Sweden

Ida B. Karlsson
Geological Survey of Denmark and Greenland (GEUS), Voldgade 10, Copenhagen 1350, Denmark
Department of Geosciences and Natural Resource Management, Copenhagen University, Rolighedsvej 23,1958 Frederiksberg C, Copenhagen, Denmark

Torben O. Sonnenborg and Jens Christian Refsgaard
Geological Survey of Denmark and Greenland (GEUS), Voldgade 10, Copenhagen 1350, Denmark

Lauren P. Seaby
Department of Environmental, Social, and Spatial Change, Roskilde University, Universitetsvej 1, Building 02,Roskilde 4000, Denmark

Karsten H. Jensen
Department of Geosciences and Natural Resource Management, Copenhagen University, Rolighedsvej 23,1958 Frederiksberg C, Copenhagen, Denmark

Dennis Trolle and Erik Jeppesen
Department of Bioscience, Aarhus University, Vejlsøvej 25, 8600 Silkeborg, Denmark
Sino-Danish Centre for Education and Research (SDC), Beijing, PR China

Anders Nielsen
Department of Bioscience, Aarhus University, Vejlsøvej 25, 8600 Silkeborg, Denmark

Department of Agroecology, Aarhus University, Blichers Allé 20, 8830 Tjele, Denmark

Jonas Rolighed, Hans Thodsen, Hans E. Andersen, Karsten Bolding, BrianKronvang and Martin Søndergaard
Department of Bioscience, Aarhus University, Vejlsøvej 25, 8600 Silkeborg, Denmark

Ida B. Karlsson and Jens C. Refsgaard
Geological Survey of Denmark and Greenland (GEUS), Øster Voldgade 10, 1350 Copenhagen, Denmark

Jørgen E. Olesen
Sino-Danish Centre for Education and Research (SDC), Beijing, PR China
Department of Agroecology, Aarhus University, Blichers Allé 20, 8830 Tjele, Denmark

K. Arnbjerg-Nielsen
Department of Environmental Engineering, Technical University of Denmark, Miljovej, Building 113, 2800 Lyngby, Denmark

L. Leonardsen
Copenhagen Municipality, Njalsgade 13, 1505 Copenhagen V, Denmark

H. Madsen
DHI, Agern Alle 5, 2970 Hørsholm, Denmark

Kirsten Halsnæs, Per Skougaard Kaspersen and Martin Drews
Climate Change and Sustainable Development Group, Department of Management Engineering,Technical University of Denmark, Building 426, Produktionstorvet, 2800 Kgs. Lyngby, Denmark

Ryuhei Yoshida
Graduate School of Science, Tohoku University, Sendai 980-8578, Japan
Faculty of Symbiotic Systems Science, Fukushima University, Fukushima 960-1296, Japan

Shin Fukui
Agro-Meteorology Division, National Institute for Agro-Environmental Sciences, Tsukuba 305-8604, Japan
Faculty of Human Sciences, Waseda Univeristy, Tokorozawa 359-1192, Japan

Teruhisa Shimada
Graduate School of Science and Technology, Hirosaki University, Hirosaki 036-8561, Japan

Toshihiro Hasegawa and Yasushi Ishigooka
Agro-Meteorology Division, National Institute for Agro-Environmental Sciences,Tsukuba 305-8604, Japan

Izuru Takayabu
Meteorological Research Institute, Tsukuba 305-0052, Japan

Toshiki Iwasaki
Graduate School of Science, Tohoku University, Sendai 980-8578, Japan

Ying Shi and Ying Xu
National Climate Center, China Meteorological Administration, Zhongguancun Nandajie 46, Haidian District,Beijing 100081, PR China

Xuejie Gao
Climate Change Research Center, Institute of Atmospheric Sciences, Chinese Academy of Sciences, Huayanli 40,Chaoyang District, Beijing 100029, PR China

Filippo Giorgi
The Abdus Salam International Centre for Theoretical Physics, PO Box 586, Trieste 34100, Italy

Deliang Chen
Department of Earth Sciences, University of Gothenburg, PO Box 460, 40530 Gothenburg, Sweden

Siqiong Luo, Di Ma, Yan Chang, Minghong Song and Hao Chen
Key Laboratory of Land Surface Process and Climate Change in Cold and Arid Regions,Cold and Arid Regions Environmental and Engineering Research Institute, Chinese Academy of Sciences, Lanzhou 730000, PR China

Xuewei Fang
Key Laboratory of Land Surface Process and Climate Change in Cold and Arid Regions,Cold and Arid Regions Environmental and Engineering Research Institute, Chinese Academy of Sciences, Lanzhou 730000, PR China
University of the Chinese Academy of Sciences, Beijing 100049, PR China

Shihua Lyu
Key Laboratory of Land Surface Process and Climate Change in Cold and Arid Regions,Cold and Arid Regions Environmental and Engineering Research Institute, Chinese Academy of Sciences, Lanzhou 730000, PR China
Chengdu University of Information Technology, Chengdu 610225, PR China

Margaret Gitau
Agricultural and Biological Engineering, Purdue University, 225 South University Street, West Lafayette, IN 47907, USA

Ulf Büntgen
Swiss Federal Research Institute WSL, 8903 Birmensdorf, Switzerland
Oeschger Centre for Climate Change Research, 3012 Bern, Switzerland
Global Change Research Centre AS CR, 61300 Brno, Czech Republic

Aleš Farda, Jan Meitner and Kamil Rajdl
Global Change Research Institute CAS, Belidla 986/4a, Brno 60300, Czech Republic

Petr Štěpánek, Pavel Zahradníček and Petr Skalák
Global Change Research Institute CAS, Belidla 986/4a, Brno 60300, Czech Republic
Czech Hydrometeorological Institute, Na Šabatce 2050/17, Praha 14306, Czech Republic

Miroslav Trnka
Global Change Research Institute CAS, Belidla 986/4a, Brno 60300, Czech Republic
Institute of Agriculture Systems and Bioclimatology, Mendel University in Brno, Zemedelska 1, 613 00 Brno, Czech Republic

Miroslav Trnka, Daniela Semerádová, Jan Balek, Petr Hlavinka and Zdeněk Žalud
Global Change Research Centre, Czech Academy of Sciences, Belidla 986/4a, Brno 60300, Czech Republic
Department of Agrosystems and Bioclimatology, Mendel University in Brno, Zemedelska 1, Brno 61300, Czech Republic

Ivan Novotný and Jan Vopravil
Research Institute for Soil and Water Conservation, Žabovřeská 250, 156 27 Praha 5 – Zbraslav, Czech Republic

Miroslav Dumbrovský
Faculty of Civil Engineering, Department of Landscape Water Management, Brno University of Technology, 60200 Brno,Czech Republic

Karel Drbal, Pavla Štěpánková and Adam Vizina
T. G. Masaryk Water Research Institute, Mojmirovo nam. 16, Brno 61200, Czech Republic

František Pavlík
State Land Office, Husinecká 1024/11a, 130 00 Praha 3, Czech Republic

Lenka Bartošová
Global Change Research Centre, Czech Academy of Sciences, Belidla 986/4a, Brno 60300, Czech Republic

Michal Belda, Eva Holtanová, Jaroslava Kalvová and Tomáš Halenka
Charles University in Prague, Dept. of Meteorology and Environment Protection, 18200, Prague, Czech Republic

Index

www.ingramcontent.com/pod-product-compliance
Lightning Source LLC
Chambersburg PA
CBHW080657200326
41458CB00013B/4891